STEREOTYPIC ANIMAL BEHAVIOUR

Fundamentals and Applications to Welfare

Second Edition

STEREOTYPIC ANIMAL BEHAVIOUR
Fundamentals and Applications to Welfare

Second Edition

Edited by

Georgia Mason

Department of Animal and Poultry Science
University of Guelph
Canada

and

Jeffrey Rushen

Pacific Agri-Food Research Centre
Agriculture and Agri-Food Canada

www.cabi.org

CABI is a trading name of CAB International

CABI Head Office
Wallingford
Oxfordshire OX10 8DE
UK

CAB North American Office
875 Massachusetts Avenue
7th Floor
Cambridge, MA 02139
USA

Tel: +44 (0)1491 832111
Fax: +44 (0)1491 833508
E-mail: cabi@cabi.org
Website: www.cabi.org

Tel: +1 617 395 4056
Fax: +1 617 354 6875
E-mail: cabi-nao@cabi.org

A catalogue record for this book is available from the British Library, London, UK.

Library of Congress Cataloging-in-Publication Data

Stereotypic animal behaviour : fundamentals and applications to welfare / edited by Georgia Mason and Jeffrey Rushen. - -2nd ed.
    p. cm.
  Includes bibliographical references.
  ISBN-13: 978-0-85199-004-0 (alk. paper)
  ISBN-10: 0-85199-004-5 (alk. paper)
  1. Animal behavior. 2. Animal welfare. 3. Stereotyped behavior (Psychiatry) I. Mason, Georgia. II. Rushen, Jeffrey. III. Title.

  QL751S712 2006
  591.5--dc22

2005020129

ISBN-10:   1-84593-042-8
ISBN-13: 978-1-84593-042-4

Typeset by SPi, Pondicherry, India
Printed and bound in the UK by Cromwell Press, Trowbridge.

The paper used for the text pages in this book is FSC certified. The FSC (Forest Stewardship Council) is an international network to promote responsible management of the world's forests.

Contents

Contributors

A.J. Badnell-Waters, *Equine Consultancy Services, Bridge Cottages, How Caple, Hereford, HR1 4SS, UK; formerly at the University of Bristol, Langford House, Langford, North Somerset, BS40 5DU, UK.*

R. Bergeron, *Département des Sciences Animales, Faculté des sciences de l'agriculture et de l'alimentation, Pavillon Paul-Comtois, Université Laval, Québec, G1K 7P4, Canada.*

S. Cabib, *Universita 'La Sapienza', Dipartimento di Psicologia, via dei Marsi 78, Rome 00185, Italy.*

R. Clubb, *Care for the Wild International, The Granary, Tickford Farm, Marches Road, Kingsfold, West Sussex, RH12 3SE, UK; formerly in the Animal Behaviour Research Group, Department of Zoology, Oxford University, South Parks Road, Oxford, OX1 3PS, UK.*

J.P. Garner, *Animal Sciences Department, Purdue University, 125 South Russell Street, West Lafayette, IN 47907, USA.*

S. Lambton, *University of Bristol, Langford House, Langford, North Somerset BS40 5DU, UK.*

J.B. Lewis, *New College of Florida, 5700 North Tamiami Trail, Sarasota, Florida 34243-2197, USA.*

M.H. Lewis, *Department of Psychiatry and Neuroscience, University of Florida and McKnight Brain Institute of the University of Florida, 100 S. Newell Drive, Gainesville, FL 32610-0256, USA.*

A. Luescher, *Animal Behavior Clinic, Purdue University, VCS, LYNN, 625 Harrison Street, West Lafayette, IN 47907-2026, USA.*

C. Lutz, *New England Primate Research Center, Harvard Medical School, One Pine Hill Road, PO Box 9102, Southborough, MA 01772-9102, USA.*

G. Mason, *Department of Animal and Poultry Science, University of Guelph, Guelph, Ontario, N1G 2W1, Canada.*

J.S. Meyer, *Department of Psychology, University of Massachusetts, Amherst, MA 01003, USA.*

D. Mills, *Department of Biological Sciences, University of Lincoln, Riseholme Park, Lincoln, LN2 2LG, UK.*

M.A. Novak, *Department of Psychology, Tobin Hall, University of Massachusetts, Amherst, MA 01003, USA; New England Primate Research Center, Harvard Medical School, One Pine Hill Road, P.O. Box 9102, Southborough, MA 01772–9102, USA.*

M.F. Presti, *Departments of Psychiatry and Neuroscience, University of Florida and McKnight Brain Institute of the University of Florida, 100 S. Newell Drive, Gainesville, FL 32610-0256, USA.*

J. Rushen, *Pacific Agri-Food Research Centre, Agriculture and Agri-Food Canada; and the Animal Welfare Program, University of British Columbia, PO 1000, 6947 Highway 7, Agassiz, BC, V0M 1A0, Canada.*

D. Shepherdson, *Oregon Zoo, 4001 SW Canyon Rd, Portland, OR, 97221 USA.*

R. Swaisgood, *Center for Conservation and Research for Endangered Species, Zoological Society of San Diego, PO Box 120551, San Diego, CA 92112, USA.*

S. Tiefenbacher, *New England Primate Research Center, Harvard Medical School, One Pine Hill Road, PO Box 9102, Southborough, MA 01772-9102, USA.*

C.A. Turner, *Mental Health Research Institute, Department of Neuroscience, 205 Zina Pitcher Place, University of Michigan, Ann Arbor, MI 48109-0720, USA.*

S. Vickery, *Animal Welfare Division, Department for Environment, Food & Rural Affairs, 1a Page Street, London, SW1P 4PQ, UK; formerly in the Animal Behaviour Research Group, Department of Zoology, Oxford University, South Parks Road, Oxford, OX1 3PS, UK.*

H. Würbel, *Institut für Veterinär-Physiologie, Justus-Liebig-Universität, Giessen, Frankfurter Str. 104, D-35392 Giessen, Germany.*

Additional contributors

J. Barber, *Prospect Park Zoo Wildlife Conservation Society, 450 Flatbush Avenue, Brooklyn, New York, NY 11225, USA.*

K.S. Bollen, *Department of Psychology, University of Massachusetts, Amherst, MA 01003, USA.*

R. van den Bos, *Ethology and Welfare Centre (EWC), Department of Animals, Science and Society, Faculty of Veterinary Medicine, Utrecht University, Yalelaan 2, NL-3584 CM Utrecht, The Netherlands.*

J. Gimpel, *Fauna Australis, Facultad de Agronomia e Ingenieria Forestal, Pontificia Universidad Católica de Chile, Vicuna Mackenna 4860, San Joaquin, Santiago, Chile.*

N. Latham, *Animal Behaviour Research Group, Zoology Department, Oxford University, South Parks Road, Oxford, OX1 3PS, UK.*

F.O. Ödberg, *Department of Animal Nutrition, Genetics, Breeding and Ethology, University of Ghent, Heidestraat 19, B-9820 Merelbeke, Belgium.*

E.M.B. Poulsen, *c/o Dr. Cam Teskey, Department of Psychology, University of Calgary, 2500 University Drive NW, Calgary, Alberta T2N 1N4, Canada; formerly Cochrane Polar Bear Habitat and Heritage Village, 1 Drury Park Road, PO Box 2240, Cochrane, Ontario, P0L 1C0, Canada.*

B.M. Spruijt, *Ethology & Welfare Centre (EWC), Department of Animals, Science and Society, Animal and Society Department, Faculty of Veterinary Medicine, Utrecht University, Yalelaan 2, NL-3584 CM Utrecht, The Netherlands.*

G. Campbell Teskey, *Department of Psychology, University of Calgary, 2500 University Drive NW, Calgary, AB T2N 1N4, Canada.*

1 A Decade-or-More's Progress in Understanding Stereotypic Behaviour

J. Rushen[1] and G. Mason[2]

[1]Pacific Agri-Food Research Centre, Agriculture and Agri-Food Canada; and the Animal Welfare Program, University of British Columbia, PO 1000, 6947 Highway 7, Agassiz, BC, V0M 1A0, Canada; [2]Department of Animal and Poultry Sciences, University of Guelph, Guelph, Ontario, N1G 2W1, Canada

Editorial Introduction

To open the book, we review the extent and nature of research into stereotypic behaviour since the first edition was published 13 years ago. We compare the numbers of recent papers on captive animals with those on human clinical subjects or research animals experimentally manipulated to produce abnormal behaviour, and also show some recent meta-analyses of trends within the former group. Contributed boxes present simple overviews of the motivational explanations typically used by ethologists, versus explanations in terms of brain function, and also review how the terms 'coping' and 'pathology' have been used. We then assess the extent to which the research questions raised by the last volume have been answered, and end by introducing this new edition's website and how the following chapters are organized.

GM and JR

1.1. Introduction

In this chapter, we review the scope and layout of the book and its accompanying website, and introduce some key concepts that recur throughout the volume, such as 'coping' and 'pathology'. We also discuss the extent to which this volume addresses the research questions highlighted at the end of the first edition (Lawrence and Rushen, 1993). This was published more than a decade ago, and set out to review what was then known about stereotypic behaviour with particular attention to the implications for animal welfare. Why is a second edition warranted? Presumably there has simply been enough new material on stereotypic behaviour of animals to convince the publishers, editors and contributors that a new edition was justified. But to what extent have the issues raised in the earlier edition been successfully resolved? What new issues have emerged? And how does this new edition resemble or differ from the old?

1.2. Research on Stereotypies since the First Edition

In the introduction to the first edition, Lawrence and Rushen (1993) reported that a total of 63 papers had been published on the stereotypic behaviour of farm animals over the preceding 27 years. Of these, nearly a third were reviews, while the majority of data-based studies were mainly descriptive, with few studies experimentally manipulating likely causal factors. Lawrence and Rushen (1993) posited an unsurprising list of potential reasons for the low volume of articles at that time. These included a lack of interest, a lack of money, a tendency to 'talk rather than do', as well as some more subtle issues such as seeing animal stereotypic behaviour as a problem to be solved rather than a phenomenon to be understood.

The second and last issues probably remain with us, but there is no doubt that scientific interest in the underlying causes of these behaviours has increased substantially. In fact, a recent search of the ISI Web of Knowledge (http://isiwebofknowledge.com/) reveals that since the last edition, an average of 16 papers a year have been published on captive animals' stereotypies – a total of 188 from the start of 1993 to the end of 2004 (see Fig. 1.1). This number is dwarfed by research on the stereotypies of human clinical subjects and of laboratory animals subjected to genetic modification, brain lesion or psychopharmacological challenge (yielding around 120 papers a year – see Fig. 1.1). However, it nevertheless represents a substantial increase in rate, and a substantial volume of new research to assimilate.

Three things are striking about the articles now being published on captive animals' stereotypies. The first is that a growing number – as we will see in this volume (especially Chapters 5–8) – now draw explicitly from those other main sources of research into stereotypic behaviour: neuroscience, clinical psychology and psychiatry. The second is that since the first edition, there is a far greater concentration on trying to understand *why* animals perform these rather bizarre-looking behaviours, i.e. more investigations that unravel the causal factors rather than simply describe what the animals are doing. Although there remains the very practical desire to prevent these behaviours from occurring wherever possible (see e.g. Chapters 9 and 10, this volume), it is thus clear that researchers have recognized the need to base such 'prevention or elimination' strategies upon a greater understanding of why such behaviours occur. Furthermore, there is also a growing realization that such behaviours have much to tell us about how 'normal' behaviour

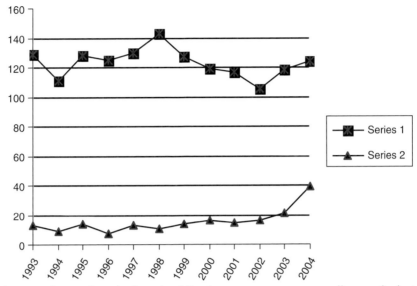

Fig. 1.1. The number of refereed publications on stereotypy, annually over the last 12 years. Records are divided into: Series 1 – papers on human patients, plus lesioned, pharmacologically treated or otherwise manipulated laboratory animals; Series 2 – papers on stereotypies emerging 'spontaneously' in captive animals.

is organized and controlled, and even about the likely psychological or neurophysiological normality of many of the millions of animals kept by humans worldwide.

The third change in the last decade or so is that a diverse array of animal species is now being studied. In 1993, most work on stereotypic behaviour focused on farm animals, or on drug-induced stereotypies in laboratory rodents. Although zoo biologists had carried out pioneering work describing these behaviours, and had raised attention (and some alarm) as to what they may mean for welfare, there were few systematic attempts either to understand or to prevent the occurrence of such behaviours in zoos. In the last 12–13 years, however, a great increase in interest in zoo animals is very evident (see e.g. Chapters 3 and 9, this volume). This has been paralleled by more research on 'spontaneous' cage stereotypies in laboratory animals (Chapters 4, 5, 7 and 8, this volume), and a growing interest in the 'problem behaviours' – including stereotypies – of the cats, dogs and horses that we keep as companion animals (Chapters 2 and 10, this volume). Indeed one reason that we added a website to this new edition (see http://www.aps.uoguelph.ca/~gmason/StereotypicAnimal Behaviour/) was to illustrate this new diversity with images and video-clips.

One result of the increased volume and taxonomic diversity of research into cage stereotypies is that it has recently allowed meta-analyses, in which the data contained in existing papers and reports are pooled for further statistical analysis to look for overall patterns or even test specific hypotheses. Such analyses are useful in revealing broad trends that would not be evident in single studies involving a small number of individuals of a single species, and can even test hypotheses that would be challenging to investigate otherwise. Two nice recent examples are presented in Chapters 3 and 9, this volume. We also give three other instances here, to further help set the scene for this new edition. First, Mason and Latham (2004) were able to estimate the total number of stereotyping animals worldwide. Although approximate, it does illustrate the vast scale of this phenomenon (see Table 1.1). The same authors also investigated the relationships between stereotypic behaviour and other measures of poor welfare. The authors pooled several hundred papers, and sorted them by the control group used as a comparison. In one type of study, animals performing stereotypic behaviour were compared with low- or non-stereotyping animals that had been raised or kept in different conditions. In the other type of study, the low- and high-stereotyping animals under comparison came from the same treatment conditions, but showed spontaneous individual differences in the behaviour. The particular treatments that led to a high incidence of stereotypic behaviour were, as we have long suspected, often linked with other signs

Table 1.1. Estimated total numbers of stereotypers for some major species/production groups (modified from Mason and Latham, 2004). Totals here represent those occurring over a period of approximately 6 months; true annual figures would be larger because of those animals generally kept for less than 12 months (e.g. laboratory mice).

Species (system)	Estimated stereotypy prevalence (% individuals)	Estimated number of stereotypers	Notes (see Mason and Latham, 2004 for details of all data sources)
Pigs (confined sows)	91.5	15,393,000	Estimated for Europe and North and Central America only
Poultry (broiler breeders)	82.6	56,498,000	Estimated for Europe and North America only
Mice (research and breeding establishments)	50.0	7,500,000	Stereotypy prevalence is a conservative guessed estimate; prevalence data are published only for a very high stereotypy strain where almost all individuals are affected; more data on common strains are therefore needed
American mink (breeding females on fur farms)	80.0	4,680,000	Prevalence estimate here ideally needs data from more farms ($N = 2$ in this estimate)
Horses (stables)	18.4	2,724,000	Population size is for the 'developed world'

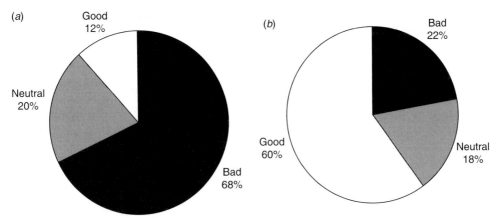

Fig. 1.2. How stereotypies and other welfare measures covary (modified from Mason and Latham, 2004). Accounts reporting additional welfare measures along with stereotypy (e.g. heart-rate changes, approach/avoidance behaviour, corticosteroid outputs) were scored as to whether they linked high stereotypy with decreased ('bad'), unchanged ('neutral') or improved ('good') welfare relative to no-/low-stereotypy controls. (*a*) Shows results from 196 reports where stereotypers and low-/non-stereotyping controls came from different treatment groups (e.g. different housing conditions or feeding regimens). (*b*) Shows results from 90 reports where stereotypers and low-/non-stereotyping controls came from within the same population/treatment group. The resulting patterns are significantly different from chance, and also significantly different from each other; see Mason and Latham (2004) for more details.

of poor welfare (see Fig. 1.2a). However, high stereotypy individuals within such environments often seemed 'better off' rather than 'worse off' than low-stereotyping animals (see Fig. 1.2b). This overview shows that the environmental factors that elicit stereotypies are not the same as the individual characteristics that predispose individual animals to develop the behaviours, an issue that recurs throughout this book. The third meta-analysis, also by Mason and colleagues, broke down abnormal behaviours (including stereotypy) by taxon, to show that different orders of mammals typically favour different types (see Fig. 1.3), and highlighting the value of research across a wide range of species.

1.2.1. Clarifying terminologies

One issue of note, as we survey these diverse pieces of work, is that researchers from different fields have, unsurprisingly, different interests, terminologies, assumptions and modes of explanation. Therefore, in an effort to help non-ethologists appreciate motivational explanations of animal behaviour (including stereotypies), both within these primary publications and in the chapters that follow, we present a box at the end of this chapter that gives a basic introduction (see Box 1.1). To follow this, and to likewise help ethologists appreciate how the vertebrate

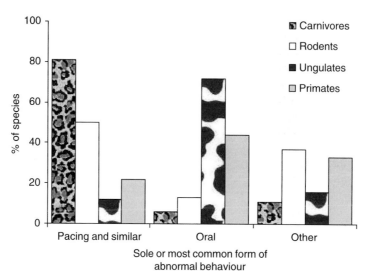

Fig. 1.3. The taxonomic distribution of different forms of abnormal behaviour (from Mason *et al.*, in press). For this survey, abnormal behaviours were defined as not known to occur in the wild, with no obvious goal or function; they thus included stereotypies, but also other behaviours, e.g. overgrooming; regurgitation-and-reingestion. Animals with apparent severe CNS dysfunction (e.g. experimentally treated with psychoactive substances, or performing self-injurious behaviours) were excluded; and husbandry differences between taxa were controlled for as much as possible (see Mason *et al.*, 2006, for details). Abnormal behaviours were categorized as: (i) pacing and similar (i.e. locomotory movements); (ii) oral (e.g. sham-chewing); and (iii) other (e.g. non-locomotory body movements like body-rocking or repetitive jumping). The 61 carnivore, 26 ungulate, 15 rodent and 19 primate species for which reports were obtained were each classified according to their sole or most commonly reported form. The frequency of different typical forms varied significantly with taxon ($\chi^2 = 51.17$ df $= 6$ $P < 0.001$).

brain generates behaviour (including abnormal behaviour), we give a very simple introduction in Box 1.2.

Furthermore, some terminology is used across multiple disciplines, but has connotations that are either rather vague or that vary between fields. Two such terms particularly relevant to discussions of stereotypy are the concepts of 'coping' and of 'pathology'. Boxes 1.3 and 1.4 therefore review these, to provide a reference definition or level of understanding that many other authors will then refer to in the chapters that follow. Again, these are given at the end of this chapter.

1.3. Issues Resolved since the Last Edition?

At the end of the last edition, Ödberg (1993) summed up some of the main issues concerning stereotypic behaviours that were then taxing the minds of investigators. He also made research recommendations for the future, as follows (his italics): '(1) Carry out as far as possible *developmental studies*, trying to induce stereotypies and investigating what changes in the organism; (2) Study *individual differences* (high/low, stereotypers/ non-stereotypers, between stereotypers, between non-stereotypers), investigating in which aspects they differ *other than in performances of stereotyping*; (3) Use increasingly *interdisciplinary approaches*, especially neuropsychological and biochemical ones, with more attention on cognitive processes; (4) Keep an open mind for different hypotheses' (p. 187). Has his advice been followed? We would have to answer with a resounding 'yes'. For example, Chapter 8 is one nice example of the first recommendation, Chapter 7 of the second, and Chapter 5 of the third; and all the chapters in the book show a desire to put forward and evaluate alternative hypotheses.

Of course, far more could be done, however, and both the overviews above and the chapters that follow reveal a number of omissions or 'blind spots'. First, much relevant literature is still not adequately used or referred to by those seeking to explain stereotypies. For example, the literature search behind Fig. 1.1 threw up a lot of hits (not included in the figure) about naturally stereotyped responding in simple neural systems (e.g. some insect movement patterns), or in what used to be termed 'fixed action patterns' such as some bird songs. These types of phenomena were discussed in the first edition (e.g. Mason, 1993) yet seem to have been forgotten in the last decade, an issue Mason returns to in Chapter 11. Second, the flow of ideas from neuroscience to ethology and related disciplines seems so far to have been one way. Third, research on avian stereotypies, even those of poultry, seems to have all but disappeared. Fourth, researchers using different types of animals have tended to also use different research approaches and techniques; this means that we quite simply do not yet have a complete understanding of the motivational, developmental and neurophysiological underpinnings of *any single model*. Chapter 11 takes this last issue further, but first, let us flag the main issues that were taxing researchers 12–13 years ago, and see to what extent they have been dealt with here.

1.3.1. What is a stereotypy?

The question of how to define or classify stereotypies was an important issue (Mason, 1993), not only because it was evident that different investigators were using different methods to decide which behaviours were included as stereotypies, but also because many of the disputes about the causes and functions of stereotypies arose partly because of a mistaken assumption that the class of stereotypies was homogeneous. Difficulties in interdisciplinary communication about stereotypies were thus thought partly to result from the fact that stereotypy meant different things in different disciplines (Ödberg, 1993). Could we assume that the same causal factors were responsible for stereotypic rocking in maternally deprived primates, stereotypic pacing in caged lions and stereotypic bar-biting by tethered sows? Should other behaviours, e.g. polydipsia in schedule-fed rodents or non-nutritive sucking by milk-fed calves, be included as stereotypies? Did the occurrence of all forms of stereotypic behaviour have the same implications for welfare across these very different animals?

The decision to classify any given behavioural pattern as a stereotypy depended, primarily, upon the way that the behaviour was performed, i.e. whether in a repetitive and invariant manner. Here, the question focused upon how much the performance of a behaviour actually had to be repetitive and invariant for it to count: detailed studies of stereotypies revealed that there was quite a deal of variability both in the timing of the performance and its repetitiveness. Clarity of the description of each instance of stereotypic behaviour is essential in comparing different studies. However, many 'normal' behaviours are also performed in a stereotypic manner, yet go unremarked. What was striking about the stereotypies being studied was that they appeared to have no function, which implied a certain degree of abnormality (however vaguely defined). It was this quality that focused the minds of researchers interested in animal welfare upon the possibility of using them to assess welfare, and to some extent a suggestion of 'abnormality' was often implicit in the decision as to whether or not any given behaviour would be included in the category of 'stereotypy'.

So, to what extent have we made progress in the way that we describe, measure and classify such behaviours? As we will see, the issue of classification and definition remains a very live one today, indeed one rendered even more important by the growing use of psychopharmacological treatments that might simply be inappropriate if we mistake the true aetiology of a particular behaviour (Chapter 10, this volume). Thus, calls for classificatory schemes are made in Chapters 2 and 4, and responses, in the form of suggested frameworks, are proposed in Chapters 5, 10 and 11. Some of the diverse behaviours that may or may not be included in future schemes are portrayed on this book's website, again to help illustrate the diverse properties of the behaviours discussed by different authors; see http://www.aps.uoguelph.ca/~gmason/StereotypicAnimalBehaviour/.

1.3.2. What causes stereotypies?

Lawrence and Rushen (1993) complained of the relative lack of systematic research to understand the causal bases of these behaviours. According to Ödberg (1993), what causal analyses had been undertaken consisted of a list of factors that might affect the occurrence of such behaviours rather than any theoretical framework based on an understanding of the organization of behaviour.

Understanding the basic causes of the behaviours is important, not only to advance fundamental scientific knowledge of how behaviour is organized, but also because the appropriate use of stereotypic behaviours to assess animal welfare requires that we understand why animals perform them. Further, it would seem likely that attempts to prevent stereotypies from occurring would benefit from a greater understanding of their causes. In the first edition of this book, the causal basis of stereotypies was approached from an analysis of the motivational (see Box 1.1), the neurophysiological and the emotional underpinning, but most approaches were essentially behavioural. A genuine theoretical model of the causes of stereotypies, however, requires that we pool our insights and adopt a more interdisciplinary approach, uniting neurophysiological, motivational and cognitive approaches (Ödberg, 1993).

To what extent have we done that? Readers must decide this for themselves as they read through the volume, but several chapters in the present volume seem to represent major advances, for instance discussing novel and hypothesis-driven cross-species comparative work on carnivores (Chapter 3, this volume), beautifully designed ethological experiments on rodents (Chapter 4, this volume), new and innovative uses on animals of the psychological tests used on human patients (Chapter 5, this volume) and sophisticated investigations of the behaviour's neurophysiological underpinnings (Chapters 7 and 8, this volume). Furthermore, one general proposal that has resurfaced in recent years is that stereotypies are pathological (see Boxes 1.2 and 1.4 for background). This term had been used in a rather loose way with respect to stereotypies for decades, but as Chapters 5, 6, 10 and 11 of this edition illustrate, it is now being used with much more precision, with uncomfortable implications for the way we currently house and use animals, and generating hypothesis-led ideas for future research.

1.3.3. How do stereotypies develop and change over time?

Ödberg (1993) drew attention to the need for more 'developmental' studies, following the history of such behaviours in individuals. This had been found useful for generating hypotheses about the causal basis of such behaviours (Cronin *et al.*, 1985), and had also shown that with time, stereotypic behaviours could sometimes become emancipated from the underlying causal factors, so that different factors affected stereotypies early in their development than when they had become fully 'established'. This finding potentially has important implications for research that tries to understand the causal basis of the behaviour, and equally for attempts to prevent or suppress stereotypic behaviour (see e.g. Mason and Latham, 2004). But has our understanding progressed? Some developmental effects are focused on in Chapter 6, which discusses the role of early social loss in primates, and Chapter 7, which discusses the role of early physical complexity. These all add substantially to our knowledge of 1993. Chapters 2, 3 and 4 also review the long-standing ideas about how developed stereotypies then change with repetition. However, the real lack of longitudinal studies of stereotypy, i.e. following individuals over time, remains striking – an issue Mason returns to in Chapter 11.

1.3.4. What is the welfare significance of stereotypies?

One question that was inadequately addressed in the first volume is a central question in animal welfare: to what extent does the occurrence of stereotypies reflect emotional suffering? Stereotypies were often assumed to occur when animals were 'stressed', an assumption undismayed by the lack of any clear definition or broadly accepted marker of stress. Since 1993, we have developed far more sophisticated, and diverse, notions of what constitutes stress, and when it is most useful to apply the concept (e.g. Moberg and Mench, 2000; Chapter 8). Has this advanced our understanding of stereotypies as part of the stress response, or as an indicator of poor welfare?

Although discussing cause rather than welfare significance was the remit of our authors, Chapters 2–8 do show how the causal factors of stereotypy are all essentially forms of stress or deprivation, while Chapter 5 presents a fascinating attempt to understand the mental world of animals performing stereotypies, from the view that they reflect some functional brain pathology. Chapter 8 even proposes a new definition of stress, central to which are the types of brain change that stereotypies may derive from. However, several chapters also provide evidence that it would be very naïve to assume simply that 'high stereotypy = bad welfare, no stereotypy = good welfare'. Mason discusses this further in Chapter 11, but let us briefly look at one potential reason for this: 'coping'.

1.3.5. What is the function of stereotypies?

Those with pedantic minds might complain of attempts to understand the function of behaviours that are, by definition, functionless. But the description of stereotypies as being apparently functionless was always more of an admission of our ignorance than a genuine description of the behaviour. In the early 1990s, particularly in the domain of farm animal welfare, there was considerable interest in the notion that the performance of stereotypic animals might actually help animals cope with stress (Dantzer and Mittleman, 1993; see also Box 1.3). For example, some of this earlier work had raised the question as to whether the performance of stereotypic behaviour helped reduce 'arousal' in animals. However, it was the finding that the performance of apparently stereotyped behaviours might be associated with a reduced hypothalamo-pituitary-adrenocortical (HPA) axis response to stress that led to the 'coping hypothesis' of stereotypies and which generated considerable research attempts to demonstrate this. By the time of the first edition of this book, however, these attempts had produced somewhat mixed results and only limited support for the coping hypothesis (Dantzer and Mittleman, 1993; Ladewig *et al.*, 1993). Have we progressed? Box 1.3 certainly helps to illustrate how sophisticated ideas have become about this; and recent evidence that some stereotypic behaviour brings with it benefits is presented in Chapters 2 and 6. It is clear, however, that as yet, more questions are raised on this topic than are answers, an issue Mason returns to in Chapter 11.

1.4. The Structure of the Book

In this book, we have selected authors who present very different perspectives on stereotypic behaviour. Chapters 2–4 (Part I) come from an ethological perspective. They discuss behaviour, including stereotypies, in terms of its motivational bases, and implicitly (or sometimes explicitly) assume that their stereotyping subjects are normal animals responding in species-typical ways (typically maladaptive ones *sensu* Box 1.4) to abnormal environments. The emphasis here is on the mismatches between the environment an animal has evolved to deal with and that into which it is placed by humans. Thus, species differences are acknowledged as an important source of variation, and even as a source of useful data with which to test hypotheses further. Furthermore, species are typically seen as important in their own right, rather than as 'models' for other organisms or conditions. The work in these chapters helps in particular to explain the form and timing of different stereotypies.

In contrast, Chapters 5–8 (Part II) have closer ties with clinical psychology, psychiatry and neuroscience. These authors typically work on three assumptions that differ from those in Part I: first, that their stereotypies of focus are the product of dysfunction, i.e. the animal is abnormal, not normal; second, that the fullest understanding of them will come from investigating the neurophysiological mechanisms involved and third, that the processes involved at this level have great cross-species generality. Thus, animals are often used as 'models' believed to display at least some of the features of the 'real' subject of interest – in practice often humans with clinical conditions. The studies reviewed here help to explain the extraordinary invariance and/or persistence we see in some forms of stereotypic behaviour.

In Part III, two chapters (Chapters 9 and 10) then illustrate how stereotypies can be tackled and reduced by those concerned about their unaesthetic appearance and/or welfare implications. These efforts are usually *post hoc* reactions to stereotypies that have emerged and been deemed problematic, typically in zoo and companion animals, whose abnormal behaviours ironically have attracted the least fundamental research. Finally, we end (Chapter 11) with a synthesis, suggestions for future research and suggestions for how terminology could perhaps improve.

Note that all chapters contain 'boxes', sometimes by the chapter's own authors, sometimes by others, that are designed to be delved into or skipped over at will. These have a number of functions. Sometimes they elaborate a self-contained 'sub-topic'; sometimes they add relevant examples, from other areas, to the topic covered in the main text; sometimes they expand on central issues that may not be familiar to all disciplines and last but not least, sometimes they flag topics of disagreement or controversy.

Box 1.1. Motivation and Motivational Explanations for Stereotypies

R. CLUBB, S. VICKERY **and** N. LATHAM

Motivational explanations of stereotypies seek to understand how they arise via an animal's internal states and responses to external stimuli. These internal and external factors are considered the same as those underlying the initiation and termination of adaptive, species-typical behaviour patterns. Motivation is thus an ethological construct used to describe why normal animals do what they do, in terms of their choice of behaviour pattern at any moment, and the time and effort they devote to performing it (e.g. Blackburn and Pfaus, 1988; Mason *et al.*, 2001; Toates, 2001).

Motivation used to be thought of in terms of drives; thus a high hunger drive would lead to increased foraging and feeding. However, drive theories have been replaced by more complex models (see e.g. Barnard, 1983) based on motivational *states* (e.g. McFarland, 1993) that are determined by an array of internal and external factors. Thus, an animal's likelihood to feed is affected by internal factors (such as an energy deficit), but also the availability and palatability of different foods, and the presence of factors eliciting competing behaviours, such as the presence of predators. Motivated behaviour often has an appetitive preparatory phase (such as searching), which culminates in a more stereotyped, species-typical consummatory phase (such as eating, mating or fighting; e.g. McFarland, 1981). In some cases, motivation is controlled by negative feedback, e.g. if consummation is successful, this reduces motivation by altering the internal state and/or eliminating the relevant environmental cues. However, motivational mechanisms often involve more complex patterns of feedback; for example, the performance of appetitive behaviours *per se* may also serve to reduce motivation, or the performance of consummatory behaviour may actually increase motivation via positive feed-back (e.g. Toates, 2001; see also Chapter 2, this volume).

The failure of such negative feedback loops is often thought to underlie stereotypies. For example, in captivity, some highly motivated consummatory behaviours (e.g. mating) may be impossible, regardless of the degree of appetitive behaviour (e.g. mate search) performed. In other cases, consummatory behaviour (e.g. feeding) may occur without the normal appetitive behaviours (e.g. grazing; Chapter 2, this volume) being possible. If such constraints leave animals in states of high motivation, this can result in frustration-related stress (e.g. Mason *et al.*, 2001), and a number of behavioural phenomena (e.g. Dawkins, 1990) including 'intention movements' (e.g. the restless escape attempts of a migratory bird confined to a cage, Dawkins, 1988), 'redirected movements' (e.g. the sucking that calves direct to pen-mates and pen furnishings in the absence of a teat; e.g. Jensen, 2003), 'vacuum activities' (e.g. the 'mimed' dustbathing movements of hens kept on bare wire floors; Lindberg and Nicol, 1997), and/or 'displacement activities' (e.g. 'out of context' ground-pecking or preening by birds during conflict situations, e.g. Tinbergen, 1952). If sustained, these activities can then develop into stereotypies – a view supported by two types of study. The first is observational, tracking in detail the development of a stereotypy from one of these 'source behaviours' (*cf.* Mason, 1991b). Ödberg (1978), for instance, described the emergence of a paw-raising stereotypy from the repeated courtship movements of a male okapi prevented from reaching a female; and Duncan and Wood-Gush (1972), the development of pacing in food-frustrated hens repeatedly trying to escape a cage. The second is experimental, and involves identifying the specific internal states or external cues key in eliciting the stereotypic behaviours. These are often the same factors that elicit normal behaviour patterns (e.g. the role of hunger in pig oral stereotypies; Chapter 2, this volume; and the role of frustrated burrow requirements in stereotypic digging by gerbils; Chapter 4, this volume). Ödberg (1978) thus asserted that 'there is one common factor to all conditions [in which stereotypies develop]: frustration. In all situations some tendency is being thwarted, some goal cannot be reached, some homoeostasis is disrupted'. Since then, much research does indeed suggest this to be the case (Chapters 2–4 and Chapter 9, this volume).

Box 1.2. A Quick Systems Sketch of Brain and Behaviour, and the Key Systems Implicated in Stereotypies

J.P. GARNER

To understand stereotypies in captive animals, we need to understand why it is that repeated motor patterns or sequences are called up repetitively within a bout, so that they occur again and again despite no obvious goal or function; and why it is these bouts themselves are repeated from one occasion to the next. In some instances, we also need to understand why they are incredibly persistent, despite costs to the animal such as self-harm. In others, we also need to understand why they are so incredibly unvarying, such that animals perform exactly the same movements each time, for example, always placing their feet in the exact same spot while repeatedly pacing or climbing. Here, I give a brief overview of the brain systems involved, to help non-neuroscientists appreciate neuroscientists' accounts of these behaviours. How does the brain turn internal and external stimuli (e.g. motivationally relevant cues) into behavioural responses? The figure below gives a simple illustration of the three main steps involved when behaviour is generated.

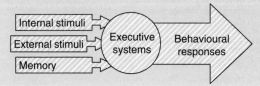

Information from the external world (via the senses) and about internal state (e.g. blood sugar levels, hormone levels) is processed and reduced to biologically meaningful information by separate systems. The cacophony of raw information arriving from the sensory organs is processed into biologically relevant representations by many dedicated perceptual systems operating in parallel, and the processing of information through these systems is called a 'stream'. For instance, visual information is first separated by the visual cortex into streams that identify form, movement, depth and colour. These visual streams are broadly integrated into an object recognition, or 'what' processing stream (located in the forebrain's temporal cortex) and a spatial and movement, or 'where' processing stream (in the parietal cortex).

All of this processing ultimately provides representations of the identity, position and movement of objects encoded in the 'cortical association areas'. The broader meaning of this information is integrated across the senses by the 'limbic system'. Associative memories are integrated and formed in interactions with the hippocampus, but the actual memories appear to be stored in the association areas that encoded the original representations. Internal physiological stimuli (*cf.* Box 1.1) are processed by the hypothalamus and relayed to other limbic areas. Further processing in the limbic system, especially the amygdala and orbito-frontal cortex, encodes emotional and value-related information (i.e. are the stimuli rewarding or aversive?) which together are termed 'affective' information.

The breadth of this information (i.e. perception, internal stimuli, memory, affect) is then processed by the brain's 'executive systems', to select and sequence both learned and instinctive behavioural responses for performance. These executive systems thus serve as a central hub for translating information into behaviour. Different executive systems have subtly different functions and are located in different areas of the brain. Probably the most important system for stereotypies is the 'contention scheduling system' involving the dorsal striatum of the basal ganglia, which selects individual movements (see Chapters 5 and 7, this volume for more information). The 'supervisory attention system' located in the prefrontal cortex may also

Continued

Box 1.2. *Continued*

play a role in some stereotypy-like behaviours, since it selects the overall 'plan' or behavioural context to be pursued (see Chapters 5 and 10, this volume for more information). Lastly, a system involving the nucleus accumbens (in the ventral striatum of the basal ganglia) may also be important since it plays a key role in determining the motivational importance accorded to a given behaviour (see Chapters 8 and 11, this volume, for more information).

The behaviours themselves are then produced by the motor systems. Motor 'programs' (the sequence of movements involved in a behaviour) are encoded in the premotor cortices, and are sequenced into individual movements by circuit loops running to the putamen (the posterior dorsal striatum) of the basal ganglia, these movements then being generated by the primary motor cortex. Thus, while sensory processing and motor functions are distinct, the executive systems are finely intertwined with the last stages of sensory and affective integration at the start of the executive processing 'stream' and the first stages of motor integration at the end of the executive 'stream'. The cerebellum modulates signals passing down the spinal cord from the motor cortex, aiding in fine feedback of movement and helping (together with executive and motor systems) coordinate the learning of highly automatic repetitive motor tasks (such as riding a bicycle). The cerebellum may thus play a role in changing some initially variable repeated movements into very predictable ones. For further details see Hubel (1988); Rolls (1994); Passingham (1995); Dias *et al.* (1996) and Brodal (2003).

Box 1.3. The Coping Hypothesis of Stereotypic Behaviour

H. WÜRBEL, R. BERGERON AND S. CABIB

Originally the term 'coping' pertained only to acutely stressful situations (e.g. electric shock; Dantzer, 1989; Wechsler, 1995), with 'coping response' referring to any behaviour apparently attenuating stressor-induced physiological responses (e.g. HPA activity, gastric ulcers). For instance, rats allowed to chew inedible objects during foot-shock showed less physiological stress than controls exposed to the same stressor but which were unable to perform these behaviours (Weiss, 1972; Tanaka *et al.*, 1985; see also Berridge *et al.*, 1999). More recently, the concept of coping has been extended to any fitness cost (e.g. Wiepkema, 1982; Wechsler, 1995) or perceived cost (i.e. negative subjective states; e.g. Lazarus and Folkman, 1984). 'Coping' then covers a broad range of responses, including both learnt behaviours (e.g. active avoidance) and unlearnt ones (e.g. hiding; fighting; displacement activities), which can either have very specific effects (e.g. reducing the HPA response to a specific stressor) or act as a 'general panacea' (i.e. attenuating negative subjective states in any adverse situation). Furthermore, some even use 'coping' to mean any behavioural attempt to control a situation, even if unsuccessful; thus Cabib (Chapter 8, this volume) presents a neurobiological conception of how unlearnt, species-typical 'coping' responses like escape attempts (which for caged animals, obviously fail) underlie stereotypies.

When used in research on stereotypic behaviour, however, 'coping' typically refers to a learnt response, which does have benefits (i.e. is at least partially successful) and may be rather specific in both cause and effects. Thus, the coping effects associated with performing a source behaviour are hypothesized to reinforce it, thereby leading to the repetitive performance typical of stereotypies. It has also been proposed that the coping effect of a stereotypy may not actually be bound to a specific situation, with the same stereotypy coming to act as a coping response in different situations, and serving as a panacea. So, what is the evidence for this idea? In the 1980s, some farm animal studies showed inverse relationships between

Continued

Box 1.3. *Continued*

stereotypy performance and gastric ulceration or heart rate (Chapter 2, this volume). Other studies found that the opioid antagonist naloxone selectively disrupted stereotypy perform- ance in sows (e.g. Cronin *et al.*, 1985), and in analogy to the 'runner's high' (e.g. Pargman and Baker, 1980), a theory attributing euphoria during long distance running to the release of endogenous opioids, these authors suggested that animals perform stereotypies to self- narcotize. Further support for coping came from laboratory rodent studies of amphetamine stereotypy (e.g. reviewed by Mason, 1991a) and human studies on the reported subjective consequences of stereotyping (e.g. reviewed by Mason and Latham, 2004). However, the 1990s saw much criticism of this idea. Evidence in favour of it was discounted as merely correlational, anecdotal or invoking 'coping' when other explanations were equally plausible (e.g. Rushen *et al.*, 1990; Mason, 1991a,b; Rushen, 1993; Garner, 1999). Thus, for example, Dantzer (1991) concluded 'there is no evidence at all that performance of stereotypies results in increased opioid activity' (see Stoll, 1997 for a similar critique of the 'runner's high' in sports science). Experimental research on rodents also yielded findings at best ambiguous in their support of this hypothesis (Chapter 4, this volume). Furthermore, several other studies yielded results inconsistent with specific predictions of the coping hypothesis (e.g. Rushen *et al.*, 1990; Terlouw *et al.*, 1991; Dantzer and Mittleman, 1993; McGreevy and Nicol, 1998).

Overall, there is no evidence that all stereotypies inevitably help animals cope or that a single mechanism (e.g. stress reduction) is always involved. Nevertheless, increasing evidence indicates that at least some stereotypies, in some species, do have some beneficial effects. Chapters 2 and 6 give some examples. Furthermore, the sheer number of correlational links between stereotypy performance and measures of improved welfare continue to intrigue, and to generate hypotheses as to how such effects might arise (Mason and Latham, 2004). However, whether any coping effects are truly *causal* in the development and continued performance of stereotypies, or merely beneficial side effects, remains unknown.

Box 1.4. Behavioural Pathology – Attempt at a Biologically Meaningful Definition

H. WÜRBEL

Pathology and disease are common words in everyday language, yet their precise biological meanings are elusive. This is mainly because 'pathology' – like 'abnormal' – is a *normative* rather than a biological term. According to *Webster's Medical Dictionary*, pathologies are 'the anatomic and physiological deviations from the normal that constitute disease or characterize a particular disease', with disease being 'an impairment of the normal state ... that interrupts or modifies the performance of the vital functions ...'. Thus, Novak and colleagues (Chapter 6, this volume) describe stereotypies as 'pathological' when they take up excessive time or cause self-harm. However, what constitutes impaired performance often depends on circum- stance, and on one's level of focus. Fever, for instance, might be classified as pathology because it has some negative side effects, yet fever is an adaptive, functional response of the organism to infection by pathogens. Thus, pathology might better be defined by its causes rather than its consequences, e.g. as 'a maladaptive phenotypic expression caused by dys- function of one or several parts of the body'. This resembles Mills' (2003) definition of 'malfunctional': 'expressions of direct disruption ... (with) no functional value in any con- text', seizures being one such example. Such responses differ from expressions caused by mismatches between an animal's phenotype and its current environment (Bateson *et al.*, 2004), or what Mills (2003) terms 'maladaptive' responses, defined for behaviours as 'attempts

Continued

Box 1.4. *Continued*

to behave in an adaptive way in an environment to which complete adaptation is not possible'.

Behavioural pathologies (or malfunctional behaviours) will typically stem from dysfunction of the nervous system. Of course, we must note that dysfunction itself is also a normative term: no sharp boundary exists between normal, healthy function and pathological dysfunction, and dysfunction can only be defined with reference to the natural variation in phenotypic expression under natural (or near-to-natural) conditions, the boundary typically needing to be defined statistically. However, overall, stereotypies might be classified as *adaptive*, if they effectively serve a coping function (see Box 1.3); *maladaptive*, if they reflect normally adaptive responses occurring inappropriately, within an 'abnormal' environment; or *pathological*, if they are caused by dysfunction, e.g. within the nervous system.

Acknowledgements

We are indebted to Alistair Lawrence for his work on the first edition. Enormous thanks also go to all the contributors to this edition, who (despite occasional bouts of rage, despair and going completely AWOL) worked incredibly hard, and with intelligence, grace and humour.

References

Barnard, C.J. (1983) *Animal Behaviour: Ecology and Evolution.* Croom Helm, London.

Bateson, P., Barker, D., Clutton-Brock, T., Deb, D., D'Udine, B., Foley, R.A., Gluckman, P., Godfrey, K., Kirkwood, T., Lahr, M.M., McNamara, J., Metcalfe, N.B., Monaghan, P., Spencer, H.G. and Sultan, S.E. (2004) Developmental plasticity and human health. *Nature* 430, 419–421.

Berridge, C.W., Mitton, E., Clark, W. and Roth, R.H. (1999) Engagement in a non-escape (displacement) behavior elicits a selective and lateralized suppression of frontal cortical dopaminergic utilization in stress. *Synapse* 32, 187–197.

Blackburn, J.R. and Pfaus, J.G. (1988) Is motivation really modulation? A comment on Wise. *Psychobiology* 16, 303–304.

Brodal, P. (2003) *The Central Nervous System: Structure and Function*, 3rd edn. Oxford University Press, Oxford, UK.

Cronin, G.M., Wiepkema, P.R. and van Ree, J.M. (1985) Endogenous opioids are in-

volved in abnormal stereotyped behaviours of tethered sows. *Neuropeptides* 6, 527–530.

Dantzer, R. (1989) *L'illusion Psychosomatique.* Editions Odile Jacob, Paris.

Dantzer, R. (1991) Stress, stereotypy and welfare. *Behavioural Processes* 25, 95–102.

Dantzer, R. and Mittleman, G. (1993) Functional consequences of behavioural stereotypies. In: Lawrence, A.B. and Rushen, J. (eds) *Stereotypic Animal Behaviour: Fundamentals and Applications to Welfare.* CAB International, Wallingford, UK, pp. 147–172.

Dawkins, M.S. (1988) Behavioural deprivation: a central problem in animal welfare. *Applied Animal Behaviour Science* 20, 209–225.

Dawkins, M.S. (1990) From an animal's point of view: motivation, fitness, and animal welfare. *Behavioural and Brain Sciences* 13, 1–9.

Dias, R., Robbins, T.W. and Roberts, A.C. (1996) Dissociation in prefrontal cortex of affective and attentional shifts. *Nature* 380, 69–72.

Duncan, I.J.H. and Wood-Gush, D.G.M. (1972) Thwarting of feeding behaviour in the domestic fowl. *Animal Behaviour* 20, 444–451.

Garner, J.P. (1999) The aetiology of stereotypy in caged animals. Ph.D. thesis, University of Oxford, Oxford, UK.

Hubel, D.H. (1988) *Eye, Brain, and Vision*, Vol. 22. Scientific American Library, Distributed by W.H. Freeman, New York.

Jensen, M.B. (2003) The effects of feeding method, milk allowance and social factors on milk feeding behaviour and cross-sucking in group housed dairy calves. *Applied Animal Behaviour Science* 80, 191–206.

Ladewig, J., de Passillé, A.M., Rushen, J., Schouten, W., Terlouw, E.C.M. and von Borell, E. (1993) Stress and the correlates of stereotypic behaviour. In: Lawrence, A.B. and Rushen, J. (eds) *Stereotypic Animal Behaviour: Fundamentals and Applications to Welfare*. CAB International Wallingford, UK, pp. 97–118.

Lawrence, A.B. and Rushen, J. (1993) Introduction. In: Lawrence, A.B. and Rushen, J. (eds) *Stereotypic Animal Behaviour: Fundamentals and Applications to Welfare*. CAB International, Wallingford, UK, pp. 41–64.

Lazarus, R.S. and Folkman, S. (1984) *Stress, Appraisal, and Coping*. Springer, New York.

Lindberg, A.C. and Nicol, C.J. (1997) Dustbathing in modified battery cages: is sham dustbathing an adequate substitute? *Applied Animal Behaviour Science* 55, 113–128.

Mason, G.J. (1991a) Stereotypies: a critical review. *Animal Behaviour* 41, 1015–1037.

Mason, G.J. (1991b) Stereotypies and suffering. *Behavioural Processes* 25, 103–115.

Mason, G.J. (1993) Forms of stereotypic behaviour. In: Lawrence, A.B. and Rushen, J. (eds) *Stereotypic Animal Behaviour: Fundamentals and Applications to Welfare*. CAB International, Wallingford, UK, pp. 7–40.

Mason, G.J. and Latham, N.R. (2004) Can't stop, won't stop: is stereotypy a reliable animal welfare indicator? *Animal Welfare* 13 (Supplement), S57–S69.

Mason, G., Cooper, J. and Clarebrough, C. (2001) The welfare of fur-farmed mink. *Nature* 410, 35–36.

Mason, G., Clubb, R., Latham, N. and Vickery, S. (2006) Why and how should we use environmental enrichment to tackle stereotypic behavior? *Applied Animal Behaviour Science*. Available online August 2nd 2006 doi: 10.106/japplanim. 2006.05.04/

McFarland, D. (1981) *The Oxford Companion to Animal Behaviour*. Oxford University Press, Oxford, UK.

McFarland, D. (1993) *Animal Behaviour*. Oxford University Press, Oxford, UK.

McGreevy, P. and Nicol, C.J. (1998) Physiological and behavioral consequences associated with short-term prevention of crib-biting in horses. *Physiology and Behavior* 65, 15–23.

Mills, D.S. (2003) Medical paradigms for the study of problem behaviour: a critical review. *Applied Animal Behaviour Science* 81, 265–277.

Moberg, G.P. and Mench, J.A. (2000) *The Biology of Animal Stress: Principles and Implications for Animal Welfare*. CABI Publishing, Wallingford, UK.

Ödberg, F.O. (1978) Abnormal behaviours: stereotypies. In: *First World Congress on Ethology Applied to Zootechnics, October 23–27, Madrid*, Industrias Grafices Espana, pp. 475–480.

Ödberg, F.O. (1993) Future research directions. In: Lawrence, A.B. and Rushen, J. (eds) *Stereotypic Animal Behaviour: Fundamentals and Applications to Welfare*. CAB International, Wallingford, UK, pp. 173–192.

Pargman, D. and Baker, M. (1980) Running high: enkephalin indicted. *Journal of Drug Issues* 10, 341–349.

Passingham, R. (1995) *The Frontal Lobes and Voluntary Action*. Oxford University Press, Oxford, UK.

Rolls, E.T. (1994) Neurophysiology and cognitive functions of the striatum. *Revue Neurologique* 150, 648–660.

Rushen, J. (1993) The coping hypothesis of stereotypic behaviour. *Animal Behaviour* 48, 91–96.

Rushen, J., de Passillé, A.M.B. and Schouten, W. (1990) Stereotypic behaviour, endogenous opioids, and post-feeding hypoalgesia in pigs. *Physiology and Behavior* 48, 91–96.

Stoll, O. (1997) Endorphine, Laufsucht und Runner's High. Aufstieg und Niedergang eines Mythos. *Leipziger Sportwissenschaftliche Beiträge* 28, 102–121.

Tanaka, T., Yoshida, M., Yokoo, H., Tomita, M. and Tanaka, M. (1985) Expression of aggression attenuates both stress-induced gastric ulcer formation and increases in noradrenaline release in the rat amygdala assessed by intracerebral microdialysis. *Pharmacology and Biochemistry of Behaviour* 59, 27–31.

Terlouw, E.M.C., Lawrence, A.B., Ladewig, J., de Passillé, A.M.B., Rushen, J. and

Schouten, W.G.P. (1991) Relationship between plasma cortisol and stereotypic activities in pigs. *Behavioural Processes* 25, 133–153.

Tinbergen, N. (1952) 'Derived' activities: their causation, biological significance, origin and emancipation during evolution. *Quarterly Review of Biology* 27, 1–32.

Toates, F. (2001) *Biological Psychology.* Pearson Education Limited, Harlow, UK.

Wechsler, B. (1995) Coping and coping strategies – a behavioural review. *Applied Animal Behaviour Science* 43, 123–134.

Wiepkema, P.R. (1982) On the identity and significance of disturbed behaviour in vertebrates. In: Bessei, W. (ed.) *Disturbed Behaviour in Farm Animals.* Ulmer, Stuttgart, pp. 1–17.

Weiss, J.M. (1972) Influence of psychological variables on stress-induced pathology. In: Porter, R. and Knight, J. (eds) *Physiology, Emotion and Psychosomatic Illness.* CIBA Foundation Symposium. Elsevier, Amsterdam.

2

Stereotypic Oral Behaviour in Captive Ungulates: Foraging, Diet and Gastrointestinal Function

R. Bergeron,[1] A.J. Badnell-Waters,[2] S. Lambton[3] and G. Mason[4]

[1]Département des Sciences Animales, Faculté des sciences de l'agriculture et de l'alimentation, Pavillon Paul-Comtois, Bureau 4131, Université Laval, Québec, G1K 7P4, Canada; [2]Equine Consultancy Services, Bridge Cottages, How Caple, Hereford, HR1 4SS, UK; formerly at the University of Bristol, Langford House, Langford, North Somerset, BS40 5DU, UK; [3]University of Bristol, Langford House, Langford, North Somerset, BS40 5DU, UK; [4]Department of Animal and Poultry Sciences, University of Guelph, Guelph, Ontario, N1G 2W1, Canada

Editorial Introduction

With millions of affected animals worldwide, ungulates are the most prevalent mammalian stereotypers. Agricultural ungulate stereotypies were also the first to attract serious scientific study. They therefore dominated the first edition of this book, and it seems probable that more individuals with stereotypies have now been studied in this taxon than in any other. Examples of the behaviours that Bergeron and co-authors consider here include crib-biting by horses, sham-chewing by sows and tongue-rolling by cattle and giraffes. Concerns about animal welfare and economic issues (e.g. stock value or productivity) have meant that many studies aimed to reduce these behaviours, rather than understand the niceties of their underlying mechanisms. Nevertheless, motivational explanations for ungulates' oral stereotypic behaviours have been developed, and to some extent tested. Ungulates are primarily herbivorous, and much evidence supports the hypotheses that their oral stereotypic behaviours derive from natural foraging. The forms of the movements are often similar, with some abnormal behaviours even involving ingestion (e.g. wood-chewing by horses); they typically peak with the delivery of food or end of a meal; and, like natural foraging, they are often reduced by factors increasing satiety. Thus, in practice, replacing captive ungulates' typically low-fibre, high-concentrate provisions with more naturalistic foodstuffs successfully reduces oral abnormal behaviour across a wide range of species. But what exactly is the link between natural foraging and oral stereotypic behaviour? This is less certain, and Bergeron and her colleagues review three principal hypotheses.

The first is that captive ungulates' diets do not fully satisfy them, because they give too little gut fill, are deficient in specific ways (e.g. too low in salt, protein or fibre), or supply too little energy (pregnant sows, for instance, are routinely fed a fraction of what they would eat *ad libitum*). Stereotypic behaviours are then proposed either to stem from unlearnt, persistent attempts to find more food, or to be learnt behaviours that help redress the animals' underlying deficits (wood-chewing to gain fibre being one possible example). The second hypothesis is that captive diets take too little time to find, chew or ruminate, leaving animals with unfulfilled motivations to perform these natural foraging activities. If natural foraging is reinforcing *per se*, quite independent of nutrient gain, then oral stereotypies can be seen as vacuum or redirected behaviours supplying at least some of the feedback normally provided by natural foraging. The third hypothesis is that oral stereotypic behaviour is not caused directly by diet quality or the minimal foraging that it requires, but instead by its consequences for gut function. Low-fibre, carbohydrate-rich foods have long been known to cause gastrointestinal dysfunction in ungulates, including gastric ulcers in horses and pigs and ruminal acidosis in cattle. More recently, experimental manipulations of both stereotypy performance and of gastrointestinal acidity have led to suggestions that oral stereotypic behaviours are a response to gut health, and perhaps even have some beneficial effects, for instance generating saliva that, if swallowed, helps to rectify gastrointestinal pH.

There is, however, evidence both for and against each of these hypotheses, and the next few years clearly need to see less *post hoc* explanation (valuable though such ideas have been) and more hypothesis-driven research, ideally combining a good physiological understanding of how various diets affect satiety and gastrointestinal function, a better understanding of the aetiology of pathologies like ulcers, and a cross-species appreciation of the different modes of ungulate foraging behaviour. Indeed different forms of oral stereotypic behaviour may well prove to have different underlying aetiologies. Some further questions posed by this chapter are as follows: How do ungulates resemble pandas, chickens and walruses? Is it ethical to physically prevent horses from stereotyping, without first tackling the underlying causes of the behaviour? And last but not least, which of the many behaviours discussed here should we actually call 'stereotypies'?

GM

2.1. Introduction

Repetitive, seemingly functionless oral and oro-nasal activities are prevalent in captive ungulates. Indeed in contrast to other taxa, they are this group's typical abnormal behaviour (see Fig. 1.2, Chapter 1, this volume). Common examples include bar-biting and sham-chewing by sows, tongue-rolling by cows and crib-biting by stabled horses. Similar behaviours also occur in exotic ungulates in zoos, for instance object-licking by bongo antelopes (Ganslosser and Brunner, 1997), dirt-eating by Przwalski's horses (e.g. Hintz *et al.*, 1976) and tongue-rolling by giraffes and okapi (e.g. Koene, 1999; Bashaw *et al.*, 2001a). These behaviours have long caused concern, for both practical and welfare reasons. Crib-biting in horses, for example, increases energy expenditure and causes tooth wear (e.g. McGreevy and Nicol, 1998a), while oral stereotypies in sows similarly increase energy-use (Cronin *et al.*, 1986), reduce weight gain (Bergeron and Gonyou, 1997) and perhaps exacerbate the effects of food restriction on hunger levels (Rushen, 2003).

The occurrence of these behaviours has also prompted more fundamental questions about their ethological origins and putative functions. Across multiple species, captive ungulates' oral behaviours often resemble species-typical feeding movements, tend to be performed at high rates around feeding and are usually affected by diet and the way that animals are fed. This suggests they share a broadly common cause relating to foraging behaviour. This chapter therefore reviews how similarities in the feeding and foraging of free-living ungulates, and in the ways they are fed in captivity, underlie these phenomena. We discuss the natural foraging biology of ungulates in Section 2.3, then review the various effects of captive diet (especially fibre levels, calorific restriction, foraging time and effects on gastrointestinal function) in Section 2.4. In Section 2.5, we consider the possible functions of these behaviours. Because of their strange appearance and apparent lack of function, stereotypic oral behaviours are often described as 'abnormal', even as 'vices' in horses (although many dislike this term, e.g. Houpt, 1993). But are ungulate stereotypies really malfunctional (in the sense of Box 1.4, Chapter 1, this volume), or are they merely maladaptive – or even adaptive? In Section 2.6, we briefly consider the possible contributory roles played by early experience (e.g. early weaning), and the physical environment (e.g. restraint), before concluding with a summary of the likely bases of ungulates' abnormal oral behaviours, a discussion of their welfare significance, and suggestions for further research. First, we look at the basics: what forms occur, how prevalent are they and why does their basic aetiology implicate foraging?

2.2. Prevalence, Nature and Possible Behavioural Origins of Stereotypic Oral Behaviour

The most common stereotypic oral behaviours in captive adult ungulates are listed in Table 2.1 and shown on this book's website. Some of these are unambiguously 'stereotypies', being repetitive, fixed in form and serving no obvious function (*cf.* e.g. Ödberg, 1978; Mason, 1991). Others, however, are less clear cut: wool-chewing by sheep and wood-chewing in horses, for instance, are relatively variable in form, have apparent goals, and so are generally not classified as stereotypies (e.g. Nicol, 1999). To reflect this diversity, here we use the term 'stereotypic behaviour' as a broad descriptive term, encompassing all repetitive unexplained behaviours, even if not highly predictable from one movement to the next (Chapter 10, this volume). Weaned infant ungulates also show oral stereotypies, but we generally do not discuss these in this chapter because they seem to relate to frustrated suckling rather than adult foraging behaviours, and their relationship with adult stereotypic behaviour is unclear.

Oral stereotypic behaviour can be very prevalent, i.e. occur in much of the population. Prevalence figures often vary between studies, but nevertheless do help give a general idea of the scale of this issue. In horses, the prevalence reported in six questionnaire surveys ranged from 0% to 8.3% for crib-biting/wind-sucking, and 5% to 20% for wood-chewing (Canali and

Table 2.1. Common abnormal oral/oro-nasal behaviours in adult domesticated ungulates.

Stereotypic behaviour	Species	Description
Crib-biting or cribbing	Horses	Grasping the edge of a horizontal surface with the incisor teeth and pulling back, while drawing air into the cranial oesophagus and emitting a characteristic grunt (e.g. McGreevy et al., 1995c; Simpson, 1998). Air is not actually swallowed during cribbing, although the short column of air that remains in the upper part of the oesophagus after cribbing could be swallowed along with food (McGreevy et al., 1995c)
Wind-sucking	Horses	Same characteristic posture and grunt as crib-biting, but without grasping a fixed object (McGreevy et al., 1995c)
Wood-chewing	Horses	Grasping wood, and at least briefly chewing it (Johnson et al., 1998). Not generally considered a stereotypy, but may precede stereotypy development (Nicol, 1999)
Tongue-rolling or tongue-playing	Cows	Swinging of the tongue outside the mouth, from one side to the other, or repetitively rolling the tongue inside the mouth (Sambraus, 1985). Tongue-playing may also occur in their water-bowls (B. McBride; personal communication, Guelph, 2005).
Object-licking	Cattle, sows, sheep	Repetitive licking of non-food objects (e.g. Bashaw et al., 2001a). Tongue movements with contact, such as licking or biting at fences, walls or the food trough, have sometimes also been termed 'para-tongue-playing' (Seo et al., 1998)
Bar-biting	Sows, sheep, cows	Taking a bar (generally a pen fixture) in the mouth and biting on it (Sambraus, 1985)
Wool-biting	Sheep	Biting off and ingesting portions of fleece (e.g. Cooper et al., 1994). Not usually termed a stereotypy
Slat-chewing	Sheep	Nibbling at edges of slats and apparent ingestion of slithers of wood and/or faecal material (e.g. Cooper and Jackson, 1996)
Sham-chewing or vacuum-chewing	Sows	Chewing with nothing in the mouth. Often accompanied by gaping, and salivary foaming from the chewing motion (Sambraus, 1985). Tongue-sucking (Whittaker et al., 1999) may also be observed
Chain-chewing or chain manipulation	Sows	Chewing on a chain that has been installed experimentally as a focus for oral activities (so that they can be logged automatically; see Terlouw et al., 1991a). Sequences vary in their degree of stereotypy
Excessive drinking (polydipsia)	Sows	Drinking or manipulation of the drinker that exceeds physiological needs (Terlouw et al., 1991a; Robert et al., 1993). Drinker activities are often performed in a ritualistic way and incorporated into sequences of stereotypic activities, such as chain-chewing or bar-biting

Borroni, 1994; McGreevy *et al.*, 1995a,b; Luescher *et al.*, 1998; Redbo *et al.*, 1998; Bachmann and Stauffacher, 2002). Waters (2002) summarized such studies to yield a median prevalence of 3.1% for cribbing/wind-sucking, and 12% for wood-chewing, figures which translate into well over a million affected individuals, given ca. 15 million horses in the developed world alone (e.g. Mason and Latham, 2004). One study of cows reported that 40 out of 95 stabled dairy cows (42%) showed stereotypies (Redbo *et al.*, 1992), mostly tongue-rolling (although this figure is probably higher than the norm; J. Rushen, personal communication; Agassiz, 2005). A survey of giraffids reported higher prevalence rates still, with 72.4% zoo animals (214 giraffes, 29 okapis and 14 unspecified individuals) showing repetitive object-licking (Bashaw *et al.*, 2001a). A few studies also report prevalence for pregnant sows. Although often based on small populations, they suggest high rates, ranging from 28% (7/25 sows; Rushen, 1984) to 100% (117/117 sows; Cronin, 1985). Mason and Latham (2004) used such papers to generate a median prevalence of 91.5%, from which they estimated that over 15 million sows across Europe and North/Central America show these behaviours.

As well as being prevalent, oral stereotypic behaviour may be time-consuming. For example, horses can spend up to 8 h crib-biting each day, performing around 8000 bites (e.g. McGreevy and Nicol, 1998b; McGreevy *et al.*, 2001a), while in other species, even the average animal may spend several hours daily in such behaviours. Thus on one site, tethered cows spent 1% to 38% of a 24-h period stereotyping (Redbo, 1990; see also Redbo, 1992), although another study put the figure far lower, at 1–2% (Bolinger *et al.*, 1997). Likewise, pregnant sows spent from 7% (Broom and Potter, 1984) to 55% (Von Borell and Hurnik, 1990) of an 8-h observation period in oral stereotypies; while one female giraffe spent more than 40% of the night-time hours licking and tongue-playing (Baxter and Plowman, 2001).

So what are the origins of such behaviours? Perhaps tellingly, forms of dietary manipulation that reduce such levels of stereotypy in farmed pigs (increased fibre and/or increased calories, as we review later), do likewise for more natural foraging behaviours directed at grass, soil and stones in sows outdoors (see Braund *et al.*, 1998; Horrell, 2000). This suggests that oral stereotypic behaviour might be related to natural foraging. Its form and timing further implicate frustrated natural foraging. It often physically re-sembles natural foraging movements, with species feeding with tongue-sweeps, such as cattle or giraffes, developing stereotypic tongue movements (e.g. Bashaw *et al.*, 2001a), but sheep, goats, horses and pigs instead showing biting/chewing behaviours (e.g. Terlouw *et al.*, 1991a; Waters *et al.*, 2002). In cattle, abnormal oral behaviours even show developmental changes that parallel natural changes in foraging mode, with young calves sucking their tongues in a manner akin to normal suckling, but adults showing the curling and uncurling tongue movements typical of grazing (Fraser and Broom, 1990). Furthermore, as we have seen, some forms of oral stereotypic behav-iour involve ingestion, e.g. of non-food solids or of water.

These behaviours also often have a close temporal association with feeding. In horses it may intersperse with food-ingestion (e.g. Kennedy *et al.*, 1993), and in many species, when food is presented in meals, stereotypic behaviour peaks around the time of delivery. However, typically it is then displayed most frequently after the food has been consumed (e.g. reviewed by Mason and Mendl, 1997). Post-feeding peaks have thus been observed in pigs (reviewed by Mason and Mendl, 1997; also Robert *et al.*, 1993, 1997; Spoolder *et al.*, 1995); giraffes (e.g. Veasey *et al.*, 1996; Tarou *et al.*, 2001); horses (e.g. Kusenose, 1992; Kennedy *et al.*, 1993; Gillham *et al.*, 1994); cattle (e.g. Sambraus, 1985; also see Fig. 2.1a); and sheep (see Fig. 2.1b). The stereotypies may be prompted by food-ingestion itself (see Terlouw *et al.*, 1993), although ingestion is not essential (see Mason and Mendl, 1997). Interestingly, in the wild, free-living giraffes also briefly show tongue-playing after feeding or drinking (Veasey *et al.*, 1996); while in wild boar housed in semi-natural enclosures, food-ingestion is also followed by rooting, and chewing at vegetation (Horrell, 2000). In Fig. 2.1b, also note the contrast in timing to the locomotor stereotypies seen pre-feeding (see Chapter 3, this volume for similar pre-feeding behaviour in captive carnivores).

Next, we discuss the experimental and epidemiological evidence for a role of frustrated foraging. We begin by discussing the natural biology of ungulates, to identify what is constrained in captivity. After all, when naturalistic foraging is impossible for caged primates or carnivores, they seldom show the extensive sham-chewing or tongue movements so typical of ungulates (e.g. Mason and Mendl, 1997; Mason, in press), suggesting that biological predispositions do play an important role.

2.3. The Natural Foraging Biology of Ungulates and How Captivity Affects It

2.3.1. The natural foraging biology of ungulates

The obvious foraging characteristic shared by ungulates is herbivory (although pigs are more correctly omnivores). Although different species vary in their relative use of grass, broad-leaved plants and/or other types of plant material (e.g. roots), and vary too in attributes like their selectivity (e.g. Van Soest, 1994), in general herbivory has several broad implications for how they naturally find and process food. The first is that because vegetation typically needs bulk-ingestion for nutrient gain, ungulates naturally spend many hours foraging. For instance, dairy cows on pasture spend nine or more hours grazing daily, and, pooling this 'prehension' with rumination, take over 72,000 bites a day (Linnane *et al.*, 2004; Newman, in press), while horses may graze for up to 16 h (e.g. Fraser and Broom, 1990). In the wild, giraffes spend 40–80% of the day browsing (Veasey *et al.*, 1996; Ginnett and Demment, 1997). In semi-natural environments, wild boars spend a quarter to a third of the day foraging and rooting (Blasetti *et al.*, 1988; Horrell, 2000); while domesticated pigs spend 22% to 28% of the day foraging, or 50% of their active time (Stolba and Wood-Gush, 1989; Buckner *et al.*, 1998).

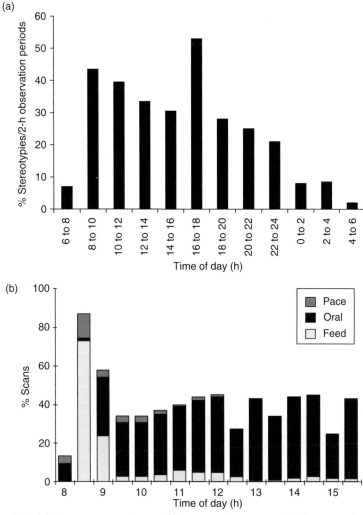

Fig. 2.1. (a) Average percentages of time spent in stereotypies by cows during 2-h observation periods across the day. Animals were fed at 06:00 and 13:30 h, and thus stereotypies increased within 2–4 h after feeding (adapted from Redbo, 1990). (b) The percentage of scans engaged in locomotor stereotypy (pacing), feeding and oral stereotypic behaviour (bar-biting, slat-chewing, wool-pulling) over the day. Data came from 30 restrictively fed lambs at 25 weeks of age. The lambs were singly housed, and received a low fibre pelleted feed at approximately 08:30 h each day and no supplementary forage. Most lambs consumed this ration within 45 min of delivery (from Cooper *et al.*, 1994). Thus oral stereotypic behaviour was relatively low pre-feeding, but pacing was relatively high.

The second implication of herbivory is that natural food is typically found in small, bite-sized portions, which may occur in clusters, e.g. one bush may be rich with leaf buds, another not. This may lead to local food-search being stimulated by ingestion: particulate food that occurs in

Box 2.1. Ungulate Ingestion and Digestion: Anatomical and Physiological Adaptations for Herbivory and their Behavioural Implications

S. LAMBTON **and** G. MASON

Ungulate digestion relies on cellulose digestion by micro-organisms in the gut. Ungulates are either pre-gastric fermenters (ruminants, e.g. cattle, sheep, goats, giraffes and camels), or post-gastric fermenters (e.g. horses, tapirs, rhinoceroses and to a lesser extent pigs). Adult ruminants are polygastric, with a three- or four-chambered stomach. In the latter species (e.g. the cow) these are the rumen, reticulum, omasum and abomasum (or true stomach), while camelids lack the omasum (e.g. Robbins, 1993; Van Soest, 1994). The rumen is the first chamber, and is a fermentation 'vat' of active bacteria, protozoa and fungi. Digesta is processed further in the reticulum, from which it is regurgitated as 'cuds'. After rumination, food passes back to the rumen for additional fermentation, before passing to the omasum for further mechanical processing, and then to the 'true stomach' or abomasum, where ruminal microorganisms are digested (e.g. Schmidt-Nielson, 1997). This type of digestive apparatus has several behavioural implications. First, non-foraging mouth movements in the form of rumination are a key part of the behavioural repertoire; e.g. occupying 6–8 h/day in cattle (Phillips, 2002); interestingly, such rumination can be accompanied by non-REM sleep (reviewed by Tobler and Schweirin, 1996). Second, the types of food selected and its intake rate affect ruminal microbial action: constraints which help shape ruminant foraging behaviour (e.g. Newman, in press). Third, because fermentation generates organic acids, ruminal pH must be controlled to protect the stomach and sustain microbial fermentation. This is largely achieved via salivation, which peaks during chewing and rumination (e.g. Meot *et al.*, 1997), the salivary bicarbonates and phosphates acting as buffers when swallowed (e.g. Sauvant *et al.*, 1999). Fourth, stomach development is itself shaped by the food ingested. In calves, for example, the digestive tract only fully develops post-weaning, the rumen not beginning to function until animals begin consuming solids (Van Soest, 1994). Thus if fed non-naturalistic foodstuffs, rumino-reticulum development is altered, e.g. zoo giraffes can show grazer-like reduced ruminal surface areas and very well-developed reticula, compared with wild, naturally browsing conspecifics (Hofmann and Matern, 1988).

Post-gastric fermenters have a simple stomach, and most mechanical processing of plant cell walls takes place in the mouth. Digestion is also initiated through chewing, by enzymes in the saliva (Pough *et al.*, 1989). Digesta then undergoes microbial fermentation in the caecum, which is enlarged to create a fermentation chamber and in horses comprises, together with the colon, around 60% of the alimentary canal (Frape, 1998). This type of digestive apparatus has two main behavioural implications: chewing the ingested food is an important part of processing, and thus these ungulates typically spend more time foraging than ruminants (e.g. Fraser and Broom, 1990) (if one excludes rumination from foraging time); and food intake rate – and passage rate – is relatively fast. Thus, in both ruminants and post-gastric fermenters, chewing-type oral movements are an important part of the behavioural repertoire.

Saliva is thus important in ungulate feeding. As well as the buffering and enzymatic functions described above, in some ungulates (notably browsers) it contains proteins that bind to plant tannins that would otherwise be detrimental (e.g. Fickel *et al.*, 1999; Clauss *et al.*, 2005). Browsers specializing in tanniferous plants also have much salivary urea recycling (Van Soest, 1994). Small wonder then that vast amounts of saliva are secreted during normal foraging: sheep may produce 6–16 l a day, and cattle, up to 100–190 l (Schmidt-Nielson, 1997). Horses also produce fairly large amounts, up to 10–12 l a day, which even in these non-ruminants helps buffer stomach acidity (Frape, 1998; reviewed by Nicol *et al.*, 2002). Finally, ungulate teeth are also adapted for herbivory, cheek teeth being high-crowned so that deep peaks and folds of enamel and the softer dentine wear differentially with use,

Continued

Box 2.1. *Continued*

forming effective grinding ridges. In some ungulates, e.g. the horse, cheek teeth are also 'open-rooted' and continuously growing (e.g. Young, 1981; Pough *et al.*, 1989), while in a few (e.g. the vicuna; Bonacic, 2005) even incisors show continual growth. Behaviourally, this means that chewing plays important roles in maintaining tooth function. Some authors even suggest that certain specialized chewing movements occur specifically to facilitate appropriate tooth wear, e.g. in sheep (Every *et al.*, 1998), although this idea is controversial (Murray and Sanson, 1998). Overall, these oral and gastrointestinal adaptations broadly mean that for ungulates, foraging movements often have functions beyond direct nutrient intake, acting also to maintain the ideal functioning of the teeth and/or gut.

patches often stimulates local search, especially if it is also rather cryptic and unlikely to flee while being searched for (Bell, 1991). Thus in wild boars, for example, stomach content analysis shows that they typically consume a lot of just one single food type at once, even though over time they eat a very diverse array of food items; and ingesting a small amount of food does promote further feeding and foraging (reviewed by Mason and Mendl, 1997; see also Horrell, 2000). To some extent, this dietary 'patchiness' may even hold for what look to us like uniform swards, because a third implication of herbivory is that ungulates are selective, responding to both specific nutrient deficits and gut functioning by carefully choosing the items they eat on the basis of fibre, sugar, mineral and/or nitrogen content (e.g. Newman, in press). Thus grazing sheep, for example, preferentially select either clover or grass at different times (Newman, in press; *cf.* Rutter *et al.*, 2004 on cattle). The fourth implication of herbivory is that ungulates have sophisticated adaptations for dealing with cellulose, tannins, silicates and other plant defences, especially specialized teeth, salivary glands and gastrointestinal tracts. These, and their behavioural implications, are reviewed in Box 2.1.

Together, it seems likely that these aspects of natural foraging are what make captive ungulates' abnormal behaviours so distinctive. Cross-species comparisons give further support to this idea (see Box 2.2). As yet, we know little about the relative roles played by ungulates' anatomical, physiological and behavioural adaptations, and indeed this may differ between species. We can, however, analyse how together these shape the likely impact of captive feeding regimes.

2.3.2. Effects of captivity on ungulate feeding, foraging and gut function

The precise feeding regime of captive ungulates varies with their natural food preferences or habits, the availability of natural foods, and economic constraints (see e.g. Henderson and Waran, 2001). In agricultural animals, it is also affected by their stage in the production cycle (e.g. pregnant sows are fed differently from lactating sows). However,

Box 2.2. Do Ungulate-like Natural Foraging Styles Lead to Ungulate-like Stereotypies in Other Animals?

G. MASON

If oral post-feeding stereotypies stem from naturally time-consuming foraging styles, from feeding on small, static, clustered food items, or even from herbivory *per se*, then they are unlikely to be unique to ungulates. We might expect them in any animal whose natural foraging behaviour has one or more of these traits, if it is fed non-naturalistically in captivity. So far, every post-feeding oral stereotypy reported outside of the ungulates does fit this pattern. Intensively farmed, trough-fed chickens thus display post-feeding spot-pecking, especially when food-deprived (reviewed by Mason and Mendl, 1997, who also review similar post-feeding spot-pecking in pigeons). Their less stereotyped, but still abnormal, 'feather-pecking' has also been reported after feeding (Blokhuis, 1986). Naturally, their wild equivalents spend much time foraging for seeds, invertebrates and vegetation (e.g. devoting 60% of their activity to ground-pecking) (reviewed by Mason and Mendl, 1997). Post-feeding oral stereotypies like paw-sucking and tongue-flicking are also fairly common in captive giant pandas, Asiatic black bears and sun bears, especially if fed non-naturalistic meals based on rice, bread or milk (Vickery and Mason, 2004; Swaisgood *et al.*, a,b in press). In the wild, giant pandas naturally spend a large proportion of their time (e.g. 14 h a day) seeking and eating shoots and leaves, while Asiatic black bears and sun bears do likewise for fruits and other vegetation, sun bears additionally consume small invertebrates (e.g. reviewed by Schaller *et al.*, 1989; Vickery and Mason, 2004). Our last case is the walrus. For decades, captive walruses have been reported performing repetitive oral behaviours like flipper-sucking, or repetitively rooting and sucking at the concrete of their pools (sometimes wearing their tusks down to stumps; e.g. Coates, 1962; Hagenbeck, 1962; Kastelein and Wiepkema, 1989; Kastelein *et al.*, 1991; see image on website); and a recent study reveals that these oral behaviours, too, peak post-feeding (D. Reiss, personal communication, New Orleans, 2004; Reiss *et al.*, in preparation; see figure).

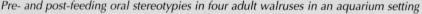

Pre- and post-feeding oral stereotypies in four adult walruses in an aquarium setting

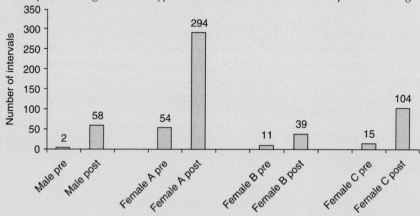

Although far from herbivorous, sure enough the walrus's natural foraging style is ungulate-like, with animals 'grazing' on patches of marine invertebrates rooted up from the seabed. This takes several hours a day, with many hundreds of small items being processed and eaten (e.g. Fisher and Stewart, 1997; Born *et al.*, 2003), and contrasts greatly with the rapidly eaten fish-based meals typical of captivity (e.g. Kastelein *et al.*, 1991).

Continued

Box 2.2. *Continued*

Adding these six species to our 'ungulate foraging story' is of course a long way from properly testing the hypothesis (*cf.* Box 3.2, Chapter 3 on cross-species comparisons), but it is intriguing. Further data are now needed (perhaps from manatees? aardwolves? rodents? the marsupials?) to see how consistently such stereotypies really do relate to natural foraging style, and to pinpoint what aspects of feeding motivation, behavioural time-budgets or even gastro-intestinal physiology are the specific predictors.

---

feeding and foraging in captivity typically differ in three ways from naturalistic situations.

2.3.2.1. *Dietary preferences and/or needs may be unfulfilled, leaving the animal motivated to feed*

Captivity often constrains the amount or composition of food that can be ingested. An extreme case is the pregnant sow, which is routinely food-restricted (to control weight gain, and maximize food intake when lactating; e.g. Cole, 1982; Mroz *et al.*, 1986). These animals are usually given just 2.5 kg of food daily: half or even a third of what they would eat *ad libitum* (Ramonet *et al.*, 1999; Bergeron *et al.*, 2000), resulting in prolonged high levels of frustrated feeding motivation (e.g. Lawrence and Illius, 1989). Brief periods of food restriction are also imposed on other captive ungulates (e.g. horses before a race; e.g. Murray, 1999; Merial, 2004), constraining natural meal patterning. Captivity often thwarts specific motivations for particular food-stuffs too. For example, both sheep and cattle select different diets (e.g. ones containing more fibre) during sub-acute ruminal acidosis (e.g. Keunen *et al.*, 2002), while free-ranging horses voluntarily select soils high in copper and iron for geophagia (e.g. McGreevy *et al.*, 2001b), and pigs with a choice select dietary protein levels in a state-dependent manner (reviewed by Lawrence *et al.*, 1993). Intensive housing conditions, however, generally prevent animals from expressing or satisfying such preferences.

2.3.2.2. *Fewer behavioural demands are made on the animal, affecting the foraging time-budget*

On farms and in zoos, homogeneous foodstuffs such as hay, browse or man-made diets (e.g. milled, low-fibre mash or pellets) are typically presented directly to the animal, in a single manger or trough. Thus food-search, and even consummatory behaviours like chewing, take a fraction of the time they would naturally. This effect is even more marked if food is restricted in quantity. Thus pregnant sows consume their daily meal of concentrate (a low-fibre food made of grain and protein-rich ingredients) in under 20 min (Ramonet *et al.*, 1999, 2000a); while concentrate-fed stabled horses may spend just 2 h feeding (Kiley-Worthington, 1983), or even as little as 20–30 minutes (Henderson and Waran, 2001). For ruminants, the situation differs further in that less time is also spent on rumination when fed concentrates, compared with diets high in natural forage (e.g. Abijaoude *et al.*, 2000;

Lindström and Redbo, 2000; Baxter and Plowman, 2001). This could be important because natural foraging activities can be intrinsically reinforcing, regardless of nutrient gain (e.g. Wood-Gush and Beilharz, 1983; Hutson and Haskell, 1990; Mason *et al.*, 2001).

2.3.2.3. Captive diets may detrimentally affect gastrointestinal function

Low fibre, high carbohydrate concentrate diets can cause gastrointestinal acidity, and thence potentially mucosal damage (especially in monogastrics' stomachs) and/or acidosis (especially in ruminants, where ruminal contents become overly acidic, impairing proper fermentation). Sub-acute acidosis seems very prevalent (e.g. 20% of dairy cows; Oetzel, 2003), and such processes are well understood for ruminants. Here, dietary concentrates decrease ruminal pH by increasing fermentation (e.g. Sauvant *et al.*, 1999; Schwartzkopf-Genswein *et al.*, 2003) and reducing chewing and rumination (e.g. Abijaoude *et al.*, 2000) thence decreasing salivation (e.g. in cattle, to around two-thirds the levels secreted when grazing; Bauman *et al.*, 1971; see also Hibbard *et al.*, 1995; Meot *et al.*, 1997). Processed, low-fibre diets also cause gastrointestinal acidity in horses and pigs. Thus in horses, grain feeding can cause hindgut acidosis (e.g. Rowe *et al.*, 1994; Johnson *et al.*, 1998), while both concentrate feeding (Rowe *et al.*, 1994; Murray, 1999) and periods of food deprivation (Murray and Eichorn, 1996) increase gastric acid secretion and can cause foregut ulcers (e.g. reviewed by Nicol, 2000; Nicol *et al.*, 2002). Such lesions are very prevalent: in some breed/management groups (e.g. racehorses) they occur in the majority of individuals (e.g. reviewed by Murray, 1999; Merial, 2004). Likewise, stomach ulceration is common in commercially kept pigs; for example, in sows, O'Sullivan *et al.* (1996) reported a mucosal lesion prevalence of 60%, and Hessing *et al.* (1992) 63%; whereas in young slaughter pigs, Ayles *et al.* (1999) report gastric ulceration in 32–100% of animals, and Hessing *et al.* (1992) 36%. Again these problems are associated with a lack of fibre, small dietary particle sizes, pellet feeding and restricted feeding (e.g. Wondra *et al.*, 1995; and reviewed by Blood and Radostits, 1989).

2.4. What Aspects of Captive Feeding Regimes Cause Oral Stereotypic Behaviour?

In the following sections, we analyse which of these aspects of captive diets underlie abnormal behaviour. We begin with a caveat: many data come from non-experimental studies, e.g. those reliant on cross-site comparisons; and even when they do come from experiments, the research goals were often practical (e.g. aiming to reduce stereotypy, or improve welfare) rather than hypothesis-testing. Furthermore, much of this work predates recent suggestions about the role of gut dysfunction. Thus it is often unclear exactly *how* effects are mediated. Consequently, we start by simply illustrating the effects of the typical high-concentrate, low-bulk diets on ungulate oral abnormal behaviour (see Section 2.4.1). We then lead on with three sections that try and tease apart the roles of: dietary deficits that leave the animal with unfulfilled motivations to ingest (*cf.* Section 2.3.2.1, above); altered time-budgets, especially, reduced foraging and rumination times (*cf.* Section 2.3.2.2) and gastrointestinal dysfunction (*cf.* Section 2.3.2.3). We do this partly by analysing the 'high-fibre diet' research in more detail, but also by drawing on further evidence from other types of manipulation.

2.4.1. The effects of high concentrate, low-fibre diets

Many authors have shown that ungulate stereotypies increase with the proportion of concentrated food in the diet. For example, lambs fed a concentrate-based diet perform more bar-biting, licking and wool-eating, than those receiving lucerne (Cooper *et al.*, 1995); in heifers, decreasing the proportion of forage, and increasing concentrates (while maintaining the energy content constant) increases the frequency of tongue-rolling, bar-biting and chain-chewing (Redbo and Nordblad, 1997); and in giraffes, feeding more fibrous forms of hay and/or adding forage (Koene, 1999; Bashaw *et al.*, 2001a; Baxter and Plowman, 2001) reduces tongue-playing.

   Turning to non-ruminants, surveys of horses show that feeding forage in large or frequent amounts, rather than more concentrated diets, is associated with a reduced prevalence of abnormal behaviours including crib-biting and wood-chewing (McGreevy *et al.*, 1995a; Redbo *et al.*, 1998). A more recent study also reveals that foals receiving concentrates are four times more likely to develop crib-biting than other foals, while the feeding of hay replacers (fermented forages that are energy dense, so fed in relatively low quantities) instead of bulkier, higher-fibre hay, significantly increases wood-chewing (Waters *et al.*, 2002). Further-more, the behaviours emerge soon after weaning, a process typically involving a switch to concentrate foods, with crib-biting initiated at a median age of 4.6 months, and wood-chewing, 7 months (see Fig. 2.2).

   In experimental studies, Dodman *et al.* (1987) also found that feeding horses grain or sweetened grain rations increased crib-biting, whereas lucerne pelleted hay had no such effect on the behaviour.

   Pigs, especially pregnant sows, have received even more attention. High-fibre diets such as those based on oat hulls reduce chain-manipulating (Robert *et al.*, 1993, 1997, 2002); pre- and post-feeding stereotypies like sham-chewing and head-waving (Ramonet *et al.*, 1999); and post-feeding vacuum-chewing and the stereotypic rubbing or biting of stall fittings (Robert *et al.*, 2002). Post-feeding chain-chewing was reduced by an oat bran diet even if lower in energy than a concentrate-based control diet (though pre-feeding chain-directed stereotypies were only reduced by an oat bran, full calorie diet) (Robert *et al.*, 1997). Furthermore, feeding gilts a restricted diet with sugarbeet pulp not only

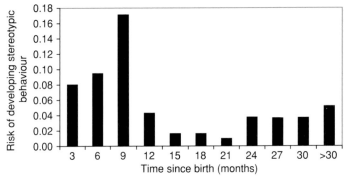

Fig. 2.2. The risk of developing a new form of stereotypic behaviour in horses at different ages. Most foals were weaned between 4 and 6 months of age (adapted from Waters *et al.*, 2002). On a finer timescale than can be seen from the figure, weaning (with its associated husbandry changes, including dietary ones), occurred at 15–35 weeks, and the emergence of new stereotypies peaked at 40 weeks. Note that the fall in risk evident from 12 months onwards thus does not indicate a decrease in the performance of existing stereotypies, merely a decline in the emergence of new ones.

reduced the incidence of post-feeding oral stereotypic licking, bar-biting and sham-chewing, but also rendered these behaviours less fixed in form (Brouns *et al.*, 1994). Figure 2.3 gives some illustrative data. Next, we move on to discuss the possible reasons as to why these low-fibre diets – and other types of dietary divergence from naturalistic foraging – promote stereotypic behaviour.

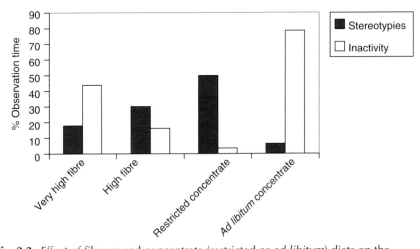

Fig. 2.3. Effect of fibrous and concentrate (restricted or *ad libitum*) diets on the percentage of time spent in stereotypic behaviours and inactive by pregnant sows in the 2-h post-feeding period. The crude fibre content was 23%, 18.2% and 5.3% for Very high fibre, High fibre and Control diets, respectively. Behaviours are expressed in median percentage of time remaining once feeding time has been removed (adapted from Bergeron *et al.*, 2000).

2.4.2. What are the roles of dietary deficits, or unfulfilled preferences, that leave animals motivated to feed?

Concentrate-based diets may fail to induce satiety, even when they meet nutritional needs, due to insufficient gut fill; thus animals fed such diets might remain motivated to eat. It is also possible that in these examples, and elsewhere too, specific appetites or food preferences are not met, playing a further role in stereotypy. Yet other cases still suggest that simple calorific deficits are also important.

2.4.2.1. Unfulfilled feeding motivations due to low satiety in low-fibre fed animals

Nicol (1999) hypothesized that in horses, hay reduces the risk of developing abnormal oral behaviour by reducing the feeding motivations via gut fill. Certainly for pigs, we know that stereotypy-reducing high-fibre diets promote short-term satiety. This is due to stomach distension (e.g. Lepionka *et al.*, 1997) plus altered nutrient absorption rates and post-meal blood concentrations of glucose, insulin and acetate (Rushen *et al.*, 1999; Ramonet *et al.*, 2000a). This satiety is often manifest in fewer postural changes, increased resting time around mealtime and reduced rooting-/foraging-like behaviours, e.g. to straw (e.g. Brouns *et al.*, 1994; Whittaker *et al.*, 1998). A negative relationship between diet bulk and feeding motivation in the post-feeding period has also been measured (Day *et al.*, 1996). In one study, high fibre fed sows even put on more weight, even though calorie intake was the same as on a control, concentrated, diet, perhaps because animals were less active (Ramonet *et al.*, 1999; though *cf.* Whittaker *et al.*, 1998). The stereotypy-reducing sugarbeet pulp, for example, causes a rapid satiety still present 2 h after the meal (Brouns *et al.*, 1997). Using operant conditioning tests for feeding motivation, Robert *et al.* (2002) also reported a lower feeding motivation both before the morning meal and after the afternoon meal, in gilts fed oat hull and lucerne diets compared to concentrate. Despite the general positive effects of high-fibre diets, however, some authors failed to find that they reduced feeding motivations (Bergeron *et al.*, 2000; Ramonet *et al.*, 2000b). Apparent discrepancies between studies may be explained by differences in the age of animals, methods of assessing hunger or the timing of measurement. For instance, reduced feeding motivations with high-fibre diets are often present only in the few hours after eating (e.g. Day *et al.*, 1996; Robert *et al.*, 1997; though *cf.* Robert *et al.*, 2002). This could explain why some high-fibre diets only reduce post-feeding stereotypies but not those that appear before the next meal.

Overall, the 'dietary fibre' studies of pigs thus suggest that hunger-reduction does correlate with stereotypy-reduction, but that fibre alone is often insufficient to achieve this around the clock: in the longer term, it seems that energy level is more important. They also suggest that motivations to ingest more nutrients are indeed important in stereotypy. Thus despite their high gut fill, high-fibre diets are less effective at reducing oral behaviour than are conventional diets served *ad libitum* (Bergeron *et al.*, 2000; see Fig. 2.3). In Section 2.4.2.2, we therefore ask – are calories the key?

2.4.2.2. Unfulfilled feeding motivations due to energy-restriction

The most striking evidence for a role of energy-restriction comes from pregnant sows: energy-deficits play a major role in these animals' oral stereotypies (e.g. Appleby and Lawrence, 1987; Terlouw *et al.*, 1991a). Thus their stereotypies are usually greatly reduced when their daily food allowance of concentrate is increased (Appleby and Lawrence, 1987; Terlouw and Lawrence, 1993; Bergeron *et al.*, 2000). Furthermore, providing sows with 1.7 times as much digestible energy – despite no more dietary bulk – significantly reduced vacuum-chewing (Bergeron and Gonyou, 1997). Restricted feeding also increases stereotypies even when time spent foraging is statistically corrected for, suggesting that hunger, not just feeding time, really is important for sows (Spoolder *et al.*, 1995). The role of food restriction in stereotypies has also been investigated in ruminants. The restricted feeding of a total mixed ration (i.e. mix of concentrate and forage) compared to *ad libitum* feeding, increased the level and prevalence of oral stereotypies in dairy cattle (Redbo *et al.*, 1996; Lindström and Redbo, 2000). Similarly, food-restricted lambs performed more slat-chewing, wool-biting and repetitive-licking (Cooper *et al.*, 1994).

2.4.2.3. Unfulfilled feeding preferences due to specific dietary deficits

Specific deficits are also implicated in some abnormal behaviour. Bar-biting and slat-chewing by lambs was specifically increased by protein restriction (e.g. Whybrow *et al.*, 1995). Anecdotally, deficits of copper, manganese or cobalt can likewise induce tongue-rolling in cattle (Sambraus, 1985). Salt blocks also anecdotally reduce crib-biting in horses (Fraser and Broom, 1990), while in dairy cows, experimentally increasing the salt (NaCl) levels in concentrate diets reduced oral stereotypies (Phillips *et al.*, 1999).

2.4.2.4. Frustrated ingestion in animals fed small amounts of motivating food

Feeding motivations can be stimulated not just by baseline internal state but also by external stimuli associated with food delivery. Although infants are not the focus of our review, this issue has been best explored in calves, where the ingestion of a small amount of milk enhances suckling motivations, and in the absence of a teat, promotes object-sucking (e.g. De Passillé *et al.*, 1993). Thus Dodman *et al.* (1987) and Gillham *et al.* (1994) proposed that sweetened grain triggered oral stereotypies in horses because it is so highly palatable (although their proposed mechanism for the link was molecular rather than motivational). A related idea was proposed by Lawrence and Terlouw (1993), who suggested that the small amount of food offered to pregnant sows is not only insufficient to reduce hunger, but actually *increases* short-term feeding motivations.

This hypothesis was inspired by findings that feeding facilitates itself by positive feedback (Wiepkema, 1971), and that food-ingestion specifically prompts stereotypies (see Terlouw *et al.*, 1993). Interestingly, as low-fibre diets are less palatable to sows (Bergeron *et al.*, 2002), this could be an additional reason why they reduce post-feeding stereotypies.

2.4.2.5. Summary and potential explanations

Nutrient deficits and feeding motivations clearly potentiate ungulate oral stereotypies. This has led to hypotheses that the behaviours represent state-dependent foraging attempts, driven by dietary deficiency and/or insufficient gut fill (e.g. Terlouw *et al.*, 1993; Cooper *et al.*, 1994; Whybrow *et al.*, 1995; Nicol, 2000; McBride and Cuddeford, 2001).

But why, then, should such behaviours be sustained hour after hour, day after day? In some instances, the behaviour may actually redress underlying deficits and so be reinforced. For example, horses' wood-chewing could be a functional response to a lack of dietary fibre (Redbo *et al.*, 1998); and the chewing of urine-soaked wood slats by sheep may even be a way of gaining nitrogenous urea when protein-deficient (e.g. Whybrow *et al.*, 1995). This last could perhaps also explain wool-chewing by sheep, since soiled wool from animals' rear ends is preferred (Sambraus, 1985), although the rectifying of salt deficiencies could be another possibility. Alternatively, it may be that it is evolutionarily adaptive to food-search until successful (Mason, in press), and that such responses are relatively hard-wired and resistant to extinction (though *cf.* e.g. Haskell *et al.*, 1996). Or such persistence may instead result from some effect of the barren environment, or abnormal early-rearing conditions, as we discuss in Section 2.6.

However, the importance of nutrient deficits in all ungulate oral stereotypic behaviour is uncertain. After all, most stereotyping ungulates are not food-restricted, so this seems unlikely to be a general explanation. Furthermore, in sows, high-fibre diets can reduce stereotypies even if they do not reduce feeding motivation any more than control diets (Bergeron *et al.*, 2000). This suggests that other factors are important too. Restrictive diets and low-fibre diets do have other features in common that represent alternative causal factors for stereotypy – they make food take less time and effort to find and process; and they can also lead to gastrointestinal dysfunction. We therefore consider these next.

2.4.3. What is the role of decreased foraging or ruminating time?

The time spent foraging and/or ruminating falls considerably when concentrated diets are offered. This may frustrate some need to perform oral behaviours and/or give ungulates 'spare time' to fill with stereotypies. But how important for stereotypies are these changes in the behavioural time-budget?

2.4.3.1. Increased foraging/rumination time and high-fibre diets

Many authors have suggested that fibrous diets act to reduce stereotypies through encouraging more naturalistic oral behaviour (e.g. Rushen *et al.*, 1993). Thus in many studies of sows, high-fibre diets double or more the time spent feeding (e.g. Robert *et al.*, 1993; Brouns *et al.*, 1994; Ramonet *et al.*, 1999), and using multiple regression, Robert *et al.* (1997) found that this increased feeding time accounted for much of the differences in stereotypy level between diets. Thus across several diets identical in calories and major nutrient levels, but different in fibre level, low chewing time *per se* emerged as the key statistical predictor of stereotypic chain-chewing post-feeding. Similarly, in giraffes, feeding hay instead of lucerne prolonged feeding time in one study (as well as reducing tongue-playing; e.g. Koene, 1999), and prolonged the time spent ruminating (Baxter and Plowman, 2001) in another, these last authors hypothesizing that opportunities to ruminate are specifically important for stereotypy-reduction. Likewise, in cattle, increasing dietary forage results in longer feeding duration, along with reduced stereotypies (Redbo and Nordblad, 1997).

Studies where foraging/rumination times are manipulated via diet quality are clearly rather hard to interpret, however, so let us look at other types of study too.

2.4.3.2. Effect of providing straw bedding or other foraging opportunities

Straw has low nutritional value and is often used as bedding. However, it can serve as a foraging substrate, since animals may manipulate, chew and even consume some of it. Correspondingly, straw also has an effect on oral stereotypies. For example, in sows, experimentally providing straw bedding or loose straw also reduces the incidence of oral stereotyped activities (e.g. Fraser, 1975). Lambton and Mason (in preparation) also found that amongst barn-housed beef cattle, individuals with tongue-playing stereotypies spent the least time manipulating their straw bedding, even though they had the same access to it as did non-stereotyping individuals. The amount of straw actually consumed was not reported in these studies, so its potential effect on gut fill cannot be assessed. However, adding straw directly to the diet itself does *not* decrease stereotypies (Fraser, 1975) or feeding motivation (Lawrence *et al.*, 1989). Thus it seems that straw as a substrate on the ground is important, perhaps because it allows naturalistic foraging (e.g. Spoolder *et al.*, 1995; Whittaker *et al.*, 1998, 1999). Conversely, reducing naturalistic foraging opportunities can induce stereotypy-like behaviour, seemingly independent of nutrient-intake. Thus when the natural rooting and stone-chewing behaviour of food-restricted outdoor sows is impaired with nose-rings which make pressing the snout against the ground painful, animals instead perform more grass-chewing, and straw- or vacuum-chewing (Horrell *et al.*, 2001).

2.4.3.3. Effects of changing the foraging behaviours required to ingest food

Other research efforts have been made to specifically manipulate time spent foraging and feeding, without adding straw, or changing the quality and/or quantity of food in any way. For example, when mesh feeders were used to force giraffes to work harder to obtain their food, this successfully decreased stereotypic licking from 13% to 2% (Bashaw *et al.*, 2001b; see also Bashaw *et al.*, 2001a). Similarly, sows fed a conventional diet in a mash form instead of pellets, which increased their feeding time, showed decreased chain-chewing after eating (Bergeron *et al.*, 2002). This latter manipulation did not reduce feeding motivation in the post-feeding period (Brouns *et al.*, 1997), perhaps unsurprisingly since it was no more calorific or bulky, and yet it still clearly had an effect on stereotypic behaviour. Finally, Lindström and Redbo (2000) used invasive techniques on cattle to dissociate the behavioural components of feeding from the nutritional consequences. They found that a 50% reduction in food allowance increased cow stereotypies (see Section 2.4.2). However, they also found that this effect vanished if the animals either received compensatory rumen content (delivered direct to the rumen), enabling a high rumination level; or if they received a high food allowance, and thus could have a long feeding time, even if their rumen content was then maintained artificially low. Together these results suggest a generally beneficial effect of oral manipulation *per se*, through feeding and/or ruminating, on stereotypy.

However, increasing the foraging behaviours required to ingest food does not always reduce stereotypic behaviour. For example, a chain-based device inside the trough that increased the time food-restricted sows spent foraging, did not have great effects on their stereotypy: it decreased post-feeding vacuum-chewing, but chain-chewing and manipulation remained high (Bergeron and Gonyou, 1997).

2.4.3.4. Summary and possible explanations

These studies suggest that expressing foraging behaviour, particularly in a complex and variable way, can *per se* reduce stereotypies, regardless of nutrient intake. This has led to hypotheses that ungulates cannot or will not completely abandon naturalistic levels of foraging, even when captivity renders this redundant. If correct, this could indicate that complete flexibility in ungulate foraging time has not been selected for by evolution, leaving ungulates unable to reduce foraging behaviour to the low level required by captivity (Mason, in press), especially if concomitantly nutrient-restricted. Alternatively, defending a certain minimum level of daily foraging could have brought with it evolutionary benefits (independent of nutrient intake) such as information gain, appropriate tooth wear and/or maintaining digestive tract function (*cf.* Box 2.1). Indeed the potential role of these last factors as *proximate* drivers of stereotypy, not just ultimate ones, has recently been suggested, as we discuss below. Furthermore, because stereotypies were typically reduced but not abolished by the treatments described above, this further suggests that issues other than foraging time do need to be considered.

2.4.4. What is the role of gastrointestinal dysfunction?

2.4.4.1. Individual differences in gastrointestinal pH and lesions, and their relationships with stereotypy

In low fibre-fed animals, individual differences in gastrointestinal acidity may predict individual differences in stereotypy. Relationships between concentrate feeding, hindgut acidity and oral activities such as grasping and wood-chewing have long been observed in horses (Willard *et al.*, 1977; Johnson *et al.*, 1998), and tooth-grinding and crib-biting have also been associated, at least anecdotally, with gastritis in these animals (Rebhun *et al.*, 1982; Blood and Radostits, 1989). A recent study by Nicol *et al.* (2002) investigated these links in more detail. Foals that had recently started to crib-bite were compared with non-stereotypic foals, their stomachs being examined via video endoscopy. The crib-biters had significantly more inflamed, dry and ulcerated stomachs, along with lower faecal pH. Similar associations have been suggested for pigs: in concentrate-fed pregnant sows, Marchant-Forde and Pajor (2003) report that a weak link between oral abnormal behaviour and gastric ulceration has been established, probably based on the findings of Dybkjær *et al.* (1994).

A different picture seems to emerge in cattle, however. Wiepkema *et al.* (1987) found that 67% of veal calves bucket-fed on milk replacer showed abomasal ulcers, the scars of past ulcers, or erosions (NB at this stage calves' stomachs are not fully developed, and thus the abomasum or 'true stomach' is their only functioning chamber). However, of those animals which developed tongue-playing, none had ulcers or scars, while those animals that did not, all had ulcers or scars. The same was not true for stereotypic biting of the crate, nor for erosions which appeared in different areas of the omasum. Canali *et al.* (2001) also found that veal calves with more abnormal oral behaviour in total had fewer abomasal ulcers, although this was not true if only strict stereotypies were looked at. Furthermore, in adult cattle, Sato *et al.* (1992) found that tongue-rolling was more common in individuals which later, at slaughter, proved not to be suffering from internal organs lesions, such as enteritis (gut inflammation) or hepatitis (liver inflammation). This finding could reflect a low incidence of ruminal acidosis, because this condition can have a range of deleterious effects throughout the body, including liver abscesses (e.g. see references cited by Keunen *et al.*, 2002, and by Hanstock *et al.*, 2004).

2.4.4.2. Individual differences in gastrointestinal motility

Crib-biting has been associated with altered gut transit time in the horse (McGreevy and Nicol, 1998a; McGreevy *et al.*, 2001a). Thus longer total gut transit times were found in crib-biters compared to control horses, although oro-caecal transit times did not differ significantly (McGreevy *et al.*, 2001a). This shows that crib-biters have reduced hindgut (but not foregut) motility. Their relative hindgut stasis suggests that the oro-caecal digestion of crib-biters is less efficient than that of non-crib-biters, perhaps because they have poor mastication and/or emulsification of food in the foregut (meaning that fibre has to be retained in their hindguts for longer), or because they have some imbalance in hindgut flora (e.g. as a result of acidosis) (Nicol, 1999). Indeed, as we will see below, the crib-biting of these individuals may actually help to reduce their gut transit times from levels which otherwise would be even slower (with the cribbing behaviour thus seeming more efficacious for foregut motility than for hindgut).

2.4.4.3. Experimental alterations of gut acidity: do these affect stereotypy?

Four studies have investigated whether altering gut pH alters oral stereotypy. Johnson *et al.* (1998) focused on hindgut acidosis in horses. They found that antibiotics controlling lactate-producing bacteria (and thence increasing faecal pH) do reduce abnormal oral behaviours (though normal eating was reduced too, making interpretation a little tricky). Turning to foregut acidity, crib-biting in horses was also reduced by oral antacids (Mills and MacLeod, 2002; Nicol *et al.*, 2002). Thus in the latter study, crib-biting foals were allocated to a control or an antacid diet for 3 months. Crib-biting foals receiving the antacid diet tended to reduce their cribbiting duration to a greater extent than foals on the control diet; and foals showing the greatest reduction in ulcer severity score with the diet also tended to show the greatest reductions in crib-biting. Finally, working with pregnant sows, Marchant-Forde and Pajor (2003) have preliminary findings suggesting that adding a bicarbonate buffer to the diet may reduce barbiting, though not all stereotypies were affected.

2.4.4.4. Summary and possible explanations

The results above have led to suggestions that stereotypy is not a response to nutrient deficits or reduced foraging time *per se*, but instead to their gastrointestinal consequences. Many of these studies are merely correlational, but the more recent experimental work does indicate that gastrointestinal acidity can play a causal role in stereotypies. Why might this be so? One possible explanation is that gastrointestinal discomfort exacerbates stereotypies by being stressful (see Chapter 8, this volume), but another is that stereotypies are an attempt to alleviate such problems via the production of buffering saliva (see Box 2.1). This idea was first suggested by Wiepkema *et al.* (1987) for calves, and broadened to horses by Nicol (1999). The apparent differences between horses and calves/adult cattle – i.e. that oral stereotypic behaviour seems to correlate positively with problems in the former, but negatively in the latter – may be because cattle produce enormous quantities of saliva (see Box 2.1): thus perhaps saliva generation has more effective results in bovids. Intriguingly, amongst crib-biting foals, those without ulcers had been cribbing for longer than those with gastric lesions (Nicol *et al.*, 2002), which would be consistent with beneficial consequences. This 'salivation hypothesis' could account for some of the discrepancies in the fibre/deficit/foraging time studies reviewed earlier, with gastrointestinal effects being the missing explanatory variable. However, the idea clearly still needs to be properly experimentally tested. Furthermore, the aetiology of ulcers needs to be more fully understood, since in calves, for example, the lack of dietary fibre does not seem to play a causal role (indeed if anything the opposite is true for these young animals with their as yet undeveloped rumens; e.g. Mattiello _et al._, 2002) and stress may be the more important factor (e.g. Dybkjær _et al._, 1994).

2.5. The Biological Significance of Oral Stereotypic Behaviour: Is It Functional?

Above we have seen how ungulate oral stereotypic behaviour may have beneficial consequences (e.g. via nutrient ingestion or saliva generation), and other researchers have further suggested that it may increase feelings of satiety (e.g. Robert _et al._, 1993 on polydipsia) and/or be generally calming (e.g. Rushen, 1984). The idea that ungulate oral stereotypic behaviour has some benefits has been supported by two further types of study, looking at within-individual changes during stereotypy-performance, or at the effects of stereotypy-prevention. (Other research has utilized individual difference within populations to compare stereotypers with non-stereotypers, but it has yielded confusing results, not least as such cross-sectional studies cannot distinguish individual differences _predisposing to_ stereotypy from those _resulting from_ stereotypy.)

Intriguingly in very young calves, during the performance of post-feeding non-nutritive sucking directed to objects like artificial teats, increases in plasma insulin and cholecystokinin are seen, which are thought to aid digestion (de Passillé _et al._, 1993). Unfortunately, no study has looked at these hormones, or at any other gastrointestinal changes during oral stereotypy in adult ungulates, but variables related to stress have been measured. In horses, plasma cortisol levels are lower after a bout of crib-biting than before (McBride and Cuddeford, 2001), and heart rates also decrease during these bouts (Lebelt _et al._, 1998; Minero _et al._, 1999). Similar analyses show that in tethered gilts, switches from non-stereotyped to stereotypic behaviour are likewise accompanied by decreases in heart rate (and vice versa) (Schouten _et al._, 2000); and the same seems true for tongue-playing heifers (Seo _et al._, 1998). These data are correlational rather than indicating cause and effect, but they are nevertheless intriguing.

Stereotypy-prevention may be attempted either to abolish an undesired behaviour (see Box 2.3), or to collect research data, and when abolition is successful, consequences sometimes ensue. For example, in the calves mentioned above, reducing their non-nutritive sucking by removing a rubber teat resulted in a decrease in their post-meal hormone release (de Passillé _et al._, 1993). In sows, in contrast, removing a chain that is stereotypically chewed did not increase heart rate or cortisol (Schouten _et al._, 1991; Terlouw _et al._, 1991b), but interpretation is hard here because the subjects did develop alternative oral behaviours, e.g. drinker-manipulation. Preventing horses crib-biting has had mixed stress physiology effects, some studies finding

Box 2.3. Is it Ethical to Physically Prevent Horses Performing Oral Stereotypies?

F.O. Ödberg

Oral stereotypies are unpopular with horse owners. They can cause incisor wear; there are beliefs – though ill founded (McGreevy *et al.*, 1995c) and based solely on correlations (e.g. Archer *et al.*, 2004; Hillyer *et al.*, 2002; Traub-Dargatz *et al.*, 2001) – that they cause colic; and overall, they can reduce a horse's commercial value (McBride and Long, 2001; Mills and McDonnell, 2005). People therefore often physically try to prevent crib-biting and wind-sucking. A horse may be fitted with a neck-strap that inflicts pressure or pain during stereotypy, an electric collar that delivers a shock during the behaviour, or a muzzle that prevents biting on to hard surfaces. Horses can also be discouraged from resting their teeth on a surface by placing sharp objects, electric wires and/or unpleasant-tasting substances there. Additionally, there are surgical approaches such as buccostomy (the creation of buccal fistulae), and various myectomies (i.e. the sectioning of specific muscles to stop the motor pattern, e.g. 'Forsell's operation').

There are several reasons to be concerned about such measures. One study (McBride and Cuddeford, 2001) showed that 'crib-straps' cause stress to both windsuckers and normal controls. Electric collars are inherently painful; and automatically triggered ones can react to non-stereotypic behaviours as well as stereotypies, thence potentially inducing learned help-lessness. Furthermore, physically preventing stereotypies could make things worse, if these behaviours actually help the animal. For instance, if oral stereotypies increase stomach pH through salivation, then surely they should not be prevented. Thus in some cases, at least, there does seem to be a decrease in arousal linked with wind-sucking, and an increase when performance is thwarted (e.g. McBride and Cuddeford, 2001; see also other studies discussed in this chapter). McGreevy and Nicol (1998a) did not find such effects, but their horses were moved from the home stable to an experimental one. Mills (personal communication, Lincoln, 2005) also found an increase in heart rate when deterrent bars were placed in the stables of weavers. Even if the link found in foals between crib-biting and gastric acidity is merely correlational (e.g. nervous individuals develop both stereotypies and gastritis/ulcers), or causal in the other direction (e.g. ulcers induce discomfort), which somehow – perhaps via the mechanisms discussed in Chapter 8 – enhances stereotypies, it still seems contra-indicated to merely prevent the stereotypy physically. As this chapter argues, ungulate oral stereotypies probably indicate thwarted foraging, and merely abolishing the symptoms does not cure this underlying problem.

Despite such concerns, when I screened seven reports on surgical responses to oral stereotypy published since 1990, all evaluated success solely or mainly by the degree of stereotypy-inhibition, and none used measurable welfare or stress parameters (Hakansson *et al.*, 1992; De Mello Nicoletti *et al.*, 1996; Jansson, 2000; Delacalle *et al.*, 2002; Fjeldborg, 1993). Only one briefly mentions that stereotypy elimination may increase stress (Schofield and Mulville, 1998), and only one enquired whether aspects of horse health improved (Brouckaert *et al.*, 2002). This probably reflects a medical education that, unfortunately, tends to focus on treating symptoms instead of understanding the underlying processes (though see Chapter 10, this volume for a more holistic veterinary view). Further, objective work is therefore needed to compare the health, condition, feed intake rates and stress levels of horses exposed to such techniques (using blind observers, and with appropriate controls such as normal horses and sham-operated windsuckers). In the interim, the ethics of such approaches remain highly questionable, especially when fundamental alternatives exist, namely improving husbandry.

nothing (e.g. McGreevy and Nicol, 1998a), others finding effects of the manipulations *per se* (e.g. cribbing collars) regardless of whether the stereo-typy was prevented (McBride and Cuddeford, 2001). However, when crib-bing is prevented, it is performed at higher levels the following day once collars are removed (McGreevy and Nicol, 1998b), a 'rebound' consistent with motivational effects. Crib-biters also eat more when deprived of crib-bing, and furthermore, their slow gut transits are reduced further (especially oro-caecal motility) if they are deprived of the opportunity to both crib-bite *and* eat hay (McGreevy and Nicol, 1998a; McGreevy *et al.*, 2001a). Relatively normal oro-caecal transit times in these animals thus seem to depend on them being able to eat fibrous food or to crib-bite.

Overall, more work is clearly needed here, but these intriguing find-ings could help explain why these activities are so time-consuming and persistent day after day. They could also perhaps explain why attempts to prevent stereotypy sometimes fail, e.g. horses may persevere with crib-biting despite preventative collars or surgery (reviewed e.g. McGreevy and Nicol, 1998a,b), while in giraffes, an attempt to reduce fence-licking by coating it with bitter substances just shifted the behaviour to new locations (Tarou *et al.*, 2003). Most importantly, they also raise concerns about the physical prevention of stereotypies that is routine in some stables (see Box 2.3) – partly since such approaches ignore the underlying problems, but partly also since they could decrease animals' welfare yet further, if these behaviours do indeed have beneficial consequences.

2.6. Other Factors Associated with Stereotypies in Captive Ungulates: Barren Environments and Early Weaning

So far we have discussed oral stereotypic behaviour as though simply strange-looking manifestations of adaptive foraging: the products of pla-cing normal animals in abnormal foraging environments. But are other aspects of husbandry important too?

2.6.1. A role for early weaning?

Agricultural ungulates are often removed from their mothers long before natural weaning age. For example, natural weaning age in pigs is esti-mated to be between 2 and 4.5 months (Newberry and Wood-Gush, 1985; Jensen and Recen, 1989), yet on farms, piglets are routinely weaned between 21 and 28 days, sometimes even earlier (Robert *et al.*, 1999; CARC, 2003). Likewise, cattle naturally wean their calves at 8–11 months (Veissier *et al.*, 1990; Reinhardt, 2002) – yet beef calves are generally taken from their mothers at around 5 months (CARC, 1991), and dairy calves are routinely separated on their first day of life, with female calves reared as replacement stock for the dairy herd then also weaned off milk at between

4 and 12 weeks (e.g. USDA, 2002). In other taxa, early maternal separation can have lasting effects on brain function and on tendencies to stereotype (Chapter 6, this volume); and within ungulates, the quality of the mother–foal relationship is a risk factor in the later development of equine stereo-typies (Waters *et al.*, 2002; Nicol and Badnell-Waters, 2005). So could the early loss of the mother, or other aspects of mother–infant relationships that are constrained by captivity, help to explain the stereotypies of adult ungulates?

The belly-nosing of early-weaned piglets anecdotally can occasion-ally persist into adulthood (see Box 6.2, Chapter 6, this volume); and the offspring of restrained sows have been reported as more stereotypic in adulthood (reviewed by Bøe, 1997). However, we simply do not know whether the quantity or quality of early maternal care influences the later persistent bar-biting and sham-chewing shown by adult pigs. Further-more, in calves, the cross-sucking common when artificially reared calves are group-housed away from the dam (e.g. Jensen, 2003) seems to increase the later risks of inter-sucking (the sucking of the teat of another animal) when young heifers, which in turn then increases the later risk of inter-sucking as adult cows (Lidfors and Isberg, 2003). However, again, whether early weaning creates a lasting predisposition towards true stereotypies in adult cattle (e.g. tongue-rolling) is unknown. Mason (Chapter 11, this volume) revisits this issue at the end of the book.

2.6.2. The physical environment

Stereotyping captive ungulates are typically physically restricted by enclos-ure and/or tethering. For instance, when cows are moved off pasture, they are not just prevented from grazing, but also kept in small stalls that prevent locomotion (e.g. Redbo, 1992). This contrasts greatly with the ranging they would show naturally: grazing cattle may travel up to 24 km daily (Fraser and Broom, 1990); while Sato *et al.* (2001) report home ranges of 2–6 km^2 for beef cattle in semi-wild conditions and Hernandez *et al.* (1999) report ranges of 14 and 47 km^2 for domestic and feral cattle, respectively.

So, could this type of physical restriction contribute to stereotypy? Some studies suggest not. When horse and giraffe stereotypies were investigated in cross-site studies, enclosure size and exercise allowance often has relatively little effect on oral behaviour (instead affecting loco-motor stereotypies, e.g. pacing by giraffes and weaving by horses; Luescher *et al.*, 1998; Bashaw *et al.*, 2001a). Furthermore, in pigs, Ter-louw and Lawrence (1993) experimentally investigated the interactive effect of food allowance and restraint. They found that sows receiving a higher food allowance (4 kg food/day) performed less drinking and chain manipulation than sows on a low food allowance (2.5 kg/day), regardless of whether loose-housed or tethered. Indeed, they even observed loose-housed sows performing *more* chain manipulation than the tethered sows (see Fig. 2.4).

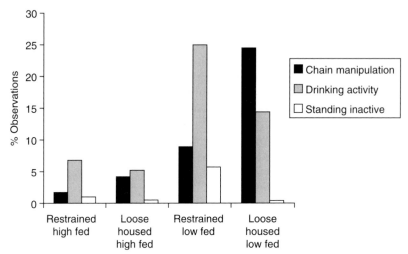

Fig. 2.4. Effect of confinement and feed level on stereotypic behaviour. The figure shows the average percentage of observations spent standing/sitting, chain-manipulating (a putative stereotypy) and drinking, in sows exposed to two levels of two different aspects of husbandry (in a 2 × 2 design): diet significantly affected oral behaviours, but the type of housing did not (adapted from Terlouw and Lawrence, 1993). Many other studies suggest that the degree of confinement does affect oral stereotypic behaviour, however (see text for discussion).

However, other work shows that oral stereotypies *are* affected by factors other than diet alone. For instance, several studies indicate that they are more frequent in pregnant sows that are confined in individual stalls or cages, compared with loose-housed females: Vieuille-Thomas *et al.* (1995) reported prevalence rates of 66% in group-housed sows, but 93% in individually stalled animals (and see also Blackshaw and McVeigh, 1984; Jensen, 1988; Broom *et al.*, 1995; Jensen *et al.*, 1995; Soede *et al.*, 1997; Pol *et al.*, 2000). Similar results are seen in cattle, with time spent in stereotypies falling greatly when animals are group-housed indoors rather than kept in small individual stalls, despite no change in feeding regime (Redbo, 1992). Likewise, when stabled horses were exercised, they showed reduced wood-chewing compared to when kept full-time in stalls (Krzak *et al.*, 1991), and other studies suggest that environmental enrichment in the form of visual contact between stabled horses also reduces this behaviour (McGreevy *et al.*, 1995a).

One possible reason for such variable findings (see Chapter 11, this volume, for an alternative) is that the treatments under comparison differ in the degree of physical restriction or freedom that they offer, which in turn might have threshold effects on behaviour. For example, Lawrence and colleagues (Terlouw *et al.*, 1991a; Lawrence and Terlouw, 1993; Haskell *et al.*, 1996) argue that high arousal and barren unvarying environments together render post-feeding foraging attempts more persistent than they would be in more naturalistic situations, and also 'channel' them into a few, repeatedly expressed behaviour patterns. In some instances, this may mean that oral or oro-nasal behaviours simply become more stereotyped in physically restrictive conditions, but not necessarily more frequent. Several studies suggest that the physical environment does not affect the total amount of oral behaviour, but does influence its form, and especially its degree of stereotypy. Thus food-restricted sows given straw on the floor manipulate this as much as they would pen fittings if straw is absent (e.g. Whittaker *et al.*, 1998), and similarly, sows held via tethers or stalls spend approximately the same amount of time chain-manipulating or bar-chewing as they would spend rooting on straw if loose housed, or chewing rocks and soil if kept outdoors (Schouten and Rushen, 1992; Dailey and McGlone, 1997). Furthermore, similarly fed outdoor sows spend roughly the same time chewing at things regardless of paddock type, but what they chew at depends on the availability of natural versus other substrates, with roots and branches being chewed if available, stones being chewed if not (Horrell, 2000). Studies comparing such conditions would thus draw different conclusions as to the effect of physical restriction or complexity, depending on how strict is their definition of stereotypy.

2.7. Conclusion and Perspectives

Overall, we have shown that in captive adult ungulates, the greater the difference between artificial and natural foraging regimes, the more abnormal behaviour is shown. Thus the greater the gap between what is possible in captivity and *ad libitum* feeding levels, and/or naturalistic fibre levels, and/or naturalistic, preferred foraging modes, the greater the degree of oral stereotypy. For instance, across a range of species, animals fed high concentrate low-fibre diets are reliably more prone to stereotypy development than animals on pasture or fed a large quantity of forage. This has clear relevance for welfare, since hunger and being unable to express preferred natural behaviour patterns are both causes of stress, while gastric or hindgut acidosis or ulceration probably causes discomfort, even pain. Management conditions that elicit oral stereotypies in ungulates are thus very likely to be sub-optimal. Such welfare considerations are particularly pertinent considering the many millions of ungulates that are fed low-fibre concentrates, and that perform stereotypic oral behaviour. More research is needed to assess the magnitude of these welfare problems, and, as we discuss further below, the best ways to alleviate them.

We have also shown that in some cases, within a given sub-optimal housing condition highly stereotypic animals may sometimes fare better than their less stereotypic conspecifics (Chapter 1, this volume); and that even where this is not known to be the case, performing bouts of stereotypy is still apparently associated with immediate benefits that are manifest as brief reductions in heart rate. This highlights how counter-productive it may be to prevent oral stereotypic behaviour physically, i.e. to merely abolish its expression without tackling its underlying causes. However, again more research is needed as to the true welfare costs of different ways of tackling these behaviours (a theme also picked up towards the end of this volume, in Chapter 10). For example, food-restricted animals may perform fewer stereotypies when given more opportunity to perform natural foraging behaviour, and thus appear to fare better, but they could still remain chronically hungry levels of food-restriction stay the same.

Overall, we thus understand fairly well the general causes of stereo-typic oral behaviour in ungulates: unnatural foraging regimes, with effects possibly exacerbated by physically restrictive environments and/or early weaning (Chapter 11, this volume). However, the precise underlying causes, and the extent to which these differ between animals of different ages, species, housing systems and preferred forms of oral stereotypy, are still the subject for much research. As we have seen, oral behaviours in captive ungulates share similarities with feeding behaviour, in their appearance, temporal distribution, and most likely in their underlying motivation. However, there are three specific means by which 'frustrated natural foraging' could give rise to stereotypies.

One such means is by leaving the animal in a state of thwarted motivation to ingest more food than is available; thus dietary deficits leave the animal with unfulfilled feeding motivations. Our food-deprived pregnant sow is one likely case in point. Furthermore, as we discussed, diet selection is naturally the principal means of modulating gastro-intestinal acidity, and herbivores also have excellent abilities to detect specific nutrient deficits and respond to them behaviourally (see e.g. Newman, in press). In captivity, in contrast, ungulates' diet selection is greatly constrained, again potentially leaving them in a state of unfulfilled feeding motivation for specific foodstuffs. However, even where such effects are demonstrably important in stereotypic behaviour, we still do not understand why deficits should result in sustained food-search hour after hour, day after day. Possible reasons include immediate consequences (such as nutrient ingestion from non-food sources); species-typical adaptations for patch-feeding (e.g. Box 2.2; Mason, in press) or even non-functional persistence resulting from stress sensitization (Chapters 8 and 11, this volume), but these issues remain unexplored. Furthermore, nor do we even always fully understand what dietary factors are needed for true satiety, or what complement of internal cues mediate this (see also Ingvartsen and Andersen, 2000). For example, a better understanding of the mechanisms by which fibrous ingredients affect digestive and metabolic processes is clearly necessary (as is assessment of the differential digestibility, net energy value, and other properties of different high-fibre diets). More consistent research across species might also help, because currently most attempts to assess hunger are done with sows; we thus know little as to whether other species are hungry when fed a large proportion of concentrate, and we also know little about polydipsia in species other than pigs, nor its possible relationship with hunger.

An alternative possibility is that ungulates have an inherent need (possibly state-dependent) to perform some foraging behaviours. Thus providing a foraging substrate such as straw, or increasing feeding time by making ingestion harder, can sometimes help to reduce stereotypies. This idea is consistent with observations that some foraging behaviours are inherently reinforcing. However, sometimes such approaches succeed in reducing stereotypic behaviour, and sometimes they do not. Does this depend on the underlying hunger levels of the animal? Or the degree to which the foraging opportunities offered are preferred and motivationally satisfying? Or instead, on the degree to which they have some beneficial physiological consequences? Or are yet other factors intervening, such as older stereotypies perhaps being harder to reverse than newly developed ones? Again we do not know. More hypothesis-driven research would help here, as currently many experiments have been designed to investigate the efficacy of methods of preventing stereotypies, without trying to understand why and how they work (or fail to). Furthermore, many studies of different species also differ in other variables too (e.g. horse studies often seem to deal with adult animals with well-developed stereotypies, while sow studies often look at young females with 'developing' stereotypies) which could be rectified in future work. Variation between species in their natural foraging biology could also be used to test hypotheses, as we see in Chapter 3, this volume. For example, perhaps browsers have a greater post-food stereotypy peak than grazers, because their food is naturally more patchy, so making ingestion-stimulated food-search more adaptive.

The third possible reason for sustained oral behaviours in captive ungulates is not because performing the behaviours is inherently important, but instead because it has useful consequences, for example, gastrointestinal health. The recent evidence linking stereotypies with gastrointestinal acidity/function opens a whole new array of research avenues. The causal relationships between dietary fibre, saliva production and gut acidity should therefore be investigated further, via hypothesis-driven experiment across a range of species. If this hypothesis *is* correct, and abnormal oral behaviours do effectively generate saliva which helps to alleviate abnormal gut pH, it also raises several further questions (e.g. Mason, in press). How do ungulates monitor their digestive tracts' pH, and does this vary with foraging niche? Do some or all ungulates monitor saliva production levels? If so, are they learned or innate responses to internal cues – or does this vary with dietary niche? How does diet interact with other factors, such as stress, in the aetiology of stomach lesions? And are other aspects of salivation important too, e.g. could adaptations for tannin-binding (see Box 2.1) play a role in some browsers' stereotypies?

Answering these questions could help to improve the fundamental understanding of ungulates, and also our abilities to husband them with good welfare. They may also apply to other taxa with somewhat similar foraging regimes, such as poultry. However, some of the questions raised by our review have even wider applicability to the other taxa discussed in the following chapters. Why do some individuals develop stereotypies, while others in the environment do not – what genetic and experiential factors are involved? How do the various forms of stereotypic behaviour interrelate? Although some attempts have been made to objectively quantify repetition and fixation (see Stolba *et al.*, 1983), most authors use subjective judgements to classify ungulate oral behaviours as stereotypic or otherwise (see e.g. Terlouw *et al.*, 1991a,b; Robert *et al.*, 1993; Bergeron and Gonyou, 1997). How should we improve this? Do we need to? Is there a real, qualitative difference between 'unambiguous' stereotypies like barbiting and 'abnormal-but-not-stereotypic' behaviours like cross-sucking or wood-chewing (e.g. as suggested by Garner in Chapter 5, this volume)? Or do they merely represent behaviours differing in their stages of development (Chapter 10, this volume), or degrees of functionality? And, last but not least, if some ungulate oral stereotypies *do* prove to be functional, then they are no longer stereotypies?

Acknowledgements

We are very grateful to the following for figures, data, feedback and discussion: Meredith Bashaw, Nick Bell, Jonathan Cooper, Naomi Latham, Brian McBride, Sebastian McBride, Daniel Mills, Jonathan Newman, Christine Nicol, Diana Reiss, Jeff Rushen, Ron Swaisgood and Sophie Vickery.

References

Abijaoude, J.A., Morand-Fehr, P., Tessier, J., Schmidely, P. and Sauvant, D. (2000) Diet effect on the daily feeding behaviour, frequency and characteristics of males in dairy goats. *Livestock Production Science* 64, 29–37.

Appleby, M.C. and Lawrence, A.B. (1987) Food restriction as a cause of stereotypic behaviour in tethered gilts. *Animal Production* 45, 103–110.

Archer, D.C., Freeman, D.E., Doyle, A.J., Proudman, C.J. and Edwards, G.B. (2004) Association between cribbing and entrapment of the small intestine in the epiploic foramen in horses: 68 cases (1991–2002). *Journal of American Veterinary Medical Association* 224, 562–564.

Ayles, H.L., Friendship, R.M., Bubenik, G.A. and Ball, R.O. (1999) Effect of feed particle size and dietary melatonin supplementation of gastric ulcers in swine. *Canadian Journal of Animal Science* 79, 179–185.

Bachmann, I. and Stauffacher, M. (2002) Prevalence of behavioural disorders in the Swiss horse population. *Schweizer Archiv fur Tierheilkunde* 144, 356–368.

Bashaw, M.J., Tarou, L.R., Maki, T.S. and Maple, T.L. (2001a) A survey assessment of variables relating to stereotypy in captive giraffe and okapi. *Applied Animal Behaviour Science* 73, 235–247.

Bashaw, M.J., Tarou, L.R., Sartor, R., Bouwens, N., Maki, T. and Maple, T.L. (2001b) Stereotypic behavior in giraffe: concerns and contingencies. In: *Proceedings of the 2001 American Zoo and Aquarium Association Conference*, St Louis, Missouri.

Bauman, D.E., Davis, C.L. and Bucholtz, H.F. (1971) Propionate production in the rumen of cows fed either a control or high-grain, low-fiber diet. *Journal of Dairy Science* 54, 1282–1287.

Baxter, E. and Plowman, A.B. (2001) The effect of increasing dietary fibre on feeding, rumination and oral stereotypies in captive giraffes (*Giraffa camelopardalis*). *Animal Welfare* 10, 281–290.

Bell, W.J. (1991) *Searching Behaviour: The Behavioural Ecology of Finding Resources*. Chapman & Hall, London.

Bergeron, R. and Gonyou, H.W. (1997) Effects of increasing energy intake and foraging behaviours on the development of stereotypies in pregnant sows. *Applied Animal Behaviour Science* 53, 259–270.

Bergeron, R., Bolduc, J., Ramonet, Y., Robert, S. and Meunier-Salaün, M.-C. (2000) Feeding motivation and stereotypies in pregnant sows fed increasing levels of fibre and/or food. *Applied Animal Behaviour Science* 70, 27–40.

Bergeron, R., Meunier-Salaün, M.-C. and Robert, S. (2002) Effects of food texture on meal duration and behaviour of sows fed high-fibre or concentrate diets. *Canadian Journal of Animal Science* 82, 587–589.

Blackshaw, J.K. and McVeigh, J.F. (1984) Stereotype behaviour in sows and gilts housed in stalls, tethers, and groups. In: Fox, M.W. and Mickley, L.D. (eds) *Advances in Animal Welfare Science*. Martinus Nijhoff Publishers, Dordrecht, pp. 163–174.

Blasetti, A., Boitani, L., Riviello, M.C. and Visalberghi, E. (1988) Activity budgets and use of enclosed space by wild boars (*Sus scrofa*) in captivity. *Zoo Biology* 7, 69–79.

Blokhuis, H.J. (1986) Feather-pecking in poultry and its relation with ground-pecking. *Applied Animal Behaviour Science* 16, 63–67.

Blood, D.C. and Radostits, O.M. (1989) *Veterinary Medicine: A Text Book of the Diseases of Cattle, Sheep, Pigs, Goats and Horses*, 7th edn. Bailliere Tindall, London.

Bøe, K. (1997) The effect of age at weaning and post-weaning environment on the behaviour of pigs. *Acta Agriculturae Scandinavica Section A* 43, 173–180.

Bolinger, D.J., Albright, J.L., MorrowTesch, J., Kenyon, S.J. and Cunningham, M.D. (1997) The effects of restraint using self-locking stanchions on dairy cows in relation to behavior, feed intake, physiological parameters, health, and milk yield. *Journal of Dairy Science* 80, 2411–2417.

Bonacic, C. (2005) Vicuña ecology and management. *Ecology Info* 27. Available at: http://www.ecology.info/vicugna.htm

Born, E.W., Rysgaard, S., Ehlmé, G., Sejr, M., Acquarone, M. and Levermann, N. (2003) Underwater observations of free-living Atlantic walruses (*Odobenus rosmarus rosmarus*) and estimates of their food consumption. *Polar Biology* 26, 348–357.

Braund, J.P., Edwards, S.A., Riddoch, I. and Buckner, L.J. (1998) Modification of foraging behaviour and pasture damage by dietary manipulation in outdoor sows. *Applied Animal Behaviour Science* 56, 173–186.

Broom, D.M. and Potter, M.J. (1984) Factors affecting the occurrence of stereotypies in stall-housed dry sows. In: Unshelm, J. and van Putten, G. (eds) *Proceedings of the International Congress on Applied Ethology in Farm Animals*. KTBL, Darmstadt, pp. 229–231.

Broom, D.M., Mendl, M.T. and Zanella, A.J. (1995) A comparison of the welfare of sows in different housing conditions. *Animal Science* 61, 369–385.

Brouckaert, K., Steenhaut, M., Martens, A., Vlaminck, L., Pille, F., Arnaerts, L. and Gasthuys, F. (2002) Windsucking in the horse: results after surgical treatment – a retrospective study (1990–2000). *Vlamms Diergeneeskundig Tijdschrift* 71, 249–255 (in Dutch).

Brouns, F., Edwards, S.A. and English, P.R. (1994) Effect of dietary fibre and feeding system on activity and oral behaviour of group housed gilts. *Applied Animal Behaviour Science* 39, 215–223.

Brouns, F., Edwards, S.A. and English, P.R. (1997) The effects of dietary inclusion of sugar-beet pulp on the feeding behaviour of dry sows. *Animal Science* 65, 129–133.

Buckner, L.J., Edwards, S.A. and Bruce, J.M. (1998) Behaviour and shelter use by outdoor sows. *Applied Animal Behaviour Science* 57, 69–80.

Canadian Agri-Food Research Council (CARC) (1991) *Recommended Code of*

Practice for the Care and Handling of Beef Cattle. Agriculture Canada Publication 1870/F.

Canadian Agri-Food Research Council (CARC) (2003) *Recommended Code of Practice for the Care and Handling of Farm Animals: Pigs. Addendum Early Weaned Pigs.*

Canali, E. and Borroni, A. (1994) Behavioural problems in thoroughbred horses reared in Italy. *Applied Animal Behaviour Science* 40, 74.

Canali, E., Ferrante, V., Mattiello, S., Gottardo, F. and Verga, M. (2001) Are oral stereotypies and abomasal lesions correlated in veal calves? In: *Proceedings of the 35th International Congress of the ISAE.* Davis, California, p. 103.

Clauss, M., Gehrke, J., Hatt, J.M., Dierenfeld, E.S., Flach, E.J., Hermes, R., Castell, J., Streich, W.J. and Fickel, J. (2005) Tannin-binding salivary proteins in three captive rhinoceros species. *Comparative Biochemistry and Physiology A – Molecular and Integrative Physiology* 140, 67–72.

Coates, C.W. (1962) Walruses and whales in the New York aquarium. *International Zoo Yearbook* 4, 10–12.

Cole, D.J.A. (1982) Nutrition and reproduction. In: Cole, D.J.A. and Foxcroft, G.R. (eds) *Control of Pig Reproduction.* Butterworths, London.

Cooper, J. and Jackson, R. (1996) A comparison of the feeding behaviour of sheep in straw yards and on slats. *Applied Animal Behaviour Science* 49, 99.

Cooper, J.J., Emmans, G.A. and Friggens, N.C. (1994) Effect of diet on behaviour of individually penned sheep. *Animal Production* 58, 441.

Cooper, J.J., McCullam, J. and Shanks, M. (1995) Effective fibre and abnormal behaviour in stall housed lambs. *Animal Production* 60, 567–568.

Cronin, G.M. (1985) The development and significance of abnormal stereotyped behaviours in tethered sows. PhD thesis, Agricultural University of Wageningen, The Netherlands.

Cronin, G.M., van Tartwijk, J.M.F.M., van der Hel, W. and Verstegen, M.W.A.

(1986) The influence of degree of adaptation to tether-housing by sows in relation to behaviour and energy metabolism. *Animal Production* 42, 257–268.

Dailey, J.W. and McGlone, J.J. (1997) Oral/nasal/facial and other behaviors of sows kept individually outdoors on pasture, soil or indoors in gestation crates. *Applied Animal Behaviour Science* 52, 25–43.

Day, J.E.L., Kyriazakis, I. and Lawrence, A.B. (1996) The use of a second-order schedule to assess the effect of food bulk on the feeding motivation of growing pigs. *Animal Science* 63, 447–455.

De Mello Nicoletti, J.L., Hussni, C.A., Thomassian, A., Gandolfi, W. and Pereira Leme, D. (1996) Estudo retrospectivo de 11 casos de aerofagia em eqüinos operados pela técnica de miectomia de Forssell modificada. *Ciência Rural* 26, 431–434.

De Passillé, A.M.B., Christopherson, R. and Rushen, J. (1993) Non-nutritive sucking by the calf and postprandial secretion of insulin, CCK and gastrin. *Physiology and Behavior* 54, 1069–1073.

Delacalle, J., Burba, D.J., Tetens, J. and Moore, R.M. (2002) Nd: YAG laser-assisted modified Forssell's procedure for treatment of cribbing (crib-biting) in horses. *Veterinary Surgery* 31, 111–116.

Dodman, N.H., Shuster, L., Court, M.H. and Dixon, R. (1987) Investigation into the use of narcotic antagonists in the treatment of a stereotypic behaviour pattern (crib-biting) in the horse. *American Journal of Veterinary Research* 48, 311–319.

Dybkjær, L., Vraa-Andersen, L., Pailey, L.G., Møller, K., Christensen, G. and Agger, J.F. (1994) Associations between behaviour and stomach lesions in slaughter pigs. *Preventative Veterinary Medicine* 19, 101–112.

Every, D., Tunnicliffe, G.A. and Every, R.G. (1998) Tooth-sharpening behaviour thegosis) and other causes of wear on sheep teeth in relation to mastication and grazing mechanisms. *Journal of the Royal Society of New Zealand* 28, 169–184.

Fickel, J., Pitra, C., Joest, B.A. and Hofmann, R.R. (1999) A novel method to evaluate the tannin-binding capacities of salivary proteins. *Comparative Biochemistry and Physiology A – Molecular and Integrative Physiology* 122, 225–229.

Fisher, K.I. and Stewart, R.E.A. (1997) Summer foods of Atlantic walrus, *Odobenus rosmarus rosmarus*, in northern Foxe Basin, Northwest Territories. *Canadian Journal of Zoology* 75, 1166–1175.

Fjeldborg, J. (1993) Results of surgical treatment of cribbing by neurectomy and myectomy. *Equine Practice* 7, 34–36.

Frape, D. (1998) *Equine Nutrition and Feeding*, 2nd edn. Longman Scientific and Technical, Harlow, Essex.

Fraser, D. (1975) The effect of straw on the behaviour of sows in tether stalls. *Animal Production* 21, 59–68.

Fraser, A.F. and Broom, D.M. (1990) *Farm Animal Behaviour and Welfare*, 3rd edn. Baillière Tindall, London.

Ganslosser, U. and Brunner, C. (1997) Influence of food distribution on behaviour in captive bongos, *Taurotragus euryceros*: an experimental investigation. *Zoo Biology* 16, 237–245.

Gillham, S.B., Dodman, N.H., Shuster, L., Kream, R. and Rand, W. (1994) The effect of diet on cribbing behaviour and plasma B-endorphin in horses. *Applied Animal Behaviour Science* 41, 147–153.

Ginnett, T.F. and Demment, M.W. (1997) Sex differences in giraffe foraging behavior at two spatial scales. *Oecologia* 110, 291–300.

Hagenbeck, C.-H. (1962) Notes on walruses *Odobenus rosmarus* in captivity. *International Zoo Yearbook* 4, 24–25.

Hakansson, A., Franzen, P. and Pettersson, H. (1992) Comparison of two surgical methods for treatment of crib-biting in horses. *Equine Veterinary Journal* 24, 494–496.

Hanstock, T.L., Clayton, E.H., Li, K.M. and Mallet, P.E. (2004) Anxiety and aggression associated with the fermentation of carbohydrates in the hindgut of rats. *Physiology and Behavior* 82, 357–368.

Haskell, M.J., Terlouw, E.M.C., Lawrence, A.B. and Erhard, H.W. (1996) The relationship between food consumption and persistence of post-feeding foraging behaviour in sows. *Applied Animal Behaviour Science* 48, 249–262.

Henderson, J.V. and Waran, N.K. (2001) Reducing equine stereotypies using an Equiball®. *Animal Welfare* 10, 73–80.

Hernandez, L., Barral, H., Halffter, G. and Colon, S.S. (1999) A note on the behaviour of feral cattle in the Chihuahuan Desert of Mexico. *Applied Animal Behaviour Science* 63, 259–267.

Hessing, M.J.C., Geudeke, M.J., Scheepens, C.J.M., Tielen, M.J.M., Schouten, W.G.P. and Wiepkema, P.R. (1992) Mucosal lesions in the pars oesophaga in pigs – prevalence and influence of stress. *Tijdschrift voor Diergeneeskunde* 117, 445–450.

Hibbard, B., Peters, J.P., Chester, S.T., Robinson, J.A., Kotarski, S.F., Croom, W.J. and Hagler, W.M. (1995) The effect of slaframine on salivary output and subacute and acute acidosis in growing beef steers. *Journal of Animal Science* 73, 516–525.

Hillyer, M.H., Taylor, F.G.R., Proudman, C.J., Edwards, G.B., Smith, J.E. and French, N.P. (2002) Case control study to identify risk factors for simple colonic obstruction and distension colic in horses. *Equine Veterinary Journal* 34, 455–463.

Hintz, H.F., Sedgewick, C.J. and Schryver, H.F. (1976) Some observations on digestion of a pelleted diet by ruminants and non-ruminants. *International Zoo Yearbook* 16, 54–62.

Hofmann, R.R. and Matern, B. (1988) Changes in gastrointestinal morphology related to nutrition in giraffes. *International Zoo Yearbook* 27, 168–176.

Horrell, R.I. (2000) Stone-chewing in outdoor pigs. Grant research report presented to the Universities Federation of Animal Welfare, UK.

Horrell, R.I., A'Ness, P.J., Edwards, S.A. and Eddison, J.C. (2001) The use of nose-rings in pigs: consequences for rooting, other functional activities, and welfare. *Animal Welfare* 10, 3–22.

Houpt, K.A. (1993) Equine stereotypies. *The Compendium* 15, 1265–1271.

Hutson, G.D. and Haskell, M.J. (1990) The behaviour of farrowing sows with free and operant access to an earth floor. *Applied Animal Behaviour Science* 26, 363–372.

Ingvartsen, K.L. and Andersen, J.B. (2000) Integration of metabolism and intake regulation: a review focusing on periparturient animals. *Journal of Dairy Science* 83, 1573–1597.

Jansson, N. (2000) Surgical treatment of windsucking in two horses younger than one year of age. *Ippologia* 11, 5–9.

Jensen, P. (1988) Diurnal rhythm of barbiting in relation to other behaviour in pregnant sows. *Applied Animal Behaviour Science* 21, 337–346.

Jensen, M.B. (2003) The effects of feeding method, milk allowance and social factors on milk feeding behaviour and cross-sucking in group housed dairy calves. *Applied Animal Behaviour Science* 80, 191–206.

Jensen, P. and Recen, B. (1989) When to wean – observations from free-ranging domestic pigs. *Applied Animal Behaviour Science* 23, 49–60.

Jensen, K.H., Pedersen, B.K., Pedersen, L.J. and Jørgensen, E. (1995) Well-being in pregnant sows: confinement versus group housing with electronic sow feeding. *Acta Agriculturae Scandinavica Section A Animal Science* 45, 266–275.

Johnson, K.G., Tyrrell, J., Rowe, J.B. and Pethick, D.W. (1998) Behavioural changes in stabled horses given nontherapeutic levels of virginiamycin. *Equine Veterinary Journal* 30, 139–143.

Kastelein, R.A. and Wiepkema, P.R. (1989) A digging trough as occupational therapy for Pacific walruses *Odobenus rosmarus* in human care. *Aquatic Mammals* 15, 9–17.

Kastelein, R.A., Paasse, M., Klinkjhammer, P. and Wiepkema, P.R. (1991) Food dispensers as occupational therapy for the walrus *Odobenus rosmarus divergens* at the Harderwijk Marine Mammal Park. *International Zoo Yearbook* 30, 201–212.

Kennedy, M.J., Schwabe, A.E. and Broom, D.M. (1993) Crib-sucking and windsucking stereotypies in the horse. *Equine Veterinary Education* 5, 142–154.

Keunen, J.E., Plaizner, J.C., Kyriazakis, L., Duffield, T.F., Widowski, T.M., Lindinger, M.I. and McBride, B.W. (2002) Effects of a subacute ruminal acidosis model on the diet selection of dairy cows. *Journal of Dairy Science* 85, 3304–3313.

Kiley-Worthington, M. (1983) Stereotypes in horses. *Equine Practice* 5, 34–40.

Koene, P. (1999) When feeding is just eating: how do farm and zoo animals use their spare time? In: van der Heide, D., Huisman, E.A., Kanis, E., Osse, J.W.M. and Verstegen, M.W.A. (eds) *Regulation of Feed Intake*. CAB International, Wallingford, UK, pp. 13–19.

Krzak, W.E., Gonyou, H.W. and Lawrence, L.M. (1991) Wood chewing by stalled horses: diurnal pattern and effects of exercise. *Journal of Animal Science* 69, 1053–1058.

Kusenose, R. (1992) Diurnal patterns of cribbing in stabled horses. *Japanese Journal of Equine Science* 3, 173–176.

Lawrence, A.B. and Illius, A.W. (1989) Methodology for measuring hunger and food needs using operant conditioning in the pig. *Applied Animal Behaviour Science* 24, 273–285.

Lawrence, A.B. and Terlouw, E.M.C. (1993) A review of behavioral factors involved in the development and continued performance of stereotypic behaviors in pigs. *Journal of Animal Science* 71, 2815–2825.

Lawrence, A.B., Appleby, M.C., Illius, A.W. and Macleod, H.A. (1989) Measuring hunger in the pig using operant conditioning: the effect of dietary bulk. *Animal Production* 48, 213–220.

Lawrence, A.B., Terlouw, E.M.C. and Kyriazakis, I. (1993) The behavioural effects of undernutrition in confined farm animals. *Proceedings of the Nutrition Society* 52, 219–229.

Lebelt, D., Zanella, A.J. and Unshelm, J. (1998) Physiological correlates associated with cribbing behaviour in horses: changes in thermal threshold, heart rate, plasma beta-endorphin and serotonin. *Equine Veterinary Journal Supplement* 27, 21–27.

Lepionka, L., Malbert, C.H. and Laplace, J.P. (1997) Proximal gastric distension modifies ingestion rate in pigs. *Reproduction Nutrition Development* 37, 449–457.

Lidfors, L. and Isberg, L. (2003) Intersucking in dairy calves – review and questionnaire. *Applied Animal Behaviour Science* 80, 207–231.

Lindström, T. and Redbo, I. (2000) Effect of feeding duration and rumen fill on behaviour in dairy cows. *Applied Animal Behaviour Science* 70, 83–97.

Linnane, M., Horan, B., Connolly, J., O'Connor, P., Buckley, F. and Dillon, P. (2004) The effect of strain of Holstein–Friesan and feeding system on grazing behaviour, herbage intake and productivity in the first lactation. *Animal Science* 78, 169–178.

Luescher, U.A., McKeown, D.B. and Dean, H. (1998) A cross-sectional study on compulsive behaviour (stable vices) in horses. *Equine Veterinary Journal Supplement* 27, 14–18.

Marchant-Forde, J.N. and Pajor, E.A. (2003) The effect of dietary sodium bicarbonate on abnormal behavior and heart rate in sows. *Journal of Animal Science* 81 (Suppl. 1), 158.

Mason, G.J. (1991) Stereotypies: a critical review. *Animal Behaviour* 41, 1015–1037.

Mason, G.J. (in press). Animal farm: food provisioning and abnormal oral behaviours in captive ungulates. In: Stephens, D.W., Ydenberg, R.C. and Brown, J.C. (eds) *Foraging*. University of Chicago Press, Chicago.

Mason, G. and Mendl, M. (1997) Do the stereotypies of pigs, chickens and mink reflect adaptive species differences in the control of foraging? *Applied Animal Behaviour Science* 53, 45–58.

Mason, G.J. and Latham, N.R. (2004) Can't stop, won't stop: is stereotypy a reliable animal welfare indicator? *Animal Welfare* 13, S57–S69.

Mason, G., Cooper, J. and Clarebrough, C. (2001) The welfare of fur-farmed mink. *Nature* 410, 35–36.

Mattiello, S., Canali, E., Ferrante, V., Caniatti, M., Gottardo, F., Cozzi, G., Andrighetto, I. and Verga, M. (2002) The provision of solid feeds to veal calves: II. Behaviour, physiology and abomasal damage. *Journal of Animal Science* 80, 367–375.

McBride, S.D. and Cuddeford, D. (2001) The putative welfare-reducing effects of preventing equine stereotypic behaviour. *Animal Welfare* 10, 173–189.

McBride, S.D. and Long, L. (2001) Management of horses showing stereotypic behaviour, owner perception and the implication for welfare. *The Veterinary Record* 148, 799–802.

McGreevy, P.D. and Nicol, C.J. (1998a) Physiological and behavioural consequences associated with short-term prevention of crib-biting in horses. *Physiology and Behaviour* 65, 15–23.

McGreevy, P.D. and Nicol, C.J. (1998b) Prevention of crib-biting: a review. *Equine Veterinary Journal Supplement* 27, 35–38.

McGreevy, P.D., Cripps, P.J., French, N.P., Green, L.E. and Nicol, C.J. (1995a) Management factors associated with stereotypic and redirected behaviour in the thoroughbred horse. *Equine Veterinary Journal* 27, 86–91.

McGreevy, P.D., French, N.P. and Nicol, C.J. (1995b) The prevalence of abnormal behaviours in dressage, eventing and endurance horses in relation to stabling. *The Veterinary Record* 137, 36–37.

McGreevy, P., Richardson, J.D., Nicol, C.J. and Lane, J.G. (1995c) Radiographic and endoscopic study of horses performing an oral based stereotypy. *Equine Veterinary Journal* 27, 92–95.

McGreevy, P.D., Webster, A.J.F. and Nicol, C.J. (2001a) Study of the behaviour, digestive efficiency and gut transit times of crib-biting horses. *The Veterinary Record* 148, 592–596.

McGreevy, P.D., Hawson, L.A., Habermann, T.C. and Cattle, S.R. (2001b) Geophagia in horses: a short note on 13 cases. *Applied Animal Behaviour Science* 71, 119–125.

Meot, F., Cirio, A. and Boivin, R. (1997) Parotid secretion daily patterns and measurement with ultrasonic flow probes in conscious sheep. *Experimental Physiology* 82, 905–923.

Merial (2004) Equine ulcers. Available at: http://gastrogard.us.merial.com/equine_ulcers.asp

Mills, D.S. and MacLeod, C.A. (2002) The response of crib-biting and windsucking in horses to dietary supplementation with an antacid mixture. *Ippologia* 13, 33–41.

Mills, D.S. and McDonnell, S.M. (2005) *The Domestic Horse: The Origins, Development and Management of its Behaviour.* Cambridge University Press, Cambridge, UK.

Minero, M., Canali, E., Ferrante, V., Verga, M. and Ödberg, F.O. (1999) Heart rate and behavioural responses of crib-biting horses to two acute stressors. *The Veterinary Record* 145, 430–433.

Mroz, Z., Partridge, I.G., Mitchell, G. and Keal, H.D. (1986) The effect of oat hulls, added to the basal ration for pregnant sows, on reproductive performance, apparent digestibility, rate of passage and plasma parameters. *Journal of the Science of Food and Agriculture* 37, 239–247.

Murray, M.J. (1999) Pathophysiology of peptic disorders in foals and horses: a review. *Equine Veterinary Journal Supplement* 29, 14–18.

Murray, M.J. and Eichorn, E.S. (1996) Effects of intermittent feed deprivation, intermittent feed deprivation with ranitidine administration, and stall confinement with ad libitum access to hay on gastric ulceration in horses. *American Journal of Veterinary Research* 57, 1599–1603.

Murray, C.G. and Sanson, G.D. (1998) Thegosis – a critical review. *Australian Dental Journal* 43, 192–198.

Newberry, R.C. and Wood-Gush, D.G.M. (1985) The suckling behaviour of domestic pigs in a semi-natural environment. *Behaviour* 95, 11–25.

Newman, J. (in press) Herbivory. In: Stephens, D.W., Ydenberg, R.C. and Brown, J.C. (eds) *Foraging*. University of Chicago Press, Chicago.

Nicol, C.J. (1999) Understanding equine stereotypies. *Equine Veterinary Journal Supplement* 28, 20–25.

Nicol, C.J. (2000) Equine stereotypies. In: Houpt, K.A. (ed.) *Recent Advances in Companion Animal Behavior Problems*. International Veterinary Information Service, Ithaca, New York.

Nicol, C.J. and Badnell-Waters, A.J. (2005) Suckling behaviour in domestic foals and the development of abnormal oral behaviour. *Animal Behaviour* 70, 21–29.

Nicol, C.J., Davidson, H.P.B., Harris, P.A., Waters, A.J. and Wilson, A.D. (2002) Study of crib-biting and gastric inflammation and ulceration in young horses. *The Veterinary Record* 151, 658–662.

Ödberg, F.O. (1978) Abnormal behaviours (stereotypies). In: *Proceedings of the First World Congress on Ethology Applied to Zootechnics*. Industrias Graficas Espana, Madrid, pp. 475–480.

Oetzel, G.R. (2003) Nutritional management and subacute ruminal acidosis in dairy herds. In: *Proceedings of the American Association of Bovine Practitioners 36th Annual Conference*, Columbus, Ohio.

O'Sullivan, T., Friendship, R.M., Ball, R. and Ayles, H. (1996) Prevalence of lesions of the pars oesophageal region of the stomach of sows at slaughter. In: *Proceedings of the American Association of Swine Practitioners*, pp. 151–153.

Phillips, C. (2002) *Cattle Behaviour and Welfare*. Blackwell Publishing, Oxford, UK.

Phillips, C.J.C., Youssef, M.Y.I., Chiy, P.C. and Arney, D.R. (1999) Sodium chloride supplements increase the salt appetite and reduce stereotypies in confined cattle. *Animal Science* 68, 741–747.

Pol, F., Courboulay, V., Cotte, J.P. and Lechaux, S. (2000) Logement en cases collectives ou en stalles individuelles en première gestation. Impact sur le bien-être des truies nullipares. *Journées de Recherche Porcine en France* 32, 97–104.

Pough, F.H., Heiser, J.B. and McFarland, W.N. (1989) *Vertebrate Life*, 3rd edn. Macmillan Publishing Company, New York.

Ramonet, Y., Meunier-Salaün, M.C. and Dourmad, J.Y. (1999) High-fiber diets in pregnant sows: digestive utilization and effects on the behavior of the animals. *Journal of Animal Science* 77, 591–599.

Ramonet, Y., Robert, S., Aumaître, A., Dourmad, J.Y. and Meunier-Salaün, M.C. (2000a) Influence of dietary fibre on digestive utilization, metabolite profiles and the behaviour of pregnant sows. *Animal Science* 70, 275–286.

Ramonet, Y., Bolduc, J., Bergeron, R., Robert, S. and Meunier-Salaün, M.C. (2000b) Feeding motivation in pregnant sows: effect of fibrous diets in operant conditioning procedure. *Applied Animal Behaviour Science* 66, 21–29.

Rebhun, W.C., Dill, S.G. and Power, H.T. (1982) Gastric ulcers in foals. *Journal of the American Veterinary Medical Association* 180, 404–407.

Redbo, I. (1990) Changes in duration and frequency of stereotypies and their adjoining behaviours in heifers, before, during and after the grazing period. *Applied Animal Behaviour Science* 26, 57–67.

Redbo, I. (1992) The influence of restraint on the occurrence of oral stereotypies in dairy cows. *Applied Animal Behaviour Science* 35, 115–123.

Redbo, I. and Nordblad, A. (1997) Stereotypies in heifers are affected by feeding regime. *Applied Animal Behaviour Science* 53, 193–202.

Redbo, I., Jacobson, K.G., van Doorn, C. and Petterson, G. (1992) A note on the relations between oral stereotypies in dairy cows and milk production, health and age. *Animal Production* 54, 166–168.

Redbo, I., Emanuelson, M., Lundberg, K. and Oredsson, N. (1996) Feeding level and oral stereotypies in dairy cows. *Animal Science* 62, 199–206.

Redbo, I., Redbo-Torstensson, P., Ödberg, F.O., Hedendahl, A. and Holm, J. (1998) Factors affecting behavioural disturbances in race-horses. *Animal Science* 66, 475–481.

Reinhardt, V. (2002) Artificial weaning of calves: benefits and costs. *Journal of Applied Animal Welfare Science* 5, 251–255.

Robbins, C.T. (1993) *Wildlife Feeding and Nutrition*, 2nd edn. Academic Press, New York.

Robert, S., Matte, J.J., Farmer, C., Girard, C.L. and Martineau, G.P. (1993) High-fibre diets for sows: effects on stereotypies and adjunctive drinking. *Applied Animal Behaviour Science* 37, 297–309.

Robert, S., Rushen, J. and Farmer, C. (1997) Both energy content and bulk of feed affect stereotypic behaviour, heart rate and feeding motivation of female pigs. *Applied Animal Behaviour Science* 54, 161–171.

Robert, S., Weary, D. and Gonyou, H.W. (1999) Segregated early weaning and welfare of piglets. *Journal of Applied Animal Welfare Science* 2, 31–40.

Robert, S., Bergeron, R., Farmer, C. and Meunier-Salaün, M.C. (2002) Does the number of daily meals affect feeding motivation and behaviour of gilts fed high-fibre diets? *Applied Animal Behaviour Science* 76, 105–117.

Rowe, J.B., Pethick, D.W. and Lees, M.J. (1994) Prevention of acidosis and laminitis associated with grain feeding in horses. *Journal of Nutrition* 124, 2742–2744.

Rushen, J. (1984) Stereotyped behaviour, adjunctive drinking and the feeding periods of tethered sows. *Animal Behaviour* 32, 1059–1067.

Rushen, J. (2003) Changing concepts of farm animal welfare: bridging the gap between applied and basic research. *Applied Animal Behaviour Science* 81, 199–214.

Rushen, J., Lawrence, A.B. and Terlouw, E.M.C. (1993) The motivational basis of stereotypies. In: Lawrence, A.B. and Rushen, J. (eds) *Stereotypic Animal Behaviour: Fundamentals and Applications to Welfare*. CAB International, Wallingford, UK, pp. 41–64.

Rushen, J., Robert, S. and Farmer, C. (1999) Effects of an oat-based high-fibre diet on insulin, glucose, cortisol and free fatty acid concentrations in gilts. *Animal Science* 69, 395–401.

Rutter, S.M., Orr, J.R., Yarrow, N.H. and Champion, R.A. (2004) Dietary preference of dairy heifers grazing ryegrass and white clover, with and without an anti-bloat treatment. *Applied Animal Behaviour Science* 85, 1–10.

Sambraus, H.H. (1985) Mouth-based anomalous syndromes. In: Fraser, A.F. (ed.) *Ethology of Farm Animals*. Elsevier, Amsterdam, pp. 391–422.

Sato, S., Kubo, T. and Abe, M. (1992) Factors influencing tongue-rolling and the relationship between tongue-rolling and production traits in fattening cattle. *Journal of Animal Science* 70 (Suppl. 1), 157.

Sato, S., Sugiyama, M., Yasue, T., Deguchi, Y. and Suyama, T. (2001) Seasonal change of land use strategy of free-ranging cattle in an agroforest setting without fencing. *Proceedings of the 35th Congress of the ISAE*, Davis, California, p. 152.

Sauvant, D., Meschy, F. and Mertens, D. (1999) Components of ruminal acidosis and acidogenic effects of diets. *INRA Productions Animales* 12, 49–60.

Schaller, G.B., Qitao, T., Johnson, K.G., Xiaoming, W., Hemind, S. and Jinchu, H. (1989) The feeding ecology of giant pandas and Asiatic black bears in the Tangjiahe Reserve, China. In: Gittleman, J.L. (ed.) *Carnivore Conservation and Evolution.* Chapman & Hall, London, pp. 212–241.

Schmidt-Nielson, K. (1997) *Animal Physiology: Adaptation and Environment*, 5th edn. Cambridge University Press, Cambridge, UK.

Schofield, W.L. and Mulville, J.P. (1998) Assessment of the modified Forssell's procedure for the treatment of oral stereotypies in 10 horses. *The Veterinary Record* 142, 572–575.

Schouten, W. and Rushen, J. (1992) Effects of naloxone on stereotypic and normal behaviour of tethered and loose-housed sows. *Applied Animal Behaviour Science* 33, 17–26.

Schouten, W., Rushen, J. and De Passillé, A.M.B (1991) Stereo-typic behavior and heart rate in pigs. *Physiology and Behavior* 50, 617–624.

Schouten, W.G.P., Lensink, J., Lakwijk, N. and Wiegant, V.M. (2000) De-arousal effect of stereotypies in tethered sows. In: *Proceedings of the 34th International Congress of the International Society for Applied Ethology*, Florianopolis, Brazil, p. 46.

Schwartzkopf-Genswein, K.S., Beauchemin, K.A., Gibb, D.J., Crews, D.H., Hickman, D.D., Streeter, M. and McAllister, T.A. (2003) Effect of bunk management on feeding behavior, ruminal acidosis and performance of feedlot cattle: a review. *Journal of Animal Science* 81 (Suppl.), E149–E158.

Seo, T., Sato, S., Kosaka, K., Sakamoto, N. and Tokumoto, K. (1998) Tongue-playing and heart rate in calves. *Applied Animal Behaviour Science* 58, 179–182.

Simpson, B.S. (1998) Behavior problems in horses: cribbing and wood chewing. *Veterinary Medicine* November 1998, 999–1004.

Soede, N.M., Helmond, F.A., Schouten, W.G.P. and Kemp, B. (1997) Oestrus and peri-ovulatory hormone profiles in tethered and loose-housed sows. *Animal Reproduction Science* 46, 133–148.

Spoolder, H.A.M., Burbidge, J.A., Edwards, S.A., Simmins, P.H. and Lawrence, A.B. (1995) Provision of straw as a foraging substrate reduces the development of excessive chain and bar manipulation in food restricted sows. *Applied Animal Behaviour Science* 43, 249–262.

Stolba, A. and Wood-Gush, D.G.M. (1989) The behaviour of pigs in a semi-natural environment. *Animal Production* 48, 419–425.

Stolba, A., Baker, N. and Wood-Gush, D.G.M. (1983) The characterisation of stereotyped behaviour in stalled sows by informational redundancy. *Behaviour* 87, 157–181.

Tarou, L.R., Bashaw, M.J., Sartor, R., Maki, T.S., Liu, S.C. and Maple, T.L. (2001) When we're not looking: nocturnal behavior of giraffe. In: *Proceedings of the 2001 American Zoo and Aquarium Association Conference*, St. Louis, Missouri.

Tarou, L.R., Bashaw, M.J. and Maple, T.L. (2003) Failure of a chemical spray to significantly reduce stereotypic licking in a captive giraffe. *Zoo Biology* 22, 601–607.

Terlouw, E.M.C. and Lawrence, A.B. (1993) Long-term effects of food allowance and housing on development of stereotypies in pigs. *Applied Animal Behaviour Science* 38, 103–126.

Terlouw, E.M.C., Lawrence, A.B. and Illius, A.W. (1991a) Influences of feeding level and physical restriction on development of stereotypies in sows. *Animal Behaviour* 42, 981–991.

Terlouw, E.M.C., Lawrence, A.B., Ladewig, J., De Passillé, A.M.B., Rushen, J. and Schouten, W.G.P. (1991b) Relationship between plasma cortisol and stereotypic activities in pigs. *Behavioural Processes* 25, 133–153.

Terlouw, E.M.C., Wiersma, A., Lawrence, A.B. and Macleod, H.A. (1993) Ingestion of food facilitates the performance of stereotypies in sows. *Animal Behaviour* 46, 939–950.

Tobler, I. and Schweirin, B. (1996) Behavioural sleep in the giraffe (*Giraffa cameleopardalis*) in a zoological garden. *Journal of Sleep Research* 5, 21–32.

Traub-Dargatz, J.L., Kopral, C.A., Hillberg Seitzinger, A., Garbes, L.P., Forde, K. and White, N.A. (2001) Estimate of the national incidence of an operation-level risk factor for colic among horses in the United Sates, Spring (1998–1999). *Journal of American Veterinary Medical Association* 219, 67–71.

United States Department of Agriculture (USDA) (2002) Dairy 2002. Part I: Reference of Dairy Health and Management in the United States. National Animal Health Monitoring System.

Van Soest, P.J. (1994) *Nutritional Ecology of the Ruminant*, 2nd edn. Cornell University Press, Ithaca, New York.

Veasey, J.S., Waran, N.K. and Young, R.J. (1996) On comparing the behaviour of zoo housed animals with wild conspecifics as a welfare indicator, using the giraffe (*Giraffa camelopardalis*) as a model. *Animal Welfare* 5, 139–153.

Veissier, I., Lamy, D. and Le Neindre, P. (1990) Social behaviour in domestic beef cattle when yearling calves are left with the cows for the next calving. *Applied Animal Behaviour Science*, 27, 193–200.

Vickery, S.S. and Mason, G.J. (2004) Stereotypic behaviour in Asiatic black and Malayan sun bears. *Zoo Biology* 23, 409–430.

Vieuille-Thomas, C., Le Pape, G. and Signoret, J.P. (1995) Stereotypies in pregnant sows: indications of influence of the housing system on the patterns expressed by the animals. *Applied Animal Behaviour Science* 44, 19–27.

Von Borell, E. and Hurnik, J.F. (1990) Stereotypic behavior and productivity of sows. *Canadian Journal of Animal Science* 70, 953–956.

Waters, A.J. (2002) Factors influencing the development of stereotypic and redirected behaviours in young horses. PhD thesis, The University of Bristol, UK.

Waters, A.J., Nicol, C.J. and French, N.P. (2002) Factors influencing the development of stereotypic and redirected behaviours in young horses: the findings of a four year prospective epidemiological study. *Equine Veterinary Journal* 34, 572–579.

Whittaker, X., Spoolder, H.A.M., Edwards, S.A., Lawrence, A.B. and Corning, S. (1998) The influence of dietary fibre and the provision on straw on the development of stereotypic behaviour in food-restricted pregnant gilts. *Applied Animal Behaviour Science* 61, 89–102.

Whittaker, X., Edwards, S.A., Spoolder, H.A.M., Lawrence, A.B. and Corning, S. (1999) Effects of straw bedding and high fibre diets on the behaviour of floor fed group-housed sows. *Applied Animal Behaviour Science* 63, 25–39.

Whybrow, J., Cooper, J., Haskell, M. and Lewis, R. (1995) Feed quality and abnormal oral behaviour in lambs housed individually on unbedded slats. In: (eds) Rutter, S.M., Rushen, J., Randle, H.D. and Eddison, J.C. *Proceedings of the 29th Congress for the ISAE*. UFAW, Potters Bar, UK pp. 251–252.

Wiepkema, P.R. (1971) Positive feedbacks at work during feeding. *Behaviour* 39, 266–273.

Wiepkema, P.R., van Hellemond, K.K., Roessingh, P. and Rombery, H. (1987) Behaviour and abomasal damage in veal calves. *Applied Animal Behaviour Science* 18, 257–268.

Willard, J.G., Willard, J.C., Wolfram, S.A. and Baker, J.P. (1977) Effect of diet on cecal pH and feeding behavior of horses. *Journal of Animal Science* 77, 87–93.

Wondra, K.J., Hancock, J.D., Behnke, K.C., Hines, R.H. and Stark, C.R. (1995) Effects of particle size and pelleting on growth performance, nutrient digestibility, and stomach morphology in finishing pigs. *Journal of Animal Science* 73, 757–763.

Wood-Gush, D.G.M. and Beilharz, R.G. (1983) The enrichment of a bare environment for animals in confined conditions. *Applied Animal Ethology* 10, 209–217.

Young, J.Z. (1981) *The Life of Vertebrates*, 3rd edn. Oxford University Press, Oxford, UK.

3

Locomotory Stereotypies in Carnivores: Does Pacing Stem from Hunting, Ranging or Frustrated Escape?

R. Clubb[1] and S. Vickery[2]

[1]Care for the Wild International, The Granary, Tickford Farm, Marches Road, Kingsfold, West Sussex, RH12 3SE, UK; formerly in the Animal Behaviour Research Group, Department of Zoology, Oxford University, South Parks Road, Oxford, OX1 3PS, UK; [2]Animal Welfare Division, Department for Environment, Food & Rural Affairs, 1a Page Street, London, SW1P 4PQ, UK; formerly in the Animal Behaviour Research Group, Department of Zoology, Oxford University, South Parks Road, Oxford, OX1 3PS, UK

Editorial Introduction

The pacing of wild carnivores was one of the first stereotypies commented on by biologists (e.g. zoo curators like Hediger, in the first half of the last century). It is also the stereotypy most familiar to the general public; indeed some forms are so notorious that the Dutch developed a verb applied to pacing, restless people, 'ijsberen': literally, 'to polar bear'. However, carnivore stereotypies attracted little in-depth research until recently, mainly because of the logistical problems of working with these animals. (As a case in point, carnivore stereotypies illustrated the front cover of this book's first edition, but little more.) Yet as Clubb and Vickery illustrate, carnivores are a taxon for which fascinating case studies have been meticulously documented; a plethora of attempts have been made to reduce their stereotypies; and perhaps most importantly, so many reports have been produced that data now exist for 30–40 separate species, allowing species comparisons to be used to formally test hypotheses about the behaviour.

As Clubb and Vickery argue, these diverse approaches differ in their support for the various hypotheses concerning this behaviour's motivational basis. Observational studies often suggest a link between the pre-feeding pacing of carnivores and natural hunting behaviour. However, these studies do not *always* indicate that foraging has a primary role, and nor do data collected when carnivores' environments are altered, often instead suggesting that thwarted escape motivations are important. Cross-species comparisons cast further doubt on a central role for frustrated hunting: no aspect of natural foraging behaviour predicts stereotypy severity, and instead natural home range size proves the key. Clubb and Vickery propose three hypotheses to account for the available evidence. The first is that multiple motivations are involved, notably ranging plus other candidates such as foraging. The second is that carnivore stereotypies represent frustrated escape attempts (to forage, range, reach a mate, or for any one of a host of reasons). The third is that non-motivational factors render naturally wide-ranging species generally prone to persistent stereotypy (e.g. because they have more stamina, or because they are rendered more dysfunctional by captivity), with motivational factors then shaping the stereotypies' timing and form. These untested hypotheses are exciting future directions for captive carnivore research. The ideas raised here about using species differences as a research tool, and about the potential multi-causal bases of stereotypies, could also be pursued in other groups too – for instance the ungulates of Chapter 2; in primates (a contributed box in the chapter describes great unexplained variance in stereotypy within this group); and in the rodents of the following chapter.

GM

3.1. Introduction

Anecdotally, captive carnivores are said to be particularly prone to stereotypy compared with other groups of animals (e.g. Boorer, 1972; Berkson, 1983; Kolter, 1995); indeed as we will see in Chapter 9, this volume, in zoos their stereotypies are much more time-consuming than those of, say, primates. These stereotypies take a range of forms (see Box 3.1 and this book's website), but typically involve locomotion of some kind, e.g. repetitive pacing. Since other forms are relatively rare, and may involve different causal factors (e.g. Carlstead and de Jonge, 1987; Mason, 1993), these locomotory stereotypies are our focus here.

The question of why carnivores stereotype has been raised ever since the first reports of the behaviour. Interest in this topic is not just scientific: the frequent performance of stereotypies can have economic consequences, for example, being associated in farmed mink (*Mustela vison*) with reduced pelt value, and potentially with reduced fertility and increased mortality (reviewed by the European Commission, 2001). These behaviours can also interfere with the aims of some captive establishments; for example, stereotypers might represent less than ideal candidates for reintroduction projects (e.g. Vickery and Mason, 2003b); have questionable educational value (Ormrod, 1987); and attract criticism of zoos by the media and the public (e.g. reviewed by Mason *et al.*, in press). Indeed the public may be justified in their concern, since some studies have shown a link between stereotypies and indicators of poor welfare (reviewed by Mason and Latham, 2004). Thus within the Carnivora, highly stereotypic individuals sometimes show greater signs of stress than those with lower stereotypy levels: for instance, highly stereotypic carnivores may excrete higher levels of the stress hormone cortisol (Bildsøe *et al.*, 1991; Wielebnowski and Brown, 2000; Shepherdson *et al.*, 2004); while seemingly stress-inducing manipulations (being housed near potential predators) and stress-relieving ones (providing hiding places) respectively increase and decrease the levels of both cortisol and pacing in leopard cats

Box 3.1. The Form of Carnivore Stereotypies

R. Clubb and S. Vickery

When carnivores develop stereotypies, these most commonly involve locomotion of some kind. A survey of the literature by Mason *et al.* (in press; see Fig. 1.3, Chapter 1, this volume) showed that locomotory stereotypies were the sole or primary form in over 80% of 61 carnivore species studied. In contrast, very few showed mainly oral stereotypies (e.g. tongue-flicking), or other forms such as repetitive jumping on the spot. The locomotory stereotypies of carnivores mainly involve pacing along a fixed route – back and forth, in circles or in a figure of eight (e.g. Clubb and Mason, 2003), but swimming in circuits (e.g. Hunter *et al.*, 2002), and weaving from side to side are also sometimes seen (e.g. Meyer-Holzapfel, 1968).

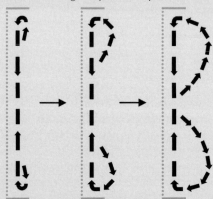

Such stereotypies can change in form over time as the behaviour develops. For instance, figure-of-eight pacing can develop from pacing back and forth along a fence line, as the turn becomes more and more pronounced (see figure, above, adapted from Meyer-Holzapfel, 1968). Meyer-Holzapfel (1968) concluded that this particular stereotypic pattern developed from frustrated escape attempts, in for instance animals wishing to reach neighbouring conspecifics, with the gradually enlarging loop resulting from a gradually diminished motivation to escape. This, and other motivational hypotheses are discussed further in our chapter.

(*Felis bengalensis*) (Carlstead *et al.*, 1993). Furthermore, the actual performance of pacing can cause direct physical harm, such as abrasions, sores or abscesses (Morris, 1964; Meyer-Holzapfel, 1968; Mason, 1991), as well as reducing social interaction (Carlstead and Shepherdson, 1994) and stunting the growth of offspring (Mason *et al.*, 1995). There are, however, studies suggesting a neutral, or even positive effect of stereotypies, such as raised productivity in mink and lower levels of baseline cortisol and other physiological stress indicators (reviewed in European Commission, 2001), making simply preventing animals from pacing a questionable tactic (see also Box 2.3, Chapter 2, this volume; Chapter 10, this volume) and emphasising the need for fundamental understanding.

Carnivores have thus attracted research attention, partly due to the high prevalence of stereotypies in this group, and also due to the charismatic nature of the species. However, carnivores are rarely housed solely for behavioural work so researchers typically work in facilities that already keep them for other purposes, e.g. zoos and fur farms. This has influenced studies in several ways, limiting controlled experimental work (due to small numbers per zoo, varied husbandry between sites and a general lack of control over these factors); restricting developmental studies (due to variable rearing histories in zoo species and some animals' relatively long lifespans); and precluding the invasive work common in laboratory animals (e.g. see Chapters 6–8, this volume). Coupled with the nature of carnivores themselves, experimental work has thus been minimal, especially on topics that could be dangerous (e.g. motivations to escape) or even illegal in some countries (e.g. motivations to hunt and kill). Furthermore, much carnivore research has been pragmatic, focusing on stopping or at least reducing stereotypies, with rather less aimed at testing specific hypotheses about the behaviours' motivational bases (*cf.* Chapter 2, this volume).

Given these limitations, three main types of research have been carried out to date: (i) behavioural observations yielding data on various characteristics of the stereotypy, the animal and its environment; (ii) manipulations revealing how stereotypies change in response to alterations of the environment; and (iii) in one case, use of the 'comparative method' to systematically compare different species' stereotypies with aspects of their behavioural ecology. Box 3.2 gives more details of the pros and cons of these approaches.

In this chapter we review such studies to analyse the evidence for each hypothesized ethological origin of carnivore pacing. As we will see, the study of carnivore stereotypies and their causes has yielded somewhat confusing and contradictory results. However, we will conclude by suggesting three main ways to explain all the evidence, and of testing these hypotheses in the future.

3.2. The Motivational Bases of Carnivore Locomotory Stereotypies

Research across a range of species suggests that stereotypies arise from the persistence and/or thwarting of highly motivated behaviours (see Box 1.1, Chapter 1, this volume). For carnivores, the most widely cited hypothesis as to the motivation underlying locomotory stereotypies implicates natural foraging – an arguably highly motivated behaviour, largely unfulfilled in captive carnivores – with pacing proposed to represent the appetitive, search phase of the hunt (Terlouw *et al.*, 1991; Mason, 1993; Mason and Mendl, 1997). Indeed, many environmental enrichment programmes aim to reduce carnivore stereotypies by providing naturalistic foraging opportunities (see e.g. Chapter 9, this volume). Some researchers have also suggested that obligate carnivores are more prone to stereotypy than carnivores with more generalist/opportunistic diets (Boorer, 1972; Kreger *et al.*, 1998).

However, others have suggested instead that species particularly prone to stereotypies are wide-ranging (Forthman-Quick, 1984; Kreger *et al.*, 1998), or naturally active (Meyer-Holzapfel, 1968; Shepherdson, 1989). Furthermore, yet other motivations still have been implicated, including: escape from specific aversive stimuli (Shepherdson, 1989; Hubrecht, 1995; Carlstead, 1998); reaching conspecifics housed nearby (Meyer-Holzapfel, 1968; Shepherdson, 1989); searching for mates during

Box 3.2. The Methods Used to Study Carnivore Stereotypies: Pros and Cons

R. CLUBB and S. VICKERY

Behavioural observations, with no experimental manipulation, have been taken primarily in zoos, and typically yield detailed descriptions (e.g. a stereotypy's form, frequency, prevalence, location, diurnal rhythm, eliciting stimuli and patterns in relation to key events), but over a rather restricted time scale and in a limited number of animals. Although unable to reveal direct causal relationships, these observations can potentially provide novel, valuable insights into possible causes of stereotypies (Mason, 1993; Carlstead, 1998). For instance, the form of a stereotypy may tell us something about the source behaviour from which it derived, and how it may be alleviated, thence generating testable hypotheses.

Manipulation studies can involve exploiting natural differences across housing systems (e.g. multi-site studies *cf.* e.g. Shepherdson *et al.*, 2004), or manipulating housing and husbandry experimentally (e.g. via 'environmental enrichments', *cf.* Chapter 9, this volume). Again, these have been primarily conducted in zoos. These studies can be time-consuming and labour intensive and, again, sample sizes are typically small (although generally larger for multi-site studies). However, a major advantage of manipulation studies is that, if designed carefully, they provide information on cause and effect.

Finally, comparative methods have recently been used in one study to investigate carnivore stereotypies. This method uses data from multiple species, allowing us to utilize differences in species' biology or behaviour to identify correlations, and even test hypotheses about biological predispositions. Here, we are thus interested in identifying fundamental relationships between two or more factors that should be evident at both a species and individual level. Importantly, these statistical analyses must take into account that some species will have characteristics that cluster together simply because they are closely phylogenetically related. If we were to simply compare species directly, without taking this into account, we may find significant relationships where none exist ((a) Type I error), or fail to find relationships that are actually there ((b) Type II error): as illustrated (below) using hypothetical data for two taxa (e.g. felids – circles and ursids – crosses) (adapted from Gittleman and Luh, 1992).

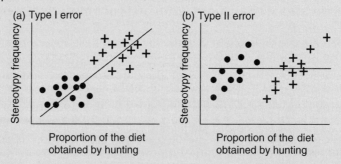

Comparative methods are designed to overcome such problems by taking phylogenetic relatedness into account, and can therefore reveal true patterns, and potentially provide general principles about the causal basis of stereotypies. Since data can often be collated from the literature, this can be a relatively cheap method. It also allows the test of ideas which would be near impossible to tackle experimentally. Downsides are that tests are restricted to available data; using data from multiple sources can introduce confounds (although such effects can be minimized with rigorous selection criteria; see Gittleman, 1989); and a reliable phylogenetic tree is needed. An overview of commonly used comparative methods and their potential role in welfare research can be found in Clubb and Mason (2004), and more general information on comparative methods in Gittleman (1989), Harvey and Pagel (1991) and Freckleton *et al.* (2002).

the natural breeding season (Carlstead and Seidensticker, 1991; Kolter and Zander, 1995) and patrolling a territory (Hediger, 1955). Below, we review the evidence for and against each of these hypotheses.

3.2.1. Do locomotory stereotypies derive from foraging behaviour?

3.2.1.1. Insights from behavioural observations

One reason for the idea that carnivore stereotypies derive from foraging is the predominance of locomotory movements over other forms of abnormal behaviour, and the way this noticeably differs from the herbivorous ungulates (Terlouw *et al.*, 1991; Mason, 1993; Mason and Mendl, 1997; Chapter 2, this volume). In most carnivores, the appetitive component of foraging would naturally take the form of roaming in search of prey, followed by active chase and then capture. Because locomotion occurs mainly during the appetitive stage of hunting, locomotory stereotypies are thought to derive from this phase, rather than from the typically more stereotyped (McFarland, 1981) consummatory phase. Repetition is then held to occur through frustration at being unable to perform the behaviour or reach the expected end-point.

Of course, several natural behaviours in carnivores involve locomotion (e.g. exploration, escape, mate search; Kolter, 1995), but additional characteristics further suggest a specific link to food-getting behaviour. For one, carnivores often pace when food is imminent. For instance, the noises and smells of food preparation or delivery elicit pacing in various species (e.g. Mason, 1993; Kolter and Zander, 1995), often in the location where food is delivered or staff delivering food can be seen to approach (Mason, 1993; Carlstead, 1998; Vickery and Mason, 2004). 'Big cats' in zoos may even pace at the sight of what in the wild would be potential prey: ponies, or even small children running past (Boorer, 1972)! Carnivores also often show increasing levels of stereotypy in the run up to feeding time (e.g. Carlstead and de Jonge, 1987; Shepherdson *et al.*, 1993; Vickery and Mason, 2004), often virtually ceasing afterwards (e.g. Carlstead and de Jonge, 1987; Mason, 1993; Vickery and Mason, 2004); indeed in a literature survey by Mason (Mason *et al.*, in press) this pre-feed peak was evident in over 70% of 21 carnivore species studied (Mason *et al.*, unpublished). Notably, in mink, pacing *per se* peaks immediately before feeding, yet other non-locomotory forms of stereotypy do not (Mason, 1993; see also Vickery and Mason, 2004). Pacing in the American black bear (*Ursus americanus*) showed a similar pattern, peaking pre-feed, but only before the natural period of hibernation when wild bears are busy foraging to build up energy reserves (Carlstead and Seidensticker, 1991). Finally, juvenile minks' interest in food, as measured by the amount of stretching to see the approaching food delivery cart, predicted their later development of pre-feed (predominantly locomotory) stereotypies as young adults (Mason, 1992).

However, there is also evidence inconsistent with the foraging hypothesis. Pacing may be accompanied by behaviour unrelated to, and even in conflict with hunting, such as vocalization in caged mink (Mason, 1993). Several authors have also found no relationship between an individual's

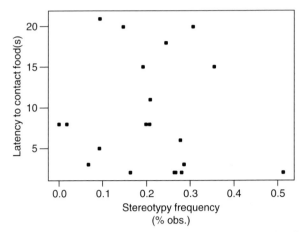

Fig. 3.1. A measure of feeding motivation – latency to contact food (in seconds) – was unrelated to stereotypy frequency (calculated as a proportion of all observations made) across 12 Asiatic black bears (5 males, 7 females) and 8 Malayan sun bears (5 males, 3 females). Data from Vickery (2003).

level of stereotypy and measures of food motivation (Mason, 1990; Vickery, 2003, see Fig. 3.1). Lastly, many carnivores pace after, as well as before feeding time (e.g. Carlstead *et al.*, 1991, 1993; Mason, 1993) and sometimes to a greater extent (e.g. Lyons *et al.*, 1997). In fennec foxes (*Vulpes zerda*), this has been attributed to frustrated post-meal caching behaviour (Carlstead, 1991); but this would not explain similar observations in other, non-caching, species. Of course, it is possible that animals stereotype after being fed because their food rations fail to relieve their hunger (*cf.* data for pigs, Chapter 2, this volume), but this would seem unlikely for animals held in zoos.

3.2.1.2. Insights from environmental manipulation

Manipulating food rations can affect stereotypy levels. Mason (1993) observed sustained high levels of stereotypy on a day when farmed mink were not fed, and Bildsøe *et al.* (1991) found that reducing minks' daily food allowance by one-third caused their stereotypy levels to triple. Zoo-housed polar bears (*Ursus maritimus*) and big cats (*Panthera* spp.) also stereotype more on days when they are not fed – so-called 'starve days' (Lyons *et al.*, 1997; Ames, 2000). Conversely, mink and bears perform less stereotypy when provided with *ad libitum* food (Houbak and Møller, 2000; Vickery, 2003), as do bears if fed more frequently (Vickery, 2003, see Fig. 3.2). Preventing access to food can also elicit pacing (e.g. Kolter and Zander, 1995), while being shut out of areas containing other valued resources does not (e.g. Lewis *et al.*, 2001; Warburton and Mason, in preparation).

Food quality and the method of presentation can also affect stereotypy levels. In zoos, foraging-based methods are probably the most widely utilized environmental enrichments (Shepherdson *et al.*, 1998; Chapter 9, this volume) and considered among the most successful (Kreger *et al.*, 1998). Thus several studies report less frequent locomotory stereotypy in

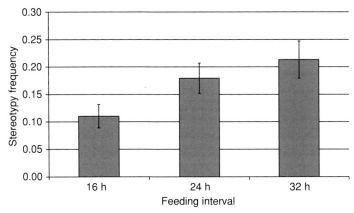

Fig. 3.2. Mean stereotypy frequency (± SEM) of 27 bears (16 Asiatic black bears [7 male, 9 female] and 11 Malayan sun bears [5 male, 6 female]) maintained on a feeding schedule with 16, 24 or 32 h between meals. All columns are significantly different from each other ($P < 0.05$). Data from Vickery (2003).

animals provided with materials that allow the performance of specific natural foraging behaviours, such as food-searching (Kastelein and Wiepkema, 1989; Shepherdson *et al.*, 1993), chasing (Markowitz and LaForse, 1987), capture (Charlton, 1995), or 'processing' such as prey-plucking (Hancocks, 1980; Forthman *et al.*, 1992; Bashaw *et al.*, 2003), as well as manipulations that simultaneously stimulate a whole range of naturalistic foraging behaviours, such as the provision of live fish or crickets (Dobberstine and Shepherdson, 1994; Bashaw *et al.*, 2003). Methods that simply make food more difficult or time-consuming to obtain, without necessarily stimulating specific natural behaviours, likewise have some success in decreasing pacing (Landrigan *et al.*, 2001; Jenny and Schmid, 2002), and offering food earlier in the day has, anecdotally, been reported to have a similar effect (Law *et al.*, 1990; although see Vickery, 2003). Also, providing various environmental enrichment items to giant pandas (*Ailuropoda melanoleuca*) caused a reduction in food anticipatory behaviour (i.e. being alert and near the area where food is delivered) as well as a reduction in stereotypic behaviour (Swaisgood *et al.*, 2001), further suggesting a link between stereotypies and the motivation to forage.

However, feeding enrichments are far from universally successful. For example, Kolter and Zander (1995) found that feeding enrichments reduced the stereotypy of one polar bear, but not that of another living in the same enclosure; while Dobberstine and Shepherdson (1994) reported that scattering low-calorie food reduced the stereotypy of three American black bears but increased it in two spectacled bears (*Tremarctos ornatus*). Furthermore, giving a stereotyping carnivore unlimited access to food does not always eliminate the behaviour, sometimes merely shifting it to less conspicuous times of day (Hansen *et al.*, 1994; Vickery, 2003). Finally, in Chapter 9, this volume, feeding-related enrichments seem no more effective than any other type, across multiple species including several Carnivora.

3.2.1.3. Insights from the comparative method

A comparative study of 33 Carnivora species, comparing the pacing levels of stereotypers with aspects of species-typical behavioural biology, found no relationship between stereotypy and a range of measurements of natural foraging behaviour (Clubb, 2001; Clubb and Mason, 2003, 2006). For instance, species that naturally invest more time in foraging did not stereotype more; and neither did species that naturally depend heavily on active hunting and killing to obtain food, compared with scavengers or more omnivorous species (see Fig. 3.3).

3.2.1.4. Summary

The strongest evidence for the foraging hypothesis comes from direct observations of carnivore stereotypies. The form, timing and circumstances of stereotypic pacing largely point to a link with feeding. This link appears somewhat weaker, however, in studies that manipulate aspects of the environment: although circumstantial evidence suggests that hunger stimulates pacing in some cases, it is difficult to draw any strong conclusions from food-related enrichments, as these have rarely been administered with a sufficient level of experimental control to demonstrate a specific motivational link. Furthermore, species differences in stereotypy levels were not explained by variation in species' natural foraging behaviour in the comparative method study, thus providing no evidence for the foraging hypothesis.

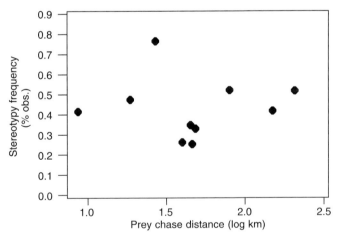

Fig. 3.3. Species that actively hunt down their prey, quantified here as the median chase distance for the main prey items, did not display more time-consuming locomotory stereotypy than their more sedentary relatives ($F_{1,9} = 0.00$, $P > 0.05$). Each point represents the median value for that species' stereotypers. Data were collated from the literature (see Clubb and Mason, 2003, 2006, for methodology).

3.2.2. Do locomotory stereotypies derive from motivations to range, or to be generally active?

Informally, wide-ranging carnivores, such as wolves and coyotes, have been identified as vulnerable to pacing in captivity (Forthman-Quick, 1984; Kreger *et al.*, 1998). Highly active species, such as bears, have also been suggested to be particularly at risk of pacing (Meyer-Holzapfel, 1968). Ranging, however, encompasses all kinds of locomotion and thus a whole range of motivations would presumably be involved, including seeking food and/or mates, patrolling a territory, migrating and exploring. Motivations to be generally active would presumably be even broader in scope, including all those mentioned above plus other activities that do not involve travel of any sort (e.g. grooming, social interaction). Because an animal could be active for much of the day yet not travel any great distance, general activity is considered distinct from ranging behaviour in this sub-section.

3.2.2.1. Insights from behavioural observations

The study of a male American black bear mentioned above suggests that pacing in this animal was caused by ranging to attain different goals: foraging before hibernation, and searching for mates during the breeding season (Carlstead and Seidensticker, 1991). Support for the suggestion that high activity levels lead to stereotypies also comes from several studies in which, within a population, the most stereotypic individuals are also the most normally active (e.g. Bildsøe *et al.*, 1990a, 1991) – although the opposite has also been reported (van Keulen-Kromhout, 1978; Hansen *et al.*, 1994; Ames, 2000). Stereotypies may also follow the same daily or seasonal rhythms as normal activity (Bildsøe *et al.*, 1990b; Hansen, 1993; Vickery, 2003).

3.2.2.2. Insights from environmental manipulation

Carnivore stereotypies are often particularly prevalent in animals housed in small enclosures that constrain normal activity (van Keulen-Kromhout, 1978), and lower in larger cages with increased opportunities for movement (e.g. Mellen *et al.*, 1998; Hansen and Jeppesen, 2000). Several studies also report that transferring individuals into more spacious enclosures reduces locomotory stereotypy (e.g. Carlstead, 1991; Kolter and Zander, 1995; Langenhorst, 1998). However, from this it is impossible to say whether stereotypy decreases because the situation fulfils some 'need' for activity *per se* – if such a thing even exists. Furthermore, some studies of carnivores transferred to more spacious enclosures found no effect (Meyer-Holzapfel, 1968; Hansen *et al.*, 1994), while mink have sometimes been reported to *reduce* pacing in smaller cages (Hansen, 1998).

Likewise with general activity, many manipulations (especially environmental enrichments) do reduce stereotypy while also increasing activity (Forthman *et al.*, 1992; Swaisgood *et al.*, 2001); but whether or not they work because they increase opportunities for various active behaviours is often impossible to say, because they may also work by meeting more specific motivations, or by other means still (*cf.* Chapter 9, this volume).

3.2.2.3. Insights from the comparative method

The strongest evidence for the ranging hypothesis comes from comparative work. Clubb and Mason (2003) tested this idea formally with their multi-species data-set, and found significant positive relationships between the extent of ranging in the wild (quantified by species' typical home range sizes and daily travel distances) and the frequency of pacing in captive individuals developing this behaviour (see Fig. 3.4). There was no activity effect, however: species that typically spend more time generally active did not show higher levels of stereotypy than less active species.

3.2.2.4. Summary

Both the ranging and activity hypotheses suffer from a lack of studies setting out to test them specifically (indeed it is probably hard to do this experimentally). Scant data – mostly observational – were available to support the hypothesis that locomotory stereotypies derive from a 'need' to be generally active, and indeed such a need is itself questionable. Comparing species yielded the strongest support for the ranging hypothesis, with naturally wide-ranging species systematically being most prone to the highest levels of pacing.

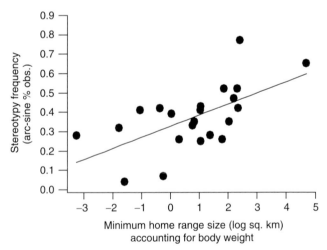

Fig. 3.4. Species that typically roam over large distances in the wild, measured by e.g. reported minimum home range sizes, were found to show higher stereotypy levels in captivity than less wide-ranging species ($F_{2,19} = 4.79$, $P = 0.01$). Each point represents the median value for a species ($n = 22$), accounting for body weight. Data were collated from the literature (see Clubb and Mason, 2003, 2006, for methodology).

3.2.3. Do locomotory stereotypies derive from motivations to explore?

Information-gathering through exploration is thought to be reinforcing in its own right (e.g. Inglis, 1983; Toates, 1983; Inglis *et al.*, 1997) and the failure of the captive environment to adequately satisfy 'needs for information' has been cited as a possible cause of stereotypies (Poole, 1998; Swaisgood *et al.*, 2001), particularly in species thought to be highly neophilic, such as bears (Ormrod, 1987; Poole, 1998).

Most insights are available from environmental manipulation studies. Stereotypies are certainly typical of animals held in barren, unchanging environments (Hediger, 1950; Morris, 1964; Carlstead, 1998) and lower stereotypy levels have been reported in animals held in, or moved to, more complex enclosures (Carlstead, 1992; Barry, 1998; Mellen *et al.*, 1998). Enrichments that introduce some element of novelty and/or promote exploratory behaviour have also successfully reduced pacing (Shepherdson *et al.*, 1993; Swaisgood *et al.*, 2001). Such manipulations also often lose their effectiveness over time, with stereotypy frequencies creeping back to baseline levels (e.g. Ödberg, 1984). Such an effect could be the result of the loss of novelty that occurs through habituation over time, prompting some to advocate programmes that frequently rotate different enrichments to maintain a high level of novelty (e.g. Eyre, 1997; Knowles and Plowman, 2001). However, whether such effects were solely due to opportunities to gather information from novel stimuli is impossible to say.

3.2.4. Do locomotory stereotypies derive from motivations to escape aversive stimuli?

3.2.4.1. Insights from behavioural observations

There are many reports of carnivores stereotyping during or after exposure to aversive stimuli, such as after aggressive encounters (e.g. Mudway, 1992; Koene, 1995; Fischbacher and Schmid, 1999); when confined in unnatural proximity to conspecifics (de Jonge and Carlstead, 1987; Kolter and Zander, 1995; Wielebnowski *et al.*, 2002a); in response to signs of potential predators (Carlstead *et al.*, 1993); during noisy cage-cleaning (e.g. Meyer-Holzapfel, 1968; Mallapur and Chellam, 2002); after unexpected or unusual loud noises (e.g. aeroplanes, or loud music: Ames, 1993; Koene, 1995); when visitors congregate in large numbers (Carlstead, 1991, see Fig. 3.5) or are very noisy (Fentress, 1976); and when locked into small indoor dens or outdoor enclosures in poor weather (Meyer-Holzapfel, 1968; Ames, 1993). In these situations, animals often stereotype against the barrier in their way (e.g. a closed door), or in places as distant as possible from the aversive stimulus (Meyer-Holzapfel, 1968; Mason, 1993).

However, as evidence against this hypothesis, factors that might be expected to elicit escape attempts, such as close human proximity or daily immobilizations, sometimes reduce rather than increase stereotypy (Bildsøe *et al.*, 1990a, 1991).

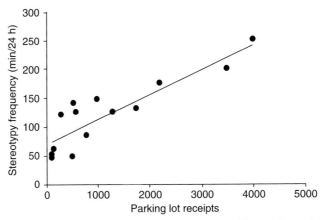

Fig. 3.5. Total stereotypic pacing frequency in a pair of fennec foxes during periods with varying visitor numbers, as measured by the number of car-parking receipts. A significant positive relationship was found (Spearman rank test = 0.72, $P < 0.001$). Data from Carlstead, published in a different form in Carlstead (1991).

3.2.4.2. Insights from environmental manipulation

Providing conditions that allow animals to hide or escape from potential dangers has been associated with lowered stereotypy levels in some carnivores. Thus small cats with more hiding places spend less time pacing (Mellen *et al.*, 1998), as do clouded leopards (Wielebnowski *et al.*, 2002b) and mink with access to enclosed nest-boxes (Hansen *et al.*, 1994), compared to individuals without such opportunities. We see similar effects within individual animals when changes are made to their environment. Four leopard cats, able to hear and smell potential predators housed nearby (lions, tigers, pumas), showed a reduction both in pacing and urinary cortisol levels when enrichments enabling the cats to hide (e.g. hollow logs, boxes) were added to their enclosures (Carlstead *et al.*, 1993), again suggesting that the motivation to escape, due to fear, was a causal factor in their pacing. Similarly, removing a stressor often reduces stereotypy. For example, the pacing of fennec foxes was reduced when the method of cleaning their enclosures was changed from noisy vacuuming to quiet sweeping (Carlstead, 1991) and, in another study, when one fennec fox was removed from the enclosure, pacing ceased completely in a remaining animal (Ödberg, 1984). Social stress and resulting motivations to escape may also explain the case of a subordinate polar bear whose pacing was unaffected by environmental enrichment but decreased when the two females with whom she shared her enclosure were confined in dens (Kolter and Zander, 1995). Lastly, the anxiety-reducing drug, Prozac, reportedly reduced the stereotypy levels of one polar bear (Poulsen *et al.*, 1996; Box 10.4, Chapter 10, this volume), possibly adding weight to the argument that stereotypies result from aversive experiences (though see Chapter 10, for caveats).

3.2.4.3. Summary

A diverse variety of aversive stimuli prompt pacing in a wide range of carnivores. Manipulations introducing stimuli judged or demonstrated to be unpleasant frequently result in increased stereotypy, while reducing such stimuli likewise decreases stereotypy. This is not, however, universally true, since not all aversive experiences have this effect (and furthermore, since we have seen, the anticipation of positive events like feeding can also elicit these behaviours).

3.2.5. Do locomotory stereotypies derive from motivations to approach conspecifics?

3.2.5.1. Insights from behavioural observations

There is some evidence that locomotory stereotypies arise from thwarted attempts to approach conspecifics. Stereotypies are often elicited when animals are housed next to conspecifics that they appear motivated to approach. Examples include a dingo (*Canis familiaris*) housed adjacent to its companions (Meyer-Holzapfel, 1968; see also Box 3.1); male mink caged next to sexually receptive females; and mink making aggressive approaches towards a neighbour (Mason, 1993). Notably, these stereotypies are often performed along the segregating boundary, suggesting that they derive from thwarted attempts to reach individuals from which they have been separated. This effect would, however, appear to be quite specific to animals housed in these types of situation.

Another related behaviour that may underlie stereotypies is the search for mates during the natural breeding season, which can involve locomotion over considerable distances. Evidence comes almost entirely from observational studies of several species of bear, although some data on mink hint at possible wider relevance. Stereotypies in these species reportedly increase during natural breeding seasons, when locomotion to seek mates would be important (Carlstead and Seidensticker, 1991; Hansen *et al.*, 1997; Ames, 2000). Thus the elevated pacing of a male black bear during the natural breeding season was proposed to derive from mate-seeking (Carlstead and Seidensticker, 1991), especially as it peaked post-feeding in timing and reduced following the placement of bear odours (from both sexes) around the enclosure – findings in marked contrast to the pattern seen during the natural foraging season. Similarly, data obtained on 14 polar bears (male and female), showed their stereotypies to peak during their natural breeding period, although the author suggests that social stress may have been involved too (Ames, 2000). Overall, the motivation to seek out mates thus has some explanatory power, but is limited to stereotypies shown at specific times of the year (for species with defined breeding seasons) and animals that usually roam in search of mates.

3.2.5.2. Insights from environmental manipulation

Stereotypies have been observed in animals that have been separated from conspecifics (in social species) or their young (in the case of adult females) (e.g. Ames, 1992; Mason, 1993; Lyons *et al.*, 1997). Again, however, these data relate to animals in a very specific situation.

3.2.6. Do locomotory stereotypies derive from motivations to patrol a territory?

3.2.6.1. Insights from behavioural observations

Some have suggested that locomotory stereotypies derive from territorial patrolling (Hediger, 1955; Morris, 1964), due to the fact that the predominant form is pacing back and forth along or around the edge of an enclosure, as if it were the animal's territory and defending the borders were important. Hediger (1950) points out that many captive species behave as if they view their enclosure as their territory, for instance by defending it against intruders. There are also reports of animals performing repetitive scent-marking in their enclosures – a territorial behaviour performed during patrolling – in combination with pacing (Boorer, 1972). Scent-marking is also reported to increase in parallel with pacing, during times of natural dispersal (Hansen *et al.*, 1997) or when placed in an enclosure previously housing other animals (White *et al.*, 2003). However, Weller and Bennett (2001) report no relationship between pacing and scent-marking in captive ocelots (*Leopardus pardalis*); pacing at enclosure edges could be for a variety of reasons (e.g. information-gathering); and there also do not appear to be any reports of other territory maintenance behaviours, such as relevant vocalizations or ground-scratching (e.g. Funston *et al.*, 1998; Sillero-Zubiri and Macdonald, 1998; Frommolt *et al.*, 2003), accompanying stereotypies in Carnivora.

3.2.6.2. Insights from the comparative method

Results of the comparative study do not provide support for this hypothesis, territorial species being no more stereotypic than non-territorial species (Clubb and Mason, 2003, 2006).

3.3. Resolving the Evidence

The evidence presented above shows patterns, but no exclusive support for any one idea. In fact, different methodological approaches tend to support different hypotheses: behavioural observations and environmental manipulations provide most evidence for frustrated foraging or escape behaviour, whereas cross-species comparisons show natural ranging to lie at the heart of carnivore pacing. Can we explain these apparent conflicts? Can we resolve these lines of evidence into one unifying motivational theory as to why carnivores exhibit locomotory stereotypies? Or could 'non-motivational' aspects of behavioural control help to further explain what is going on? We tackle these questions below.

3.3.1. Could methodological problems be behind the conflicting evidence?

One possibility is that the data we have discussed are 'noisy' and/or over-interpreted – because manipulations were aimed at reducing stereotypy for practical reasons rather than specific hypothesis testing, and control data were not collected so that alternative explanations cannot be ruled out. For instance, a reduction in stereotypy caused by enlarging an enclosure could be evidence for the ranging hypothesis, yet the change may instead stem from some other alteration, such as the novelty of the new space, or the increased distance it allows from human visitors. Likewise, increasing environmental complexity provides more opportunities for exploration (consistent with the 'frustrated exploration' hypothesis), but it also typically increases usable space, hiding places, opportunities for more varied behaviours and may additionally make the public seem more distant (Chapter 7, this volume, similarly discusses the diverse properties often offered by 'environmental enrichment').

As well as these sorts of uncontrolled-for independent variables, enrichment studies might be misleading if they work by means other than reducing the underlying motivation (Chapters 9 and 10, this volume). For example, manipulations may reduce stereotypies simply by occupying the animal's time. In this case, a reduction would only be evident if the time the manipulation occupies (e.g. time spent interacting with an enrichment item) is not statistically controlled for (*cf.* Swaisgood *et al.*, 2001; Vickery, 2003). Such time-occupying manipulations often work best when they are novel, and gradual increases in stereotypy might be observed over time as the animal either gets better at utilizing the enrichment (e.g. Ings *et al.*, 1997) or loses interest in it. There may also be other reasons for observed apparent changes in behaviour. Taking the example of cage enlargement, locomotion may simply *look* less stereotyped in the larger enclosure rather than the manipulation having tackled the root motivation: after all, there are only so many ways an animal can walk around in a small square cage (*cf.* Chapter 2, this volume, on 'channelling').

Conversely, a failure to reduce stereotypy via environmental manipulation does not necessarily mean that the behaviour's underlying motivation was not met. Instead, the duration of the manipulation may simply have been insufficient to reverse the stereotypy (as shown in parrots: Meehan *et al.*, 2001), particularly if mechanisms other than motivational frustration are also involved (see below).

3.3.2. Could a combination of motivations, or one 'umbrella' motivation, be the key?

Leaving aside these concerns about experimental design, data interpretation and potential confounds in many observational and manipulation studies, perhaps one alternative explanation is that there are genuinely multiple motivations are at work, inducing superficially

similar stereotypies for very different reasons across a range of species –
or even in the same species or individual. However, this would not
explain why there appear such strong patterns in terms of the factors
that influence stereotypies across carnivores: these do suggest that common mechanisms are involved. Furthermore, the 'multiple, equally important motivations' idea would not explain why different approaches
seem fairly consistently to point in different directions.

Instead, our apparently contradictory results could be complementary. We suggest two motivational explanations that could account for
them. The first is that the comparative study identified the predominant
causal motivational factor (ranging), while observational and manipulation studies identify lesser, albeit influential factors (e.g. frustrated foraging). Thus, frustrated ranging may lie at the heart of the matter,
correlating as it does with stereotypy severity, but other motivational
factors explain some of the remaining variation in the data.

The second way the evidence can be drawn together in a motivational
explanation was first pointed out by Shepherdson (1989): perhaps all
locomotory stereotypies derive from motivations to escape the enclosure,
with this occurring for a variety of different reasons. Thus carnivores may
be motivated to retreat from aversive stimuli (as we reviewed above), and/
or to escape to obtain some desired resource that is lacking (e.g. food,
novelty, a larger range or potential mates), and/or to reach some resource
that can be perceived but not obtained (e.g. conspecifics housed next
door; food being prepared nearby). A wide range of manipulations that
alter these factors (e.g. by providing the goals of such behaviours or an
opportunity to perform them) would then lead to a reduction in stereotypy. This might at first sight appear merely an 'umbrella term' encompassing a multitude of different motivations, but we propose something
more specific. In this scenario, locomotory stereotypies do not derive
from specific, separate natural behaviour patterns (such as actively
searching for and chasing down prey, or roaming in search of a mate),
but instead all represent one form of behaviour: thwarted escape. This
could explain why carnivore stereotypies can be accompanied by behaviours that seem more related to general arousal than to specific implicated
motivations (e.g. hunting), such as vocalization in a pacing mink (Mason,
1993). This could also explain why thwarting quite different motivations
(e.g. foraging, mate seeking) leads to the development of such similar-
looking stereotypies. It could also explain why the behaviour often occurs
at enclosure edges. One prediction of this idea is that, as well as being
affected by factors that diminish specific motivations, escape – and hence
locomotory stereotypies – should be affected by any factor that alters the
perceived aversiveness of the cage environment; thus any enrichment, or
indeed drug, that generally lowers stress levels should be effective, even
within a single individual (and any generally stress-inducing factor
should have the opposite effect).

So how does the comparative study implicating ranging behaviour fit
in? Perhaps the captive environment is more aversive to wide-rangers,
resulting in a greater desire to escape; or they may instead be strongly

inclined to move to pastures new when finding themselves in a sub-optimal environment, rather than 'sticking it out' until conditions improve (Clubb and Mason, in press). However, further possible explanations involve non-motivational aspects of behavioural control, as we discuss further below.

3.3.3. Could 'non-motivational' factors be at work?

Mechanisms other than, or in addition to, the frustration of specific motivations may play a role in carnivore stereotypies. These mechanisms may be quite normal. For example, in being adapted to travel over greater distances, wide-rangers may have higher levels of physical endurance and hence simply be able to stereotype for longer periods of time in captivity than other carnivores (Clubb and Mason, 2006). Thus range size here simply yields a non-motivational predisposition to more sustained pacing.

Instead of, or in addition to, frustrated motivation, stereotypies may represent habits that have developed from being repeated time and time again (e.g. Dickinson, 1985; Mason and Latham, 2004). Such mechanisms might potentially help to explain some confusing results, such as why stereotypies are seen in diverse situations and elicited by many different stimuli (e.g. Fox, 1971; Fentress, 1976; Ödberg, 1978; Chapters 4 and 10, this volume, for more on 'establishment' and 'emancipation'). Furthermore, it could also perhaps explain why carnivore stereotypies are often not reduced (Meyer-Holzapfel, 1968; Vickery and Mason, 2003a) or even increase (Ames, 1992) following enrichment (see also Bildsøe *et al.*, 1991). One study on mink did suggest their stereotypies to be habit-like, animals being less responsive to a noise during stereotypy than when they were moving around normally (Clubb, 2001, see Fig. 3.6). A similar study on Asiatic black bears (*Ursus thibetanus*) and Malayan sun bears (*Helarctos malayanus*) also found that animals were slower to respond to a food-rewarded operant task when they were stereotyping than when they were generally active (Vickery, 2003).

Alternatively, carnivore stereotypies may not stem from normal behavioural control but instead arise from changes making animals abnormally persistent in all types of behaviour (e.g. 'perseverative', as discussed in Chapters 5 and 10, this volume). This could be due to some early deprivation (e.g. Sandson and Albert, 1984; *cf.* Chapter 6, this volume) or simply to time spent in captivity (*cf.* Chapter 7, this volume). This would further help to explain why carnivore stereotypies often continue after stimuli that should terminate them (e.g. Ödberg, 1978; Templin, 1993); and why some individuals' stereotypies can persist to such an extent that offspring are neglected (e.g. Mason *et al.*, 1995), the animal becomes injured or, despite a change of environment, the animal paces the exact dimensions of a previous smaller enclosure (Meyer-Holzapfel, 1968). Thus anecdotal evidence does suggest that such mechanisms are involved in the stereotypies of at least some carnivores (Mason and Latham, 2004; Mason *et al.*, 2006), yet there have been few empirical tests. Just one study, on Asiatic black bears

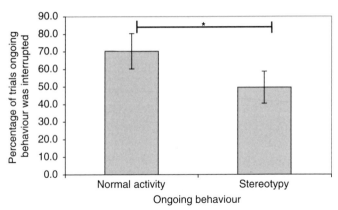

Fig. 3.6. Mean proportion of trials (± SEM) in which adult female mink ($n = 11$) stopped their ongoing activity in response to the sound of a bell (Clubb, 2001). Fewer mink were interrupted when engaged in a bout of stereotypic behaviour than when they were moving normally around their cage. *$P < 0.05$.

and Malayan sun bears, has investigated this; and highly stereotypic individuals were indeed more generally perseverative than their less stereotypic counterparts, continuing to perform a previously rewarded operant response for longer when it became unrewarding (Vickery and Mason, 2003b, 2005; see Fig. 5.2, Chapter 5, this volume). Thus overall, the stereotypies observed and manipulated in at least some carnivores may not represent symptoms of a current frustrated motivation. Instead, they may simply be ingrained habits, or even symptoms of neural changes that have made the animal very persistent in everything it does. In the latter case, naturally large range sizes may thence predispose carnivores to more severe stereotypy by predisposing them to greater perseveration (see also Box 5.4, Chapter 5, this volume).

3.4. Summary, Conclusions, and Suggestions for Future Work

Captive carnivores are informally known to be very prone to stereotypies. Their most common forms are locomotory in nature, consisting primarily of pacing back and forth and variations thereof. Frustration of a strongly motivated behaviour is generally thought to be the underlying cause of stereotypies, and a diverse array of source behaviours has been implicated for captive carnivores. The most widely cited hypothesis focuses on foraging – a potentially highly motivated behaviour that remains largely unfulfilled in captive carnivores – with pacing back and forth being proposed as equivalent to the appetitive search phase of the hunt. However, other motivational explanations include: escape from aversive stimuli; attempts to reach conspecifics or mates; and patrolling a territory. Some researchers have also highlighted species particularly prone to stereotypies, such as those that are wide ranging, naturally active or that have generalist/opportunistic lifestyles, suggesting further motivations behind stereotypy development. In this chapter, we therefore reviewed much of the existing research on carnivore locomotory stereotypies with a view to identifying the motivation(s) underlying their development. Unfortunately, studying carnivore stereotypies carries with it many restrictions, imposed primarily by the availability and nature of the study animals; and this may partly explain why the evidence does not point exclusively in any one direction. Furthermore, many of the studies we review did not set out to test the specific hypotheses considered here, so it should come as no surprise that methodologies are not always stringent and results sometimes open to interpretation. However, despite these caveats, patterns did emerge, with different methodological approaches tending to support different hypotheses. Thus behavioural observations and environmental manipulation studies provided the most evidence for the frustration of foraging behaviour and escape from aversive stimuli, while the cross-species study supported only the ranging hypothesis.

We suggest three possible explanations for this, for future testing. The first is that multiple motivations are involved in the behaviour, in a complementary rather than exclusive manner, collectively explaining stereotypy development. Thus, a motivation to hunt may explain the variance that remains between species after ranging behaviour has been taken into account. Testing this might require more comparative work in the future, once more field data on natural behavioural biology have accumulated, allowing us to run multivariate regressions across the species (so far, our multi-species data-set only allows us to run univariate regressions with any statistical power; Clubb and Mason, 2006). The prediction would then be that species that invest a lot of time in hunting, for instance, stereotype more than expected given how far they range in the wild.

Our second hypothesis is that the behaviour arises for one reason – to escape – but for varying purposes, such as to forage, find a mate or escape from some aversive event. In this scenario, all stereotypic pacing thus arises from repeatedly thwarted escape attempts. As such, it should be reduced by any manipulation that makes the captive environment a more desirable place to be, but increased by any manipulation that increases its perceived aversiveness. One might also test this idea via experiments similar to those conducted on mice to show that stereotypic bar-gnawing is motivated by escape (Nevison *et al.*, 1999; Lewis and Hurst, 2004; Chapter 4, this volume). For example, using a small carnivore such as mink, one would predict that all locomotory stereotypies should be performed next to a regularly opened exit (e.g. providing access to a familiar alternative enclosure), rather than in other locations.

Our third hypothesis is that non-motivational factors are involved, with natural home range size leading to some general predisposition to sustained stereotypy, the form and timing of which is then shaped by other, motivational factors. For instance, stereotypy levels may reflect 'exercise physiology'-related abilities to perform sustained locomotion (Clubb and Mason, 2006); represent ingrained habits, that are then triggered by multiple cues; or be due to perseveration, perhaps because past deprivation has led an animal to become abnormally persistent in all behaviours (Chapter 5, this volume). Investigating species differences in such attributes would thus be useful. For example, adapting methods commonly used in human psychology has allowed the investigation of perseverative tendencies, and measurement of habit-like qualities of stereotypies, in bears and mink (e.g. Vickery and Mason, 2003b; Chapter 5, this volume); similar studies could thus be conducted on zoo-housed carnivores with careful experimental design (see Mason *et al.*, 2006), to test how such factors relate to species differences in ranging.

It might also be beneficial to incorporate other methods, to date primarily used to study stereotypies in other taxa. For instance, rigorous developmental studies (*cf.* e.g. Cronin, 1985 on sows) would be welcome, shedding light on the potentially varying causal factors involved as stereotypy emerges and matures. Thus if stereotypies become more habit-like over time, developmental studies following individual animals would allow this to be tested directly and enable the effect of environmental manipulations to be investigated at varying stages. It is also clear that more well-controlled, hypothesis-driven experimental work is required (*cf.* e.g. research on rodents, as reviewed in Chapter 4, this volume) to further test the hypotheses discussed in this chapter, perhaps helped by a greater use of existing laboratory-housed carnivores (e.g. dogs, cats, ferrets), fur farms, kennels or catteries, where large sample sizes and more standardized housing could facilitate such rigorous approaches. Finally, well thought out epidemiological studies could also help, such as via the use of multiple zoos (*cf.* e.g. Mellen *et al.*, 1998; Shepherdson *et al.*, 2004; *cf.* Chapters 2 and 6, this volume).

Conversely, it should be noted that the study of stereotypies in other taxa could well benefit from some of the approaches described here. In particular, the comparative method used fruitfully in carnivores could now be used with other groups (e.g. ungulates), to see whether similar relationships between ranging and pacing are found, or to test hypotheses relating to other types of stereotypy (see e.g. Chapter 2, this volume). There is also much unexplained, well-documented species variation in abnormal behaviour within primates (see Box 3.3 by Novak and Bollen), and rodents (as discussed in the following chapter), potentially representing further valuable material for hypothesis-testing, especially for ideas it would be hard to tackle experimentally.

To conclude, results from further research on carnivore stereotypy would be of inherent interest, but also have practical implications too. Identifying why carnivores pace is vital if we are to accurately infer the welfare of these stereotyping animals, and also to properly identify measures that can be taken to prevent or reduce the development of these stereotypies successfully. After all, as we will see in Chapter 9, current enrichment-use in zoos very rarely abolishes stereotypies that have already emerged. Of course, it may prove not actually feasible to provide captive environments for all carnivores that eliminate pacing; thus in the future, we may perhaps have to be content viewing some species on the big screen, and not in the zoo.

Box 3.3. Differences in the Prevalence and Form of Abnormal Behaviour Across Primates

M.A. NOVAK and K.S. BOLLEN

Within primates, just as in the carnivores of this chapter, the incidence, severity and form of stereotypic behaviour varies greatly across species. Thus in one study, prosimians displayed virtually none, while the incidence was 4.6% for New World monkeys, 6.7% for Old World monkeys and 6% for apes (Trollope, 1977). We recently conducted a survey of 3465 primates in zoological facilities, on the prevalence of abnormal behaviour (Bollen *et al.*, submitted for publication). Self-injurious behaviour occurred in all groups, from prosimians to apes, but the level was very low (approximately 2.2% of subjects) and did not vary significantly with taxon. In contrast, stereotypic behaviour did show taxon effects, again indicating that prosimians, and also New World monkeys exhibit the least (7%), followed by Old World monkeys (13.8%) and apes (39.6%). The greater incidence of abnormal behaviour in our study compared to Trollope's may stem from differences in assessment method, sample size and/or our inclusion of additional kinds of abnormal behaviour, e.g. appetitive disorder. Form also varied with taxon: in our survey, prosimian stereotypic behaviour comprised whole-body movements such as pacing and bouncing, whereas New and Old World monkeys showed both whole-body and self-directed stereotypies (Chapter 6, this volume for examples), and apes, in contrast, most commonly showed coprophagy or regurgitation/reingestion. Levels of abnormal behaviour in primates in zoological gardens are substantially lower than in laboratories (*cf.* Chapter 6, this volume), but its presence at all, in these animals which are typically normally socially reared and housed, suggests that factors additional to social environment are important (*cf.* Chapter 6, this volume). It is not known, however, what such factors are.

References

Ames, A. (1992) Managing polar bears in captivity. In: Partridge, J. (ed.) *Management Guidelines for Bears and Raccoons.* The Association of British Wild Animal Keepers, Bristol, UK, pp. 41–50.

Ames, A. (1993) The behaviour of captive polar bears. UFAW Animal Welfare Research Report No. 5, Universities Federation for Animal Welfare, Hertfordshire, UK.

Ames, A. (2000) The management and behaviour of captive polar bears. Ph.D. thesis, The Open University, UK.

Barry, E. (1998) Polar bears at Dublin zoo: the effect of enclosure modification and environmental enrichment on behaviour. In: Hofer, H., Pitra, C. and Hofman, R.R. (eds) *Second International Symposium on Physiology and Ethology of Wild and Zoo Animals.* Blackwell Wissenschafts-Verlag, Berlin, 33, 116.

Bashaw, M.J., Bloomsmith, M.A., Marr, M.J. and Maple, T.J. (2003) To hunt or not to hunt? A feeding enrichment experiment with captive large felids. *Zoo Biology* 22, 189–198.

Berkson, G. (1983) Repetitive stereotyped behaviors. *American Journal of Mental Deficiency* 88, 239–246.

Bildsøe, M., Heller, K.E. and Jeppensen, L.L. (1990a) Stereotypies in adult ranch mink. *Scientifur* 14, 169–176.

Bildsøe, M., Heller, K.E. and Jeppesen, L.L. (1990b) Stereotypies in female ranch mink; seasonal and diurnal variations. *Scientifur* 14, 243–247.

Bildsøe, M., Heller, K.E. and Jeppesen, L.L. (1991) Effects of immobility stress and food restriction on stereotypies in low and high stereotyping female ranch mink. *Behavioural Processes* 25, 179–189.

Bollen, K.S., Well, A. and Novak, M.A. (Submitted for publication). A survey of abnormal behaviour in captive zoo primates.

Boorer, M.K. (1972) Some aspects of stereotypic patterns of movement exhibited by zoo animals. *International Zoo Yearbook* 12, 164–168.

Carlstead, K. (1991) Husbandry of the fennec fox *Fennecus zerda:* environmental conditions influencing stereotypic behaviour. *International Zoo Yearbook* 30, 202–207.

Carlstead, K. (1992) Stress, stereotypic pacing, and environmental enrichment in leopard cats (*Felis bengalensis*). *American Association of Zoological Parks and Aquariums Annual Conference Proceedings*, pp. 104–111.

Carlstead, K. (1998) Determining the causes of stereotypic behaviours in zoo carnivores: towards developing appropriate enrichment. In: Shepherdson, D.J., Mellen, J. and Hutchins, M. (eds) *Second Nature: Environmental Enrichment for Captive Mammals.* Smithsonian Institution Press, Washington, DC, pp. 172–183.

Carlstead, K. and de Jonge, G. (1987) Stereotypy in ranch mink (*Mustela vison*). In: *20th International Ethology Congress National Zoological Park*, Smithsonian Institute, Washington DC.

Carlstead, K. and Shepherdson, D. (1994) Effects of environmental enrichment on reproduction. *Zoo Biology* 13, 447–458.

Carlstead, K. and Seidensticker, J. (1991) Seasonal variation in stereotypic pacing in an American black bear *(Ursus americanus).* *Behavioural Processes* 25, 155–161.

Carlstead, K., Seidensticker, J. and Baldwin, R. (1991) Environmental enrichment for zoo bears. *Zoo Biology* 10, 3–16.

Carlstead, K., Brown, J.L. and Seidensticker, J. (1993) Behavioural and adrenocortical responses to environmental changes in leopard cats (*Felis bengalensis*). *Zoo Biology* 12, 321–331.

Charlton, N. (1995) An investigation into the behaviour of captive jaguars (*Panthera onca*) and an environmental enrichment study. B.Sc. thesis, University of London, London.

Clubb, R.E. (2001) The roles of foraging niche, rearing conditions and current husbandry on the development of stereotypies in carnivores. Ph.D. thesis, University of Oxford, UK.

Clubb, R. and Mason, G. (2003) Captivity effects on wide-ranging carnivores. *Nature* 425, 473–474.

Clubb, R. and Mason, G. (2004) Pacing polar bears and stoical sheep: testing ecological and evolutionary hypotheses about animals welfare. In: Kirkwood, J.K., Roberts, E.A. and Vickery, S. (eds) *Proceedings of the UFAW International Symposium Science in the Service of Animal Welfare, Edinburgh 2003. Animal Welfare* 13 (Supplement), S33–S40.

Clubb, R. and Mason, G. (2006) Natural behavioural biology as a risk factor in carnivore welfare: how analysing species differences could help zoos improve enclosures. *Applied Animal Behaviour Science.* Available online August 2nd 2006. doi: 10.1016/japplanim.2006.05.033.

Cronin, G.M. (1985) The development and significance of abnormal stereotyped behaviours in tethered sows. Ph.D. thesis, University of Wageningen, The Netherlands.

de Jonge, G. and Carlstead, K. (1987) Abnormal behaviour in farm mink. *Applied Animal Behaviour Science* 17, 375.

Dickinson, A. (1985) Actions and habits: the development of behavioural autonomy. *Philosophical Transactions of the Royal Society of London. Series B: Biological Sciences* 308, 67–78.

Dobberstine, J. and Shepherdson, D.J. (1994) Food-scattering enrichment for zoo bears: does it really work? *Shape of Enrichment* 3, 9–10.

European Commission (2001) *The Welfare of Animals kept for Fur Production.* Commission of the European Communities. Brussels, Belgium. Report of the Scientific Committee on Animal Health and Animal Welfare. Adopted on 12–13 December 2001.

Eyre, S. (1997) The effectiveness of environmental enrichment in preventing and curing stereotypic behaviour in a sand cat (*Felis margarita harrisoni*). *Ratel* 24, 156–165.

Fentress, J.C. (1976) Dynamic boundaries of patterned behaviour: interaction and self-organization. In: Bateson, P.P.G. and Hinde, R.A. (eds) *Growing Points in Ethology*. Cambridge University Press, Cambridge, UK, pp. 135–169.

Fischbacher, M. and Schmid, H. (1999) Feeding enrichment and stereotypic behavior in spectacled bears. *Zoo Biology* 18, 363–371.

Forthman, D.L., Elder, S.D., Bakeman, R., Kurkowski, T.W., Noble, C.C. and Winslow, S.W. (1992) Effects of feeding enrichment on behaviour of three species of captive bears. *Zoo Biology* 11, 187–192.

Forthman-Quick, D.L. (1984) An integrative approach to environmental enrichment. *Zoo Biology* 3, 65–77.

Fox, M.W. (1971) Psychopathology in man and lower animals. *Journal of the American Veterinary Medical Association* 159, 66–77.

Freckleton, R.P., Harvey, P.H. and Pagel, M. (2002) Phylogenetic analysis and comparative data: a test and review of evidence. *American Naturalist* 160, 712–726.

Frommolt, K.-H., Goltsman, M.E. and Macdonald, D. (2003) Barking foxes, *Alopex lagopus*: field experiments in individual recognition in a territorial mammal. *Animal Behaviour* 65, 509–518.

Funston, P.J., Mills, M.G.L., Biggs, H.C. and Richardson, P.R.K. (1998) Hunting by male lions: ecological influences and socioecological implications. *Animal Behaviour* 56, 1333–1345.

Gittleman, J.L. (1989) The comparative approach in ethology: aims and limitations. In: Bateson, P.P.G. and Klopfer, P.H. (eds) *Perspectives in Ethology*. Plenum Press, New York, pp. 55–76.

Gittleman, J.L. and Luh, H.-K. (1992) On comparing comparative methods. *Annual Review of Ecology and Systematics* 23, 383–404.

Hancocks, D. (1980) Bringing nature into the zoo: inexpensive solutions for zoo environments. *International Journal for the Study of Animal Problems* 1, 170–177.

Hansen, S.W. (1993) Circadian and annual rhythm in the activity of captive beech marten (*Martes foina*). *Scientifur* 17, 95–106.

Hansen, S.W. (1998) The cage environment of the farm mink – significance to welfare. *Scientifur* 22, 179–185.

Hansen, C.P.B. and Jeppesen, L.L. (2000) Short term behavioural consequences of denied access to environmental facilities in mink. *Agricultural and Food Science in Finland* 9, 149–155.

Hansen, S.W., Hansen, B.K. and Berg, P. (1994) The effect of cage environment and *ad libitum* feeding on the circadian rhythm, behaviour and feed intake of farm mink. *Acta Agriculturae Scandinavica* 44, 120–127.

Hansen, S.W., Houbak, B. and Malmkvist, J. (1997) Does the 'solitary' mink benefit from having company? Seminarium Nr. 280, 116. NJF Utredning, Helsingfors, Finland.

Harvey, P.H. and Pagel, M.D. (1991) *The Comparative Method in Evolutionary Biology*. Oxford University Press, Oxford, UK.

Hediger, H. (1950) *Wild Animals in Captivity*. Butterworths, London.

Hediger, H. (1955) *Studies of the Psychology and Behaviour of Animals in Zoos and Circuses*. Butterworths, London.

Houbak, B. and Møller, S.H. (2000) Activity and stereotypic behaviour in mink dams fed *ad libitum* or restricted during the winter. *Scientifur* 24, 146–150.

Hubrecht, R.C. (1995) Enrichment in puppyhood and its effects on later behavior of dogs. *Laboratory Animal Science* 45, 70–75.

Hunter, S.A., Bay, M.S., Martin, M.L. and Hatfield, J.S. (2002) Behavioral effects of environmental enrichment on harbour seals (*Phoca vitulina concolor*) and gray seals (*Halichoerus grypus*). *Zoo Biology* 21, 375–387.

Inglis, I.R. (1983) Towards a cognitive theory of exploratory behaviour. In: Archer, J. and Birke, L. (eds) *Exploration in Animals and Humans*. Van Nostrand Reinhold, Wokingham, UK, pp. 72–116.

Inglis, I.R., Forkman, B. and Lazarus, J. (1997) Free food or earned food? A review and fuzzy model of contrafreeloading. *Animal Behaviour* 53, 1171–1191.

Ings, R., Waran, N.K. and Young, R.J. (1997) Effect of wood-pile feeders on the behaviour of captive bush dogs (*Speothos venaticus*). *Animal Welfare* 6, 145–152.

Jenny, S. and Schmid, H. (2002) Effect of feeding boxes on the behavior of stereotyping Amur tigers (*Panthera tigris altaica*) in the Zurich Zoo, Zurich, Switzerland. *Zoo Biology* 21, 573–584.

Kastelein, R.A. and Wiepkema, P.R. (1989) A digging trough as occupational therapy for Pacific Walruses (*Odobenus rosmarus divergens*) in human care. *Aquatic Mammals* 15, 9–17.

Knowles, L. and Plowman, A. (2001) Overcoming habituation in an enrichment programme for tigers. *Federation Research Newsletter* 2, 2.

Koene, P. (1995) Bear behaviour in large enclosures. In: Koene, P. (ed.) *International Workshop on Captive Bear Management*. International Bear Foundation, Ouwehands Zoo, Rhenen, The Netherlands, pp. 43–50.

Kolter, L. (1995) Control of behaviour and the development of disturbed behaviour patterns. In: Gansloβer, U., Hodges, J.K. and Kaumanns, W. (eds) *Research and Captive Propagation*. Finlander Verlag, Furth, pp. 248–255.

Kolter, L. and Zander, R. (1995) Potential and limitations of environmental enrichment in managing behavioural problems of polar bears. In: *Proceedings of the Second International Conference on Environmental Enrichment*. Copenhagen Zoo, Copenhagen, DK Denmark, pp. 131–141.

Kreger, M.D., Hutchins, M. and Fascione, N. (1998) Context, ethics, and environmental enrichment in zoos and aquariums. In: Shepherdson, D.J., Mellen, J.D. and Hutchins, M. (eds) *Second Nature: Environmental Enrichment for Captive Mammals*. Smithsonian Institution Press, Washington, DC, pp. 59–82.

Landrigan, D., Dalziel, F., Luscri, J. and Metzer, J. (2001) Stereotypic pacing: alternatives and outcomes for two Malayan sun bears. *Shape of Enrichment* 10, 7–10.

Langenhorst, T. (1998) Behavior of a brown bear group (*Ursus arctos*) with behavioural enrichment at the Hellbrunn/Salzburg zoo. *Zoologische Garten* 68, 167–186.

Law, G., Boyle, H., Johnson, J. and MacDonald, A. (1990) Food presentation. Part 1: Bears. *Ratel* 17, 44–46.

Lewis, R.S. and Hurst, J.L. (2004) The assessment of bar-chewing as an escape behaviour in laboratory mice. *Animal Welfare* 13, 19–25.

Lewis, L., Cooper, J.J. and Mason, G.J. (2001) The behavioural responses of mink (*Mustela vison*) to deprivation of highly valued resources. In: Garner, J.P., Mench, J.A. and Heekin, S.P. (eds) *Proceedings of the 35th Congress of the International Society for Applied Ethology*. Center for Animal Welfare, UC Davis.

Lyons, J., Young, R.J. and Deag, J.M. (1997) The effects of physical characteristics of the environment and feeding regime on the behavior of captive felids. *Zoo Biology* 16, 71–83.

Mallapur, A. and Chellam, R. (2002) Environmental influences on stereotypy and the activity budget of Indian leopards (*Panthera pardus*) in four zoos in southern India. *Zoo Biology* 21, 585–595.

Markowitz, H. and LaForse, S. (1987) Artificial prey as behavioural enrichment devices for felines. *Applied Animal Behaviour Science* 18, 31–43.

Marriner, L. and Drickamer, L. (1994) Factors influencing stereotyped behaviour of primates in a zoo. *Zoo Biology* 13, 267–275.

Mason, G.J. (1990) Individual variation in the stereotypies of caged mink. *Applied Animal Behaviour Science* 28, 300–301.

Mason, G.J. (1991) Stereotypies and suffering. *Behavioural Processes* 25, 103–115.

Mason, G. (1992) Individual variation in the stereotypies in caged mink. Ph.D. thesis, University of Cambridge, Cambridge, UK.

Mason, G.J. (1993) Age and context affect the stereotypies of caged mink. *Behaviour* 127, 191–229.

Mason, G.J. and Latham, N.R. (2004) Can't stop, won't stop: is stereotypy a reliable animal welfare indicator? In: Kirkwood, J.K., Roberts, E.A. and Vickery, S. (eds) *Proceedings of the UFAW International Symposium Science in the Service of Animal Welfare, Edinburgh 2003. Animal Welfare* 13 (Supplement), S57–S69.

Mason, G.J. and Mendl, M. (1997) Do the stereotypies of pigs, chickens and mink reflect adaptive species differences in the control of foraging? *Applied Animal Behaviour Science* 53, 45–58.

Mason, G., Leipoldt, A. and de Jonge, G. (1995) Why do female mink with high stereotypy levels have slow-growing offspring. In: Rutter, S.M., Rushen, J., Randle, H.D. and Eddison, J.C. (eds) *Proceedings of the 29th International Congress of the International Society for Applied Ethology.* UFAW, Exeter, UK Potter's Bar, pp. 133–134.

Mason, G., Clubb, R., Latham, N. and Vickery, S. (2006) Why and how should we use environmental enrichment to tackle stereotypic behaviour? *Applied Animal Behaviour Science.* Available online August 2nd 2006 doi: 10.1016/japplanim. 2006.05.04/

McFarland, D. (1981) *The Oxford Companion to Animal Behaviour.* Oxford University Press, Oxford, UK.

Meehan, C.L., Mench, J.A. and Garner, J.P. (2001) Environmental enrichment prevents the development of abnormal behaviors and modifies fear responses in young orange-winged Amazon parrots. In: Garner, J.P., Mench, J.A. and Meekin, S.P. (eds) *Proceedings of the 35th International Congress of the ISAE.* Center for Animal Welfare at UC Davis, 42.

Mellen, J., Hayes, M. and Shepherdson, D. (1998) Captive environments for small felids. In: Shepherdson, D., Mellen, J. and Hutchins, M. (eds) *Second Nature: Environmental Enrichment for Captive Animals.* Smithsonian Institution Press, Washington, DC, pp. 184–201.

Meyer-Holzapfel, M. (1968) Abnormal behaviour in zoo animals. In: Fox, M.W. (ed.) *Abnormal Behaviour in Animals.* Saunders, London, pp. 476–503.

Morris, D. (1964) The response of animals to a restricted environment. *Symposium of the Zoological Society, London* 13, 99–120.

Mudway, G. (1992) The territorial behaviour of pine martens (*Martes martes*) during the breeding season at the Welsh Mountain Zoo. *Ratel* 19, 148–151.

Nevison, C.M., Hurst, J.L. and Barnard, C.J. (1999) Why do male ICR(CD-1) mice perform bar-related (stereotypic) behaviour? *Behavioural Processes* 47, 95–111.

Ödberg, F.O. (1978) Abnormal behaviours: stereotypies. In: *First World Congress on Ethology Applied to Zootechnics.* Industrias Graficas Espana, Madrid, pp. 475–480.

Ödberg, F.O. (1984) The altering of stereotypy levels: an example with captive fennecs (*F. zerda*). In: Unshelm, J., Van Putten, G. and Zeeb, K. (eds) *Proceedings of the International Congress on Applied Ethology in Farm Animals.* KTBL, Kiel, Dalmstadt, pp. 299–301.

Ormrod, S.A. (1987) Standards for modern captive animal management. In: Gibson, T.E. (ed.) *Proceedings of the Animal Welfare Foundation's Fourth Symposium: The Welfare of Animals in Captivity.* BVA Animal Welfare Foundation, pp. 22–27.

Poole, T. (1998) Meeting a mammal's psychological needs: basic principles. In: Shepherdson, D.J., Mellen, J.D. and Hutchins, M. (eds) *Second Nature: Environmental Enrichment for Captive Animals.* Smithsonian Institution Press, Washington, DC, pp. 83–94.

Poulsen, E.M.B., Honeyman, V., Valentine, P.A. and Teskey, G.C. (1996) Use of fluoxetine for the treatment of stereotypical pacing behavior in a captive polar bear. *Journal of the American Veterinary Medical Association* 209, 1470–1474.

Sandson, J. and Albert, M.L. (1984) Varieties of perseveration. *Neuropsychologia* 22, 715–732.

Shepherdson, D. (1989) Stereotypic behaviour: What is it and how can it be eliminated or prevented? *Ratel* 16, 100–106.

Shepherdson, D.J., Carlstead, K., Mellen, J.D. and Seidensticker, J. (1993) The influence of food presentation on the behaviour of small cats in confined environments. *Zoo Biology* 12, 203–216.

Shepherdson, D.J., Mellen, J.D. and Hutchins, M. (eds) (1998) *Second Nature: Environmental Enrichment for Captive Mammals.* Smithsonian Institution Press, Washington, DC.

Shepherdson, D., Carlstead, K. and Wielebnowski, N. (2004) Cross institutional assessment of stress responses in zoo animals using longitudinal monitoring of faecal corticoids and behaviour. In: Kirkwood, J.K., Roberts, E.A. and Vickery, S. (eds) *Science in the Service of Animal Welfare. Proceedings of the UFAW International Symposium.* Universities Federation for Animal Welfare, Edinburgh, UK, Animal Welfare 13.

Sillero-Zubiri, C. and Macdonald, D. (1998) Scent-marking and territorial behaviour of Ethiopian wolves *Canis simensis.* *Journal of Zoology* 245, 351–361.

Swaisgood, R.R., White, A.M., Zhou, X., Zhang, H., Zhang, G., Wei, R., Hare, V.J., Tepper, E.M. and Lindburg, D.G. (2001) A quantitative assessment of the efficacy of an environmental enrichment programme for giant pandas. *Animal Behaviour* 61, 447–457.

Templin, R. (1993) Stereotypic movements in zoo animals. In: *Proceedings of the 27th International Congress of the International Society for Applied Ethology.* Humboldt University, Berlin, Germany Berlin/KTBL, Darmstadt, pp. 54–59.

Terlouw, E.M.C., Lawrence, A.B. and Illius, A.W. (1991) Influences of feeding level and physical restriction on development of stereotypies in sows. *Animal Behaviour* 42, 981–991.

Toates, F.M. (1983) Exploration as a motivational and learning system: a cognitive incentive view. In: Archer, J. and Birke, L. (eds) *Exploration in Animals and Humans.* Van Nostrand Reinhold, Wokingham, UK, pp. 55–71.

Trollope, J. (1977) A preliminary survey of behavioural stereotypies in captive primates. *Laboratory Animals* 11, 195–196.

van Keulen-Kromhout, G. (1978) Zoo enclosures for zoo bears *Ursidae:* their influence on captive behaviour and reproduction. *International Zoo Yearbook* 18, 177–186.

Vickery, S. (2003) Stereotypy in caged bears: individual and husbandry factors. PhD thesis, University of Oxford, Oxford, UK.

Vickery, S. and Mason, G. (2003a) Understanding stereotypies in captive bears: the first step towards treatment. *Proceedings of the Fifth Annual Symposium on Zoo Research.* Marwell Zoo, Hampshire, UK.

Vickery, S.S. and Mason, G.J. (2003b) Behavioural persistence in captive bears: implications for reintroduction. *Ursus* 14, 35–43.

Vickery, S. and Mason, G.J. (2004) Stereotypic behavior in Asiatic black and Malayan sun bears. *Zoo Biology* 23, 409–430.

Vickery, S.S. and Mason, G.J. (2005) Stereotypic behaviour in bears and its relationship with general behavioural persistence: further data. *Applied Animal Behaviour Science* 91, 247–260.

Warburton, H., Clarebrough, C., Cooper, J. and Mason, G. (in preparation) Drink or swim: the nature and cost of local resources differentially affect preference and frustration in mink.

Weller, S.H. and Bennett, C.L. (2001) Twenty-four hour activity budgets and patterns of behavior in captive ocelots (*Leopardus pardalis*). *Zoo Biology* 71, 67–79.

White, B.C., Houser, L.A., Fuller, J.A., Taylor, S. and Elliott, J.L.L. (2003) Activity-based exhibition of five mammalian species: evaluation of behavioral changes. *Zoo Biology* 22, 269–285.

Wielebnowski, N. and Brown, J. (2000) Assessment of adrenal activity combined with husbandry and behavioral evaluations of the North American clouded leopard population. In: Pukazhenthi, B., Wildt, D. and Mellen, J. (eds) *Felid Taxon Advisory Group Action Plan.* American Zoo and Aquarium Association, Wheeling, Virginia, pp. 37–38.

Wielebnowski, N.C., Ziegler, K., Wildt, D.E., Lukas, J. and Brown, J.L. (2002a) Impact of social management on reproduction, adrenal and behavioural activity in the cheetah (*Acinonyx jubatus*). *Animal Conservation* 5, 291–301.

Wielebnowski, N.C., Fletchall, N., Carlstead, K., Busso, J.M. and Brown, J.L. (2002b) Non-invasive assessment of adrenal activity associated with husbandry and behavioral factors in the North American clouded leopard population. *Zoo Biology* 21, 77–98.

4   The Motivational Basis of Caged Rodents' Stereotypies

H. WÜRBEL

Institut für Veterinär-Physiologie, Justus-Liebig-Universität, Giessen, Frankfurter Str. 104, D-35392 Giessen, Germany

Editorial Introduction

With their small body sizes and rapid reproductive rates, rodents are ideal research animals. They are thus ideal models for conducting the type of controlled experiments on stereotypies that can be challenging with, say, carnivores or ungulates. Furthermore, since mice and rats already play a central role in behavioural neuroscience, they also open up worlds of sophisticated techniques and data on central nervous system (CNS) functioning that are simply unavailable for most other animals. Despite this, the detailed description and ethological analysis of rodents' cage stereotypies is relatively recent. Indeed for mice, their nocturnal active periods meant that their stereotypies went largely unnoticed by the scientific community until Würbel's own work in the last decade. Here, he builds on the previous two chapters to describe what is known of the motivational bases of rodent stereotypies. He first discusses the wide variety of stereotypic behaviours seen across different species, which range from forms that resemble natural behaviours, such as stereotypic digging, through to energetic and bizarre somersaulting or 'looping the loop'. Some of these forms are common to several species, but others appear more species-typical; and species also differ greatly – and, so far, puzzlingly – in their degree of stereotypy. Würbel then describes some well-designed, hypothesis-led ethological research investigating the causation of two examples: stereotypic digging in the gerbil and bar-mouthing in the laboratory mouse. These experiments carefully manipulated specific environmental stimuli, to show that motivations to escape from the cage and/or to seek appropriate shelter underlie these stereotypies.

However, while these motivational accounts can explain why animals repeat certain 'source behaviours', it is not clear that they explain why these activities become so time-consuming or ritualistic in appearance. So, do motivational explanations suffice? Würbel shows convincingly that they do not. The basic behavioural biology of laboratory mice and rats, for instance, is similar, and they are also kept in similar types of laboratory housing. We might therefore expect them to have equivalent motivations to hide or escape – and yet the stereotypies so common in mice are more or less absent in rats. Furthermore, once rodent stereotypies develop they can become astonishingly persistent, and even unresponsive to the types of environmental enrichment that effectively prevent them appearing in younger animals. Würbel therefore argues that other behavioural processes – perhaps even pathological ones – need to be invoked, and he then carefully analyses the evidence that the behaviours might represent repeated habits; that they are reinforced by some beneficial consequences; or that they stem from CNS dysfunction – a theme developed further in subsequent chapters. Finally, Würbel argues that we need more sophisticated definitions for these behaviours than are currently used, perhaps even a classificatory scheme – a point that is illustrated nicely by a contributed box on rodent wheel-running: is it a stereotypy or exercise, or even both?

GM and JR

4.1. Introduction

Like most other mammals and birds, most rodents (see Box 4.1) develop abnormal stereotypic behaviours when they are kept in barren cages, such as the standard laboratory cages used in biomedical research. Research on stereotypic behaviour in rodents generally differs, however, from that on ungulates and carnivores described in the preceding two chapters, with less interest in finding practical ways of reducing the behaviour, and more focus on fundamental experimental and developmental work. This is perhaps due to the ease with which these animals can be kept in large

---

Box 4.1. Rodents – Their Diversity and Adaptability

H. WÜRBEL

The order Rodentia forms the largest and most diverse group of mammals, comprising approximately 1700 species: 40% of all known mammalian species (e.g. Hurst, 1999). The main feature that links them is their unique gnawing action, provided by their masseter jaw muscles and ever-growing, sharp incisor teeth. This gnawing action allows rodents to feed on the toughest nuts and seeds and to gnaw through wood and roots in search for food and shelter. They are typically small-bodied, although the largest species, the capybara (*Hydrochoerus hydrochaeris*), may be as tall as 60 cm high and 130 cm long. Although most rodents are frugivorous/herbivorous, some are omnivorous (e.g. most of the genus *Rattus*) or even strictly insectivorous (e.g. the duprasi, *Pachyuromys duprasis*). Rodents inhabit virtually every type of terrestrial habitat. Some species are arboreal (e.g. arboreal squirrels, New World porcupines); while others dig extensive burrow systems under ground (e.g. gerbils, mole-rats, ground squirrels). Most rodents are nocturnal and depend on shelters against predators (e.g. mice, rats, voles, hamsters, gerbils), while those living in less predated areas are diurnal and nest above ground (e.g. guinea pigs, chinchillas). However, individual species vary greatly in the range of habitats they occupy, reflecting different degrees of adaptability between them. Among the extreme generalists are the house mouse (*Mus musculus*) and the Norway rat (*Rattus norvegicus*) (e.g. Latham and Mason, 2004). Their adaptability may have facilitated adaptation to human habitation, allowing them to exploit rich food sources. In turn, this may have predisposed them for use as laboratory animals. Today mice and rats derived from these two species account for about 85% of all animals used in research worldwide, and each year about 27 million rodents are used in animal experiments in the USA and EU alone (Moore, 2001). Many of these animals develop abnormal stereotypic behaviour in standard laboratory cages, which has raised concerns about their welfare as well as about the validity of the research they are being used for (e.g. Knight, 2001; Würbel, 2001; Sherwin, 2004).

numbers in controlled conditions, but also because such research is easier to conduct in smaller and shorter living species. In this chapter, I review ethological work on rodent stereotypies to assess their motivational bases, and discuss these in relation to other factors that may be involved in their development.

While an estimated 50% of all laboratory mice reliably develop stereotypies in standard laboratory cages (Würbel and Stauffacher, 1994), there are only a few anecdotal reports of stereotypies in laboratory rats, despite their being housed under the same conditions. Similarly, while stereotypies are often observed in standard housed gerbils (Wiedenmayer, 1997a) and bank voles (Ödberg, 1986), they seem to be uncommon, if not absent, in guinea pigs. In Section 4.2, I therefore describe the diversity of rodent stereotypies, including differences in the incidence and prevalence of different forms of stereotypies across different species and strains, and explore how this diversity might relate to underlying genetic differences.

Like stereotypies in other species (e.g. Mason, 1991), caged rodents' stereotypies do not appear suddenly in full form. Instead, they develop, originating mainly from behavioural responses to thwarting or motivational conflict such as intention movements, displacement activities or redirected behaviours (e.g. Ödberg, 1986; Würbel et al., 1996; see also Box 1.1, Chapter 1, this volume). Therefore, the nature and form of the behaviour from which a stereotypy develops, i.e. its 'source behaviour pattern', might tell us something about the underlying motivation (e.g. Rushen et al., 1993). In Section 4.3, I therefore review the literature to examine how behavioural expressions of thwarting and motivational conflict in rodents relate to stereotypy development, and whether the form of a stereotypy reflects the underlying motivational problem. I also examine whether species differences in the incidence and form of stereotypies can be explained in terms of differences in the animals' behavioural biology.

While specific causal factors often explain the occurrence of particular source behaviour patterns and their repetition (Mason and Turner, 1993), they usually fail to account for all the changes in form and performance typically observed when stereotypies develop. Indeed, the most characteristic features of stereotypies are arguably dynamic, developmental changes, rather than the static – and partly subjective – features (repetitive, invariant, no function) that form the basis of the conventional definition. Thus, over time, stereotypies normally increase in frequency and duration while becoming more and more fixed in form and orientation (Ödberg, 1986; Würbel et al., 1996; Wiedenmayer, 1997a). Furthermore, their performance may become less dependent on the original eliciting circumstances (Würbel et al., 1996; Wiedenmayer, 1997a), and they may sometimes persist even when animals are placed in an environment where stereotypies would not normally develop (Cooper et al., 1996). Interpretations of these developmental changes are dominated by two alternative views, namely that they reflect the acquisition of a behavioural strategy to cope with adverse conditions (e.g. Cronin *et al.*, 1985; Wiepkema *et al.*, 1987) or, alternatively, that they result from pathological changes at the neural level, leading to a disruption of normal brain functioning (Dantzer, 1986, 1991; Garner and Mason, 2002). In Section 4.4, I therefore examine how different explanations for these developmental changes could relate to underlying motivational processes, and then whether motivational processes alone are sufficient to explain stereotypy development in rodents.

The welfare implications of caged rodents' stereotypies may vary greatly, depending on the exact mechanisms underlying them, and the consequences of their development. Stereotypies have always been considered as a sign of impaired welfare (Ödberg, 1978; Mason, 1991), yet the nature, duration and extent of this impairment has remained uncertain. In the best case, stereotypies may merely reflect an earlier frustration, being mental 'scars' (Mason, 1993). In the worst case, they reveal acquired brain disorders and/or chronic suffering (Wemelsfelder, 1993; Garner and Mason, 2002). In Section 4.5, I will therefore discuss the welfare implications of caged rodents' stereotypies based on my conclusions about their causation and consequences. I will also discuss some possible avenues for future research, and the definitional problems that still plague research into stereotypies.

4.2. Diversity in Caged Rodents' Stereotypies

4.2.1. Species differences in the incidence of stereotypies

Few of the estimated 1700 rodent species are common in zoos, laboratories or as pets, and very few have actually been studied systematically for cage-induced stereotypies. There are scientific reports on stereotypies for field voles (*Microtus agrestis*) (e.g. Fentress, 1976), bank voles (*Clethrionomys glareolus*) (Ödberg, 1986), chinchillas (*Chinchilla lanigera*) (Kersten, 1997), gerbils (*Meriones unguiculatus*) (Wiedenmayer, 1997a), deer mice (*Peromyscus maniculatus*) (e.g. Powell *et al.*, 1999), black rats (*Rattus rattus*) (Callard *et al.*, 2000), African striped mice (*Rhabdomys pumilio*) (e.g. Schwaibold and Pillay, 2001), golden hamsters (*Mesocricetus auratus*) (Vonlanthen, 2003) and various strains of laboratory mice (Würbel and Stauffacher, 1994; Würbel *et al.*, 1996) (see Table 4.1).

However, no reports on stereotypies were found for guinea pigs, *Cavia porcellus*, nor, surprisingly, laboratory rats. Hurst *et al.* (1999) observed bar-chewing (a major stereotypy in mice, e.g. Würbel *et al.*, 1996; Nevison *et al.*, 1999a) in standard-housed laboratory rats, but did not call it a stereotypy due to its apparent function (escape attempts), its variability in form and orientation even within individuals, and its low levels of performance (0.37% and 1.63% of active time in males and females, respectively). Both rats and guinea pigs do show wheel-running when provided with a running wheel, and rats, like mice, may develop excessive levels of wheel-running even at the expense of food intake (reviewed Sherwin, 1998). But is

Table 4.1. Names and definitions of rodent stereotypies and their occurrence among different rodent species according to the literature and personal observations.

Name	Definition	Species
Bar-mouthing (also known as bar-gnawing; bar-chewing; wire-gnawing)	Hanging on the cage lid (from all paws or the fore paws only) or standing on the hind legs while chewing on a bar. The bar is held in the gap between the incisors and molars (the diastema). May be performed on the spot or by moving along the bar while chewing	Mouse (see book's website), gerbil, bank vole, golden hamster
Jumping (also known as jack-hammering)	Jumping up-and-down (on all four legs or on the hind legs only) at a cage wall or, more commonly, in a cage corner	Mouse (see book's website), bank vole, deer mouse, African striped mouse
Digging	Excessive digging in the cage corner	Gerbil
Looping	Climbing up to the cage lid and dropping down by releasing the fore paws first. May develop into backflipping	Mouse, bank vole
Backflipping (also known as somersaulting)	Backward flip from one cage wall or the food rack towards the opposite cage wall, with or without touching the cage lid and/or the opposite cage wall during the flip	Mouse, black rat, bank vole, deer mouse, African striped mouse
Running to-and-fro	Running back and forth along a cage wall. May develop into figure of eight or 'windscreen wiper' stereotypy	Mouse, bank vole
'Windscreen wiper' (also known as weaving)	Oscillating like a windscreen wiper, with the fore paws moving against the wall but the hind paws stationary	Bank vole, African striped mouse
Figure of eight	Running to-and-fro along a cage wall while turning away from the wall on each turn	Bank vole
Cage top twirling	Spinning around while hanging on the cage lid from the fore paws	Mouse
Circling	Running in tight circles on the floor	Mouse
Patterned running	Running about the cage along fixed routes	Bank vole, deer mice, African striped mouse
Rearing	Rearing up against a cage wall	Bank vole

Box 4.2. Wheel-running: a Common Rodent Stereotypy?

N. Latham and H. Würbel

Pet owners have long recognized that running wheels or discs occupy rodents in otherwise non-stimulating environments, while in research laboratories, they have proved useful for studying chronobiology and phenomena like 'activity-based anorexia'. Recently, they have also become widespread 'environmental enrichments' (Chapter 7, this volume). Caged rodents can run extraordinary distances in running wheels (although this does not translate into equal distances of normal locomotion, due to the reduced frictional forces acting in wheels; Sherwin, 1998): rats average 4–8 km per night, and can run up to 43 km (Richter, 1927 cited by Sherwin, 1998; Werme *et al.*, 2002a,b), and depending on genetic background, mice average 2.8–8.1 km at night, with a maximum of 31 km (Kavanau, 1967 cited by Sherwin, 1998; Johnson *et al.*, 2003). When voluntary and moderate, wheel-running has physiological and cognitive benefits (see e.g. Van Praag *et al.*, 1999; Johnson *et al.*, 2003), and is psychologically rewarding. For example, rodents will work to reach a wheel, and develop conditioned place preferences to environments associated with the after effects of wheel-running (reviewed by Sherwin, 1998; Werme *et al.*, 2002a). However, other properties give cause for concern. When 'excessive', wheel-running may model addiction or compulsive behaviours (see e.g. Altemus *et al.*, 1996; Werme *et al.*, 2002a), perhaps through its effects on ΔFosB levels in striatal neurons (see e.g. Werme *et al.*, 2002b; Nestler, 2004). With food restriction, the rapid weight loss it engenders is also used to model anorexia nervosa (Morrow *et al.*, 1997; Hebebrand *et al.*, 2003). Furthermore, some suggest that wheel-running is stereotypic (e.g. Kuhnen, 2002). So, is this common behaviour a stereotypy?

  Sherwin (1998) argued that wheel-running has properties that he took to differ from those of 'true' stereotypies: it can occur instantly upon provision of a wheel; remains responsive to external influences; can occur even in enriched environments; and, as we saw above, may be rewarding. Yet these features do not conflict with what we see in some accepted stereotypies; for example, some stereotypies can develop and/or persist in 'enriched' environments (see e.g. Meyer-Holzapfel, 1968; Powell *et al.*, 1999), and be potentially rewarding (see e.g. Box 1.3, Chapter 1, this volume). Furthermore, wheel-running certainly exhibits stereotypy's central defining features, in being repetitive, unvarying and apparently functionless or goalless. In addition, it shares other similarities. First, home range size predicts both stereotypies and wheel-running in some carnivores (Clubb, 2001). Second, food restriction can induce both behaviours (stereotypy: Chapters 2 and 3, this volume; wheel-running: see e.g. Altemus *et al.*, 1996; Morrow *et al.*, 1997; Hebebrand *et al.*, 2003). Third, the selective serotonin reuptake inhibitor fluoxetine (Prozac) reduces both behaviours (stereotypy: Chapter 10, this volume; wheel-running: Altemus *et al.*, 1996). Finally, like drug-induced stereotypies (and consistent with Chapters 5 and 7, this volume), wheel-running increases with ΔFosB-mediated increased excitability in striatonigral projections, while decreasing with increasing activation of the inhibitory striatopallidal pathway (Werme *et al.*, 2002b) (see Chapters 5 and 7, this volume, for the relationships and effects of these pathways).

  Overall, it is likely that wheel-running occurs in stereotypic and non-stereotypic forms. Unfortunately, this subject has little been studied, but possible future research directions include further investigating whether factors that predict or elicit stereotypies do likewise for wheel-running. For instance, do early weaning and isolation-rearing (Chapter 6, this volume) increase wheel-running as well as stereotypies? And does wheel-running correlate with perseveration (Chapter 5, this volume)? In terms of the current definition, however, we would argue that wheel-running is no less a stereotypy than the bar-chewing, somersaulting and jumping of many laboratory rodents. The current reluctance to define it as such perhaps stems from an unwillingness to suggest that stereotypy,

Continued

Box 4.2. *Continued*

a behaviour typically associated with poor welfare, is produced by an 'environmental enrich-ment'. Future research may provide more definite evidence of stereotypic properties of wheel-running, and enable us to distinguish stereotypic from non-stereotypic forms of the behaviour.

wheel-running a stereotypy? The simple answer is that we do not know (see Box 4.2), and indeed the answer may very well vary from one form of the behaviour to the next. In the absence of conclusive evidence I will not consider wheel-running further in this chapter, and instead refer to the review by Sherwin (1998) and the extensive primary literature on this behaviour. Furthermore, even if rats and guinea pigs do develop stereotypic wheel-running when provided with a running wheel, a barren cage on its own still does not seem to induce any stereotypic behaviour in these two species. Nor is this lack of reported cage stereotypy due to a lack of research: both species have been extensively studied, and under many housing conditions, ranging from the barren to the enriched (e.g. Sachser *et al.*, 1994; Hurst *et al.*, 1996, 1997, 1998, 1999; Sachser, 1998). Thus, it appears that there are indeed significant species differences within rodents in the propensity to develop cage-induced stereotypies.

4.2.2. Species differences in the form of stereotypies

Despite the lack of detailed comparative studies on stereotypy develop-ment in rodents, there is good evidence that species also differ in the main forms of stereotypies they develop or, at least, in the prevalence of par-ticular forms (Table 4.1). The main stereotypy in mice is bar-mouthing (Würbel *et al.*, 1996; Nevison *et al.*, 1999a), yet they are also found to perform most of the other rodent stereotypies (Würbel and Stauffacher, 1994). Bank voles mainly develop jumping, but also show a wide variety of other forms (Ödberg, 1986; Cooper *et al.*, 1996; Garner and Mason, 2002). African striped mice appear to develop the same stereotypies as bank voles, though their main stereotypy is backflipping (Schwaibold and Pillay, 2001). Similarly, deer mice mainly show jumping and backflip-ping (Powell *et al.*, 2000), and backflipping was also the main stereotypy in black rats (Callard *et al.*, 2000). Gerbils preferably develop stereotypic digging in the cage corners (Wiedenmayer, 1997a), but, like golden ham-sters (Vonlanthen, 2003), also show some bar-mouthing (Wiedenmayer, 1997b; Waiblinger, 2003). Barbering, i.e. the repetitive plucking of an animal's own fur or that of a cage mate, is another abnormal repetitive behaviour shown by several rodents. However, I did not include this in this review as it lacks essential stereotypical qualities, and possibly also differs from stereotypies in its underlying neural basis (Chapter 5 and Box 10.2, Chapter 10, this volume).

Thus, with the possible exception of primates (Chapter 6, this volume), rodent stereotypies appear much more diverse than those in other taxa (e.g. carnivores, ruminants), comprising oral and locomotor stereotypies as well as other (often idiosyncratic) motor patterns. Furthermore, several behaviours are sometimes combined into complex stereotypic sequences. For example in mice, bar-mouthing may occur as part of a complex stereotypic movement pattern, whereby the mouse climbs up along the food rack to the far corner of the cage lid, where it performs a bout of bar-mouthing before dropping and running back to the food rack to start the sequence all over again (personal observation).

Variation in the exact form of a particular stereotypy may occur between as well as within species (e.g. bar-mouthing), while some stereotypies (e.g. bar-mouthing, jumping, backflipping) may be performed identically even by individuals from different species. It is tempting to attribute variation in the incidence and form of stereotypies to variation in the general behavioural biology between species. However, since there is also considerable variation in the incidence and form of stereotypies within species, I will first examine the genetic basis of such individual variation.

4.2.3. Genetic differences: within-species studies

The genetic basis of caged rodents' stereotypies has not been extensively studied. However, studies within species do point to genetic differences in the propensity to develop cage stereotypies. In bank voles, Ödberg (1986) and Schönecker and Heller (2000) found a higher incidence of stereotypies in offspring of stereotyping mothers compared to offspring of non-stereotyping mothers. Similarly, in black rats it was found that the level of stereotypic backflipping runs in families (Callard *et al.*, 2000). These results might be explained by non-genomic maternal effects (e.g. maternal activity affecting the propensity of offspring to develop stereotypies). However, Schwaibold and Pillay (2001) recently demonstrated in African striped mice that fostering pups of stereotyping females on to non-stereotyping females did not alter the higher incidence of stereotypies in these offspring. Thus, individual differences in the propensity to develop stereotypies seem at least partly genetically determined. However, nothing is yet known about which genes might be involved in these differences, and to exclude the possibility of non-genomic prenatal maternal effects (e.g. prenatal stress), these results would also need to be replicated using embryo transfer.

Other work has looked at strain differences. Studies in laboratory mice revealed strain differences both in the form and level of stereotypy performance (Würbel and Stauffacher, 1994; Würbel *et al.*, 1996; Nevison *et al.*, 1999b). Differences in the level of performance appear to be related to differences in general activity, with a higher incidence of stereotypy development in the more active strains (Würbel and Stauffacher, 1994). This parallels the correlation found between individual levels of stereotypy performance and activity levels within populations of mice and bank voles (Ödberg, 1986; Cooper and Nicol, 1996; Würbel *et al.*, 1996;

Garner and Mason, 2002). (Strain differences in mouse stereotypy are also discussed further in Chapter 8, this volume, especially in relation to stress responses.)

Concerning the form of the behaviour, Schönecker and Heller (2000) found in a correlational study that bank voles tend to develop the same stereotypies as their mothers, whereas a cross-fostering study of African striped mice revealed no evidence for maternal transmission of the form of stereotypy (Schwaibold and Pillay, 2001). Furthermore, in mice there are no clear strain differences in the form of stereotypy. Different stereotypies were found within single strains, while the same stereotypies were found across different strains. For example, bar-mouthing and jumping were found in both outbred Zur:ICRs and the nude athymic mutants Zur:ICR nu/nu (Würbel and Stauffacher, 1994). In one study, a strain difference in the prevalence among two forms of stereotypy, jumping and bar-mouthing, appeared to be related to differences in physical development, with poor physical condition at the onset of stereotypy development favouring one over another behavioural response from which these two stereotypies develop (Würbel et al., 1996, Fig. 4.1).

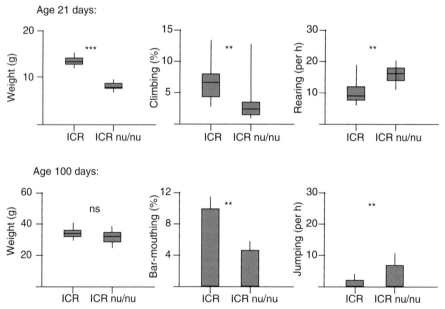

Fig. 4.1. Stereotypy development in two strains differing in physical development. ICR and ICR nu/nu mice differ in early physical development. ICR nu/nu are physically retarded at 21 days (weaning age) as indicated by their lower body weight. However, they catch up and the difference in body weight has disappeared by 100 days of age, when mice are fully adult. Early physical retardation is associated with altered preferences for two different escape strategies – climbing at the cage lid with attempts to squeeze or gnaw through the bars and rearing at the cage wall with attempts to jump out of the cage. Initially, climbing and jumping are not stereotypic, but develop into stereotypic bar-mouthing and jumping, respectively. Note that the relative difference in prevalence among the two escape strategies at 21 days of age is still reflected in the relative difference in prevalence amongst the two stereotypies in the adult mice (from Würbel et al., 1996).

While none of these studies were designed to answer questions about species differences in rodent stereotypy development, they indicate that both the form and level of stereotypy performance may depend on subtle differences in physical condition and general activity levels, with both genetic and environmental factors playing important roles. However, to relate species differences in the incidence and form of stereotypies to variation in behavioural biology, we first need to know more about the causal factors underlying stereotypy development.

4.3. Origin and Motivational Bases of Caged Rodents' Stereotypies

4.3.1. The thwarting of highly motivated behaviours: escape and/or the search for shelter

Based on developmental studies, both jumping in voles (Ödberg, 1986) and bar-mouthing and jumping in mice (Würbel *et al.*, 1996; Würbel and Stauffacher, 1998), have been suggested to originate from attempts to escape the home cage. Mice start to develop stereotypies right after weaning, which typically happens at 21 days of age (Würbel *et al.*, 1996), and which involves rehousing the offspring in same sex groups of either litter mates or mixed litter groups in new cages. Thus, compared to the gradual process of natural weaning, artificial weaning is immediate and involves complete loss of contact with the mother as well as many of the litter mates, and sometimes regrouping with strangers. Artificial weaning elicits a persistent stress response, probably due to both the novelty of the new cage and the separation of the young mice from their mothers. The importance of this last factor is supported by the fact that elevated stress levels persist longer in prematurely weaned mice (18 days of age) as well as in standard weaned mice (21 days) of low body weight, for which the loss of the mother is presumably more detrimental. Furthermore, both of these groups show more escape attempts (which may be aimed, at least in part, at returning to the mother) following weaning than mice of high body weight weaned at standard age (Würbel and Stauffacher, 1997, 1998; see Fig. 4.2). Mice use two different escape strategies. Some rear up at the cage wall and attempt to climb up or jump over it. Others climb up along the food rack to the edge of the cage lid and attempt to squeeze through the gaps between the bars of the cage lid and eventually gnaw their way out through the bars (Würbel *et al.*, 1996). Over time, these two escape strategies gradually develop into stereotypic jumping and bar-mouthing, respectively.

In two elegant studies on laboratory mice, Hurst and colleagues confirmed that such bar-related behaviours do indeed represent escape attempts. They replaced part of one sidewall of the cages by a sliding door

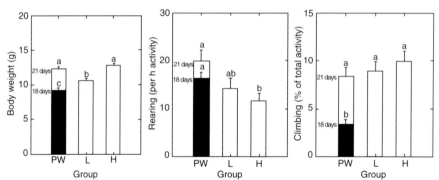

Fig. 4.2. Physical condition at weaning affects the prevalence of different escape strategies in laboratory mice. Mice weaned prematurely at 18 days of age (PW) showed more rearing and less climbing on the day of weaning than mice weaned at normal weaning age of 21 days (L, H). Furthermore, among the mice weaned at normal weaning age, those of low body weight (L) tended to show more rearing and less climbing at 21 days of age than those of high body weight (H). Physical condition may affect the ability of the mice to climb up to the cage lid, thereby altering the prevalence among the two escape strategies. Note that PW mice still tended to show more rearing and less climbing than H and L mice at 21 days of age, even though their body weight was similar to that of H mice of the same age. However, in contrast to the strain differences presented in Figure 4.1, this difference in the prevalence among the two escape strategies did not result in a differential expression of the two stereotypies – bar-mouthing and jumping – in the adult mice. This might have been due to the smaller effect sizes of both the weight differences and the differences in performance of climbing and rearing compared to the strain differences shown in Figure 4.1 (from Würbel and Stauffacher, 1998).

made of the same bars as those of which the cage top is made. Using these cages they found that bar-related behaviour was preferably oriented towards those bars (cage top or side) that were regularly opened for cage maintenance (Nevison *et al.*, 1999a) or that gave the mice temporary access to a large arena (Lewis and Hurst, 2004). In the latter experiment, the orientation of bar-related behaviour was even more skewed to the opening door, suggesting that actual 'escape' had a reinforcing effect on the orientation of the behaviour.

Correlational evidence suggests that the choice between the two different escape strategies, and thence stereotypies, might depend on physical condition at weaning, with heavier or later weaned mice showing relatively more climbing/bar-mouthing, and less rearing/jumping than lighter or early weaned mice (Würbel and Stauffacher, 1997, 1998; Nevison *et al.*, 1999a; Fig. 4.2; see also Fig. 4.1). However, attempts to experimentally manipulate performance of different source behaviour patterns in mice and bank voles in order to test the prediction that this would affect later prevalence among the corresponding forms of stereotypies, have as yet failed (Würbel and Stauffacher, 1998; Garner, 1999).

Besides returning to their mother, there may be many other reasons for mice to attempt to escape, such as aversion to the home cage. Other possible motivations include increasing chances of reproductive opportunity, to explore the environment or the smells and sounds of mice in neighbouring cages, or to search for shelter or additional food sources (Würbel and Stauffacher, 1998; Nevison *et al.*, 1999a). Indeed, a shelter in the form of a simple cardboard roll reduced stereotypy levels by 50% (Würbel *et al.*, 1998b), indicating that the need for shelter may contribute to the motivation to escape the home cage, eventually leading to stereotypy development. Access to shelter was also proposed to mediate the escape motivation of bank voles, since providing cover in the form of twigs and leaves reduced stereotypy levels in this species (Ödberg, 1987; Cooper and Nicol, 1996). The most convincing case for this specific motivation to underlie stereotypy development was, however, made by Wiedenmayer (1997a) in an ingenious experimental study in gerbils.

Gerbils readily develop high levels of stereotypic digging when kept in barren laboratory cages (see Table 4.1). Stereotypic digging starts to develop at 24 days of age, when corner-digging becomes progressively excessive compared to substrate-digging (Wiedenmayer, 1997a). Adult gerbils may spend more than 20% of their active time digging stereotypically in cage corners (Wiedenmayer and Brunner, 1993), and cage size does not seem to matter (Wiedenmayer, 1996). Wiedenmayer (1997a) demonstrated that the motivation to hide in a shelter, rather than to actually dig, underlies stereotypic digging in gerbils (Fig. 4.3). However, like bar-mouthing and jumping in mice and voles, digging in the corners of barren cages may actually represent an attempt to escape (possibly in search of shelter).

4.3.2. Role of exposure to eliciting stimuli

Exposure to external eliciting stimuli may complement internal causal factors such as the motivation to return to the mother or to find shelter as discussed above (e.g. see Box 1.1, Chapter 1, this volume). Except in cases where habituation is a counteracting force (McFarland, 1989), constant or intermittent exposure to a stimulus leads to frequent performance of the corresponding response (Mason and Turner, 1993). This may well play a role in rodent stereotypies. Laboratory gerbils are constantly exposed to the edges of the barren cages that arguably resemble structures in the wild (i.e. a tunnel at an early stage of construction) that would normally orient the beginning of a digging bout to facilitate burrow construction (Wiedenmayer, 1997a).

Similarly, laboratory mice are constantly exposed to the gaps between the bars of the cage lid which may be perceived by the mice as potential exit routes and thus elicit escape attempts (Würbel *et al.*, 1998a). Nevison *et al.* (1999a) have argued that habituation of escape attempts would be maladaptive since 'the reproductive potential of a wild mouse that

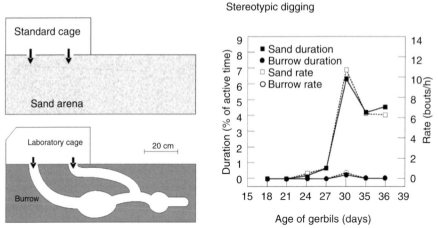

Fig. 4.3. Stereotypic digging in the gerbil: is it reduced by naturalistic digging opportunities? Early development of stereotypic digging in gerbils depending on whether they had access to an artificial burrow system without any substrate to dig or to an arena filled with dry sand, which allowed them to dig but without the digging resulting in a stable burrow. Gerbils with access to the sand arena rapidly developed stereotypic digging, whereas access to the burrow completely prevented the development of stereotypic digging, confirming that the motivation to hide in a shelter, rather than the motivation to dig, underlies this stereotypy in gerbils (from Wiedenmayer, 1997a).

gave up a response to flee if trapped … may be curtailed'. The same may apply to gerbils that stopped seeking shelter. Furthermore, cages for most laboratory rodents are stored on racks surrounded by other cages and so are constantly exposed to odour cues of conspecifics. These odour cues are perceived by the animals through the grid cage-top. The orientation of stereotypic jumping and bar-mouthing in mice and voles corresponds with sites where they are most likely to perceive odour cues from neighbouring animals (Würbel *et al.*, 1996). However, the site of perception of stimuli from neighbouring mice is confounded with the site that regularly opens for cage maintenance. Nevison *et al.* (1999a) (see also Lewis and Hurst, 2004) nicely dissociated these two factors by providing bars both on the top and side of the cage, whereby half of each was covered by Perspex to withhold airborne odour cues. In half of the cages, the top bars were regularly opened for cage maintenance, while in the other half the side bars were opened. As discussed above, mice directed most bar-related behaviour towards the bars that were regularly opened for cage maintenance, indicating their motivation to escape. However, the mice consistently directed more bar-related behaviours towards the uncovered portions of both sites, indicating an additional motivation to explore airborne odour cues.

4.3.3. The role of lack of consummation

Once a sequence of behaviour has been initiated, repetition may be further facilitated by a lack of consummation. Thus, animals may get stuck in an appetitive sequence of behaviour (Dantzer, 1986; Box 1.1, Chapter 1, this volume) when the behavioural response does not have consequences that meet the animal's expectations.

Again, work on gerbils may illustrate this possibility. After Wiedenmayer (1997a) had found that digging in cage corners was abolished by the provision of an artificial burrow system, he attempted to substitute the complex burrow system with a simpler tool that could be implemented in laboratory husbandry practice. Based on the suggestion that the experience of moving into an enclosed area from an exposed one is a critical factor in gerbil flight and concealment responses (Clark and Galef, 1977, 1981), he showed that a simple chamber with a tunnel-shaped entrance was able to prevent the development of stereotypic digging completely, while the same chamber without the tunnel was not (Wiedenmayer, 1997a; see Fig. 4.4). This finding indicates that gerbils do not recognize a dark chamber as a shelter unless they experience moving into it through a narrow tunnel. This underscores just how specific are the environmental needs of animals that must be satisfied to guarantee the normal development of behaviour. This may be particularly true for behaviours that are crucial for survival and reproduction in the wild and hence resistant to habituation or extinction.

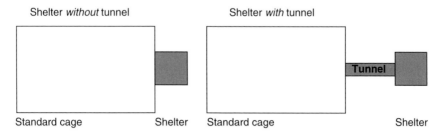

Fig. 4.4. Burrow stimuli that switch off stereotypic digging in the gerbil. Gerbils were given access to a shelter through a hole in the back of the cage (56 × 34 × 19 cm) starting at 16 days of age. The shelter (13 × 13 × 10 cm) was either attached directly to the cage or separated from the cage by a tunnel (20 cm, 5 cm diameter). At 36 days of age, gerbils in cages with a tunnel spent twice as much time inside the shelter compared to gerbils in cages without a tunnel. Furthermore, the tunnel completely prevented the development of stereotypic digging, while 9 out of 12 gerbils in cages without a tunnel developed stereotypic digging. Whether or not this system would also abolish established stereotypies has not been studied (from Wiedenmayer, 1997a).

4.3.4. Do these motivational explanations account for all the properties of rodent stereotypy?

Overall, these examples suggest that premature or sudden weaning, lack of shelter and the inability to explore cues from outside the cages may cause problems to at least some rodents, and that their persistent attempts to solve these problems may trigger stereotypy development. Thereby, internal causal factors (e.g. motivation to hide) and external causal factors (salient cues; e.g. edges of the cage) may interact, with the continued performance of the elicited responses perhaps further facilitated by a lack of consummation, and lack of habituation.

However, these examples further suggest that different motivations (e.g. for shelter, for access to neighbouring conspecifics) may elicit the same behavioural response (e.g. bar-mouthing), thereby leading to the same stereotypy (see Chapter 3, this volume for a similar idea regarding carnivore stereotypies). On the other hand, the same motivation (e.g. for shelter) may also underlie a variety of behavioural responses (e.g. bar-mouthing, jumping, digging), thereby leading to different forms of stereo-typy. This means that caution is needed when attempting to infer the underlying motivational problem from the behavioural content or form of a stereotypy.

Species differences further suggest that motivational explanations are not the whole story. As outlined above, species comparisons reveal differences as well as similarities in the incidence and form of stereotypies. These may, at least in part, depend on features of the animals' general behavioural biology, e.g. the factors that motivate them, and the species-typical responses they then show. Standard weaning practice, lack of shelter, and confinement may pose problems to most rodent species. However, while some responses to these problems seem to be shared by several species (e.g. bar-mouthing, jumping, locomotion), others may be more species-specific (e.g. cage-corner digging in gerbils). Self-made burrows may be more crucial to survival in gerbils than in mice. Unlike gerbils that rely exclusively on access to self-made burrows (Ågren *et al.*, 1989), free-living mice can use alternative behavioural strategies as they often live in areas where other forms of shelter are present. If no burrow is available, and digging one is not possible, they may engage in alternative strategies in search for shelter. Thus, one possibility is that mice are more variable in their responses to a lack of shelter than gerbils. However, in the absence of systematic studies on species differences using comparative approaches (*cf.* Chapter 3, this volume), such hypotheses remain essentially speculation. Furthermore, non-motivational explanations could also help account for species differences in stereotypy.

As we have seen, despite showing excessive wheel-running when provided with a running wheel, rats and guinea pigs seem to have a very low propensity to develop cage stereotypies or may not develop them at all. This is particularly puzzling in laboratory rats, since the general behavioural biology of both Norway rats and house mice suggests that they need largely similar environmental conditions. Moreover, like mice, rats show bar-related escape behaviours when housed in standard laboratory cages (Hurst *et al.*, 1996, 1999). Thus, both mice and rats may be faced with a similar thwarting of highly motivated behaviours and in response to these may display similar behavioural responses. However, in rats, in contrast to mice, these responses do not seem to develop into stereotypies. This suggests that species differences in the incidence of stereotypies may reflect differences related to the mechanisms by which behavioural responses to thwarting develop into established stereotypies. Such differences do not necessarily reflect differences in the behavioural biology of the animals and may therefore say little about their environmental needs. Instead, they may depend on the neurophysiological mechanism underlying behaviour control.

Furthermore, fully developed stereotypies have a number of properties that apparently distinguish them from normal, flexible motivated behaviour (e.g. decrease in variation, establishment and emancipation). These characteristics further suggest that thwarted motivation alone is not the sole cause of rodent stereotypy. In Section 4.4, I therefore review concepts about additional mechanisms potentially underlying stereotypy development, and discuss the available evidence from studies in rodents.

4.4. From Repetition to Stereotypy: Developmental Processes Underlying Caged Rodents' Stereotypies

The motivational factors discussed in the previous section can, to varying degrees, explain the initial performance and continued repetition of specific behavioural responses by caged rodents. However, as I have just reviewed, these factors do not seem to fully account for all the within- and between-species differences seen in rodent stereotypy. Furthermore, they would not account for the developmental changes attributed to true stereotypies by most authors (Fentress, 1976; Ödberg, 1978; Dantzer, 1986, 1991; Mason, 1991, 1993; Lawrence and Terlouw, 1993; Mason and Turner, 1993; Rushen *et al.*, 1993; Garner, 1999; see also Chapter 10, this volume). It is therefore important to distinguish between the causal factors underlying the repeated performance of potential source behaviours, and the developmental processes by which these initial behavioural responses are transformed into established and apparently purposeless stereotypies.

There are four developmental changes that cannot be easily explained by the motivational factors discussed above. First, these factors do not explain the dramatic *increase in frequency and duration* of the behaviours, which is commonly observed during stereotypy development (e.g. Kennes *et al.*, 1988; Cooper and Nicol, 1991; Würbel *et al.*, 1996; Wiedenmayer, 1997a; Nevison *et al.*, 1999a). In bank voles, for example, some individuals perform up to 45,000 jumps a day, day after day (Ödberg, 1986). Second, they do not explain the qualitative changes in form that are often seen, whereby the behaviour becomes increasingly *unvarying.* Würbel *et al.* (1996) described how initially variable bouts of climbing on the cage lid, pushing the nose here and there between the bars of the lid, while intensely sniffing and accidentally biting at bars, gradually developed into unvarying and rigid bar-mouthing. Third, they fail to account for *emancipation*, i.e. the stereotypy becoming increasingly elicited by a wider range of stimuli or motivational states than is seen early in development. Finally, they fail to account for *establishment*, i.e. the increasing persistence of stereotypies under conditions under which they would not normally develop. The gradual *emancipation* and *establishment* of stereotypies were shown by Cooper *et al.* (1996), who found that environmental enrichment abolished stereotypies completely in young (2 months) bank voles, while it was much less effective in mid-aged voles (6 months) and had hardly any effect on aged voles (14 months) (see also Kennes *et al.*, 1988 for pharmacological evidence; and Chapter 7, this volume, for recent similar data on deer mice). All four aspects suggest that additional mechanisms need to operate on repeatedly initiated source behaviour patterns in order for them to develop into true stereotypies. In this section, I will discuss these mechanisms and examine the literature on caged rodents' stereotypies for evidence in support of each of them.

4.4.1. The coping hypothesis

As discussed in more detail in Box 1.3, Chapter 1, it has been hypothesized that stereotypies might develop and be sustained because of rewarding properties (such as stress reduction, release of endogenous opioids, or other changes that might improve subjective well-being) that are associated with the behavioural responses that animals exhibit during conflict or thwarting. These rewarding consequences would then act as reinforcers, thereby increasing the animals' motivation to perform the same behavioural pattern on subsequent occasions. Such a process might thus account for the seemingly compulsive, addiction-like nature of some established stereotypies (e.g. their resistance to environmental enrichment). Furthermore, if the same rewarding properties were also associated with the response being performed in other, similarly aversive situations, response generalization could also explain emancipation. Here, I review the evidence that stereotypies help rodents to cope with adverse environmental conditions.

4.4.1.1. Is there evidence for a role of endogenous opioid peptides?

In Box 1.3, Chapter 1, I and others challenge the idea that stereotypies cause endogenous opioid release, and certainly in rodents, experimental tests have failed to find convincing evidence for such a mechanism. Thus, Kennes *et al.* (1988) found that blocking the effects of endogenous opioids by the opioid antagonist naloxone reduced stereotypic jumping in young bank voles, but it had no effect on performance of established stereotypies in adult animals. Although these results appear to support the idea that the rewarding properties of endogenous opioids might be involved in the early development of stereotypies, they do not provide support for an involvement in the continued performance of stereotypies. More importantly, however, the same developmental pattern can be explained by the sensitizing effect of endogenous opioids on dopaminergic pathways, and these were indeed found to mediate the performance of established stereotypies (Kennes *et al.*, 1988; see also Box 8.1, Chapter 8, this volume).

4.4.1.2. Is there evidence for coping with stress?

Another version of the coping hypothesis considered the rewarding properties to be associated with a stress-reducing effect of stereotypy performance. Most evidence for this hypothesis stemmed from studies in farm animals, and was largely correlational (see also Box 1.3, Chapter 1, this volume). To obtain more convincing evidence, experiments in rodents were undertaken to examine whether the selective prevention of stereotypy performance would lead to elevated stress levels, as predicted by the hypothesis.

Kennes and de Rycke (1988) found that lowering the cage ceiling to prevent stereotypic jumping in bank voles led to a higher and more sustained increase in plasma corticosterone levels of high stereotypers compared to low stereotypers. At first sight, these results seem to provide evidence in favour of the coping hypothesis. However, they might simply reflect the fact that the interference with familiar behavioural routines by lowering the cage ceiling was more pronounced in the high stereotypers. A similar effect was found in laboratory mice, where an acute increase in stress levels in response to selective prevention of stereotypic barmouthing could be explained without invoking a stress-reducing effect of stereotypy performance: stress levels rapidly returned to baseline despite the continued prevention of the stereotypy, and without alternative stereotypies developing (Würbel and Stauffacher, 1996; see Fig. 4.5).

4.4.1.3. Is there other evidence for a coping effect?

Another approach left the mechanism by which coping might act unspecified, but instead looked for behavioural changes that were predicted by the hypothesis that stereotypies have beneficial consequences of some form. For example, if stereotypies help animals to cope with adverse environments, stereotypers should find such environments less aversive than non-stereotypers. In bank voles, Cooper and Nicol (1991) found a negative relationship between stereotypy level and preference for an enriched environment, and the more the stereotypies developed, the more the voles preferred the barren cages where the stereotypies were performed. This effect was even significant when time in each cage was corrected for the time spent stereotyping (Garner, 1999). These findings seem to support the prediction that stereotyping voles found barren cages less aversive than non-stereotyping voles. However, alternative explanations have been proposed, e.g. that stereotyping voles found enriched

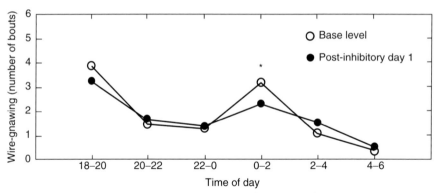

Fig. 4.5. Stereotypic bar-mouthing in ICR mice across the 12 h dark period on the day before (open circles) and after (filled circles) selective prevention of this behaviour. Bar-mouthing was selectively prevented by reducing the distance between the bars of the cage lid. Contrary to a prediction of the coping hypothesis, no post-inhibitory rebound of bar-mouthing was detected, and peak levels on post-inhibitory day 1 were even reduced compared to the day before prevention. This indicates that the motivation to perform bar-mouthing did not increase during prevention. Data are collapsed for three groups of mice selectively prevented from bar-mouthing for 1, 5 or 10 days, respectively (from Würbel *et al.*, 1998a).

and barren cages equally aversive (Rushen, 1993), or that they were impaired in reversal learning (Garner, 1999). The latter explanation is based on the fact that the positions of the cages were reversed between each choice test. In order to choose the preferred cage type, the voles had to alternate their direction at the choice point from test to test. Therefore, a deficit in their ability to learn such positional reversals may have interfered with the expression of their preference for cage type.

Alternatively, evidence for a coping effect of stereotypies might come from finding that the internal motivation to perform stereotypic behaviour increases with time since stereotypies were last performed. If stereotypies help animals attenuate detrimental effects that accumulate over time under conditions of chronic adversity, the motivation to perform the stereotypy should increase with increasing time since last performance. This can be studied by selectively preventing the behaviour for a certain time, and assessing whether performance is increased following prevention, a phenomenon called post-inhibitory rebound (Kennedy, 1985; Nicol, 1987). Post-inhibitory rebound was found following prevention of wheel-running (e.g. in rats; Sherwin, 1998; see also Box 4.2). In contrast, Würbel *et al.* (1998a) failed to find a similar effect in response to selective prevention of bar-mouthing in laboratory mice (Fig. 4.6). However, from a theoretical point of view it is still unclear whether a rise of internal motivation, and hence post-inhibitory rebound, is a necessary prediction for behaviours that have beneficial consequences under such conditions. Bridging theories of instrumental conditioning (e.g. Dickinson, 1994) and motivation (e.g. Toates and Jensen, 1991) might help to generate less ambiguous predictions to test the coping hypothesis in the future.

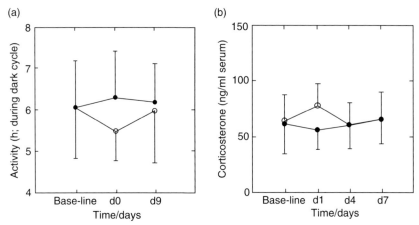

Fig. 4.6. The effects of selectively preventing stereotypic bar-mouthing by mice. Selective prevention of stereotypic bar-mouthing by reducing the distance between the bars of the cage lid was associated with a short-term increase in plasma corticosterone (a) and a short-term decrease in total activity (b). Within 3 days, however, both plasma corticosterone levels and activity were back at pre-treatment basal levels, and prevention of bar-mouthing did not affect chronic measures of stress after 10 days of prevention. The short-term increase in acute stress levels could thus reflect the animals' response to being prevented from carrying out familiar behavioural routines. As soon as behavioural organization was re-established, physiological signs of stress had disappeared even without new stereotypies being developed (from Würbel and Stauffacher, 1996).

4.4.1.4. Conclusions

Overall, evidence in support of the coping hypothesis is still sparse and inconsistent for rodents. The coping hypothesis considers stereotypies to be adaptive, functional responses and stereotypy development to be based on purely motivational processes. However, after reviewing the evidence above, the question arises as to whether motivational processes are sufficient to explain stereotypy development, and whether stereotypies are indeed functional responses. In the following sections, I therefore examine what other, non-motivational processes might account for stereotypy development and, in particular, whether stereotypies might reflect pathology rather than successful coping.

4.4.2. Maladaptive or pathological processes?

Some of the changes that are seen as motivated behaviours develop into stereotypies may be the result of maladaptive or even pathological processes. Maladaptive processes are those that are adaptive in naturalistic situations, but which may cause non-adaptive or even counterproductive outcomes in captivity; while pathologies are the product of dysfunction, for instance within the CNS (see Box 1.4, Chapter 1, this volume, for more detail).

The view that stereotypy is a pathology caused by an environment that overtaxes the animals' capacity to adapt is far less popular – at least among ethologists – than the view that it is a normal response of a normal animal to an abnormal environment (Garner, 1999): although stereotypies are commonly considered to reflect impaired welfare, they are rarely seen as truly pathological. If stereotypy reflects pathology, it is likely to be one of the dysfunctions of the CNS. In the following sections, I will therefore briefly examine hypotheses that consider stereotypies to reflect CNS dysfunction (ideas that are developed further in the following chapters), as well as the possibility that stereotypies are maladaptive.

In contrast to stereotypies in caged animals, the neurophysiological basis of stereotypies in human mental disorder (e.g. schizophrenia, autism), and stereotypies induced by drugs (e.g. amphetamine) or brain lesions, is rather well understood. These stereotypies are thought to result from impaired basal ganglia function, primarily in terms of dopaminergic regulation (Chapters 5, 7 and 8, this volume). For example, psychomotor stimulant drugs such as amphetamine and apomorphine can induce stereotypic behaviour through activating dopaminergic systems in the basal ganglia. The behaviour will become more stereotyped, i.e. more intense and rigid, with increasing doses of the drug (Lyon and Robbins, 1975), but a similar shift in form and intensity can also be induced by repeated administration of a constant dose, indicating behavioural 'sensitization' to the drug (Robbins and Sahakian, 1981; see also Chapter 8, this volume), or drug-induced behavioural disinhibition.

Dantzer (1986, 1999) provided the first formal application of behavioural sensitization to stereotypy development in captive animals. He argued that stereotypies result when animals perform appetitive behaviours that cannot become consummatory, for instance due to a lack of suitable features in the environment. When normal satiety mechanisms via negative feedback fail, the positive sensory feedback of the appetitive behaviour on the neural systems controlling it may then lead to an increased probability of this behaviour being performed on subsequent occasions. In addition, he argued that progressive sensitization of the neural elements of the repeatedly activated pathways would explain not only the increase in performance of stereotypies, but also their *ritualization, emancipation* and *establishment*.

The exact mechanism (neural sensitization) proposed by Dantzer (1986) was stated in rather vague terms and so did not generate specific predictions that could be tested. Also, it is unclear whether the neural sensitization meant by Dantzer (1986) would be a product of the repetition of behaviour *per se*, or whether it would depend on the arousal (or stress) induced by the chronic thwarting of highly motivated behaviour. The former possibility has been invoked in some hypotheses considering 'channelling' (Lawrence and Terlouw, 1993; *cf.* Chapter 2, this volume), lack of behavioural competition (Hinde, 1962; Mason and Turner, 1993) and habit formation (e.g. Hinde, 1970; Fentress, 1976; Mason and Turner, 1993) to underlie stereotypy development. In contrast, the latter possibility has been invoked in the hypothesis that stereotypies develop due to stress-induced behavioural sensitization (Cabib and Bonaventura, 1997).

These hypotheses are developed in more detail below. Note that for all of them, it is unknown whether their effects should best be defined as maladaptive or pathological, partly because we still know so little about their underlying mechanisms (see below), and partly because 'pathological' is a normative term, thus one whose definition relies on having data on the normal biological range of variation (see Box 1.4, Chapter 1, this volume for more detail).

4.4.2.1. Channelling, lack of competition and routine formation

The barren nature of laboratory cages provides only few incentives for the animals to interact with. Environmental restrictions placed on the variable expression of highly motivated behaviour have been proposed to 'channel' the behaviour into a few simple behavioural elements (see similar discussions in Chapters 2 and 3, this volume), thus contributing to stereotypy development. Channelling, together with a lack of behavioural competition as a result of a barren environment, may then lead to a higher level of performance of fewer behavioural elements less variably.

This in turn may facilitate the establishment of so-called central or closed-loop control, i.e. the process by which once flexible patterns become less and less variable with repetition and progressively independent of environmental guidance (Hinde, 1970; Fentress, 1976; Mason and Turner, 1993; Mason and Latham, 2004). This same process under normal circumstances facilitates routine formation (e.g. grooming behaviour in mice; Fentress, 1976). However, in captivity it might act on a behaviour that would not normally be repeated over and over again, thereby leading to the fixation of functionless behaviour.

It is important to note that channelling, lack of behavioural competition and routine formation do not necessarily invoke conflict or thwarting as a precondition for stereotypies to develop. Thus, while the reduction or abolition of stereotypies in rodents by environmental enrichment is normally attributed to the satisfaction of specific environmental needs (e.g. shelter; Ödberg, 1987; Cooper and Nicol, 1996; Wiedenmayer, 1997a; Würbel *et al.*, 1998b), it might as well be effective by counteracting channelling or increasing behavioural competition, via reducing the repetitiveness of behaviour and increasing its variability (see Chapter 3, this volume for a similar discussion and Chapter 9, this volume for more detail on how environmental enrichments could act). However, to date the role of these processes has not been experimentally investigated in rodents, nor indeed any other captive animal.

4.4.2.2. Stress-induced behavioural sensitization

Evidence for stress-induced behavioural sensitization mainly stems from studies on drug-induced stereotypies in rats and mice (see Chapter 8, this volume for a fuller discussion). Behavioural sensitization to stimulant drugs is not only obtained by the repeated administration of the drugs, but also by repeated exposure to various stressors, such as mild tail pinch, inescapable foot shock, food deprivation or immobilization. During such stressful experiences, endogenous opioid peptides are massively released (Amir *et al.*, 1981) and these have been shown to exert modulatory effects on dopaminergic pathways (e.g. Akil *et al.*, 1984; Cabib *et al.*, 1984, 1989). Thus, endogenous opioids, instead of acting through rewarding properties (see above), might play an important role in sensitizing dopamine systems in the basal ganglia, which also mediate opioid-induced stereotypies (Longoni *et al.*, 1991).

If cage-induced stereotypies depend on behavioural sensitization mediated by stress, individual stress levels at the onset of stereotypy development should correlate with later levels of stereotypy performance. Support for this was found in laboratory mice, where adult stereotypy levels correlated with corticosterone levels 48 h after weaning (Würbel and Stauffacher, 1997). These results parallel the stress-induced enhancement of apomorphine-induced stereotypic climbing in mice (Cabib *et al.*, 1984). As discussed above, adult stereotypy levels did not correlate with the performance of the source behaviour pattern immediately after weaning (Würbel and Stauffacher, 1997). This suggests that weaning stress affected later stereotypy development not by affecting performance levels of the source behaviour pattern, but rather by affecting its tendency to become stereotypic. At the behavioural level, a similar observation was made in mink, where early weaning was found to predispose young mink to higher later stereotypy levels in the absence of a difference in performance levels of the source behaviour patterns (Mason, 1996; see also Chapter 6, this volume, for more on early social experience and its long-term effects on abnormal behaviour).

Further evidence comes from pharmacological manipulations of cage-induced stereotypic jumping in bank voles. Endogenous opioids, which also play a crucial role in the development of stress-induced behavioural sensitization to stimulant drugs (see above), were found to be critically involved in the early, but not later, stages of stereotypy development (Kennes *et al.*, 1988), as predicted by this hypothesis.

4.4.2.3. *General disinhibition of behavioural control mechanisms*

Stereotypies in human mental disorder, and stereotypies induced by stimulant drugs or brain lesions, are associated with a range of other characteristic behavioural changes, including enhanced rates of behavioural initiation, impulsivity (i.e. making behavioural choices without first integrating the available information), impaired response suppression (i.e. the inability to withhold responding even at the cost of delaying or missing reward) and perseveration (e.g. in extinction tasks; see Chapters 3 and 5, this volume), all of which have been attributed to a general disinhibition of behavioural control mechanisms of the dorsal basal ganglia (Garner, 1999; Garner and Mason, 2002). Thus, if caged animals' stereotypies were a sign of the same basal ganglia dysfunction, stereotypy performance should correlate with these other symptoms (Garner, 1999; Garner and Mason, 2002). Using bank voles to test this prediction, Garner found that stereotypic bar-mouthing in voles correlated with increased rates of behavioural activation, hyperactivity, impairments of response timing and slower extinction learning, and that all of these signs were intercorrelated (Garner and Mason, 2002). These findings suggest a single underlying deficit consistent with impaired response selection caused by basal ganglia dysfunction.

Given the conditions under which cage stereotypies typically develop, it is tempting to take these results as evidence that stereotypies do indeed reflect acquired brain pathology caused by inappropriate housing conditions. However, strictly they merely demonstrate that individual variation in stereotypy performance correlates with individual variation in other behavioural signs of basal ganglia function, and this variation might simply reflect the normal range of variation of a population of healthy individuals. To demonstrate that stereotypy reflects a pathological insult to this brain system, it still needs to be demonstrated that these signs change together as stereotypy develops, and back again when the stereotypy is cured, e.g. by environmental enrichment (Garner, 1999; see also Chapter 5, this volume).

4.4.2.4. Conclusions

Taken together, present evidence suggests that established rodent stereotypies could indeed reflect impaired brain function, although, as discussed above, the possibility that some rodent stereotypies may serve a coping function has not been ruled out. Stereotypies generally seem to develop from adaptive behavioural responses aimed at coping with the chronic thwarting of highly motivated behaviour. However, in those cases where the behaviour develops further into an established stereotypy, present evidence suggests that coping has failed, resulting in abnormal behaviour that is maladaptive, or even pathological – possibly reflecting some form of impaired brain function.

The precise nature and aetiology of this potential pathology has to date remained elusive. However, it could explain why wild-caught voles placed in barren cages as adults did not develop stereotypies, while their offspring exposed to barren housing from birth did (Cooper and Nicol, 1996; see also Box 7.1, Chapter 7, this volume). It is well established that impoverished environments can impair normal brain development (Würbel, 2001). This impairment is characterized, among others, by fewer neurons and reduced synaptic connectivity between them (Van Praag *et al.*, 2000). Perhaps, an impoverished brain architecture induced by an impoverished rearing environment is another important prerequisite for stereotypies to develop (Würbel, 2001; see also Chapter 7).

Furthermore, stereotypy development may depend on a degree of brain plasticity that is no longer present in adult animals. In rodents, not many brain areas that show little plasticity in adulthood remain highly plastic beyond weaning, when stereotypies normally develop.

One specific brain area that does show plasticity until late adolescence, but not in adulthood, is the frontal cortex (e.g. Fuster, 2002). Importantly, the frontal cortex plays a key role in the inhibitory control of behaviour via extensive projections (mainly dopaminergic) to the basal ganglia (e.g. Passingham, 1995). This may provide an interesting addition to the hypothesis of basal ganglia dysfunction (see also Box 5.4, Chapter 5, this volume) that might be worthy of further investigation with a developmental angle.

Clearly, there is a great need for studies which integrate both behavioural and neurobiological approaches, if we are to uncover all the mechanisms involved in stereotypy development. Based on their long-standing tradition as laboratory animals, rodents provide ideal subjects for this research.

4.5. General Conclusions

4.5.1. The motivational basis of caged rodents' stereotypies

So what are the main conclusions that can be drawn from this review about the motivational basis of caged rodents' stereotypies? Current evidence indicates that stereotypies mainly develop in response to the thwarting of behaviours that are crucial for survival and reproduction in the wild. In rodents, the motivation to seek shelter and to explore odour cues perceived from outside the cage seem to be important, but other factors, such as the motivation to return to the mother following weaning may be additional forces. The thwarting of these behaviours, along with salient environmental cues, triggers sustained appetitive behaviours and/or intention movements that seem highly resistant to habituation and hence are repeated over and over again. Some of these responses may be fairly species-specific (e.g. cage-corner digging in gerbils), while others are common to many rodent species (e.g. bar-mouthing, jumping, running).

However, for such behavioural responses to develop into established stereotypies, I argue that additional conditions need to be met. While environmental restrictions placed on the variable expression of behaviour (i.e. channelling) and/or a lack of behavioural competition might facilitate further stereotypy development, the barren nature of the environment may primarily act through the impairment of normal brain development, resulting in impoverished brain architecture which may predispose animals to stereotypy development. This idea is consistent with the observation that fully developed adult rodents do not seem to develop stereotypies when newly placed into barren housing conditions. Furthermore, the arousal (or stress) engendered by the initial thwarting of highly motivated behaviour may further facilitate stereotypy development through sensitizing the neural systems involved. Over time, this may lead to dysfunction of the neural systems involved in behaviour control, leading to a range of correlated behavioural changes. Thus, as proposed by Garner (1999), stereotypy may be just one symptom of a general disruption of inhibitory control of behaviour.

However, it should be clear from this chapter that there is much we still do not understand about rodent stereotypy. Despite decades of extensive research, our understanding of the causation and consequences of stereotypies, as well as the mechanisms underlying their development, has remained elusive. Fundamental questions are as yet unanswered and these need to be addressed if facts are truly to be separated from fiction. In the following, final sections, I therefore highlight some promising areas for future research, and last but not least, I discuss the possible implications of stereotypy for rodent welfare.

4.5.2. Issues to be resolved by future research

4.5.2.1. Establishing a biologically meaningful classification system

Part of our lack of understanding of stereotypies may be inherent in their definition, which has repeatedly given rise to controversy (e.g. Dantzer, 1991; Mason, 1991; Duncan *et al.*, 1993; see also Box 4.2, this volume on wheel-running). The defining properties *'repetitive, unvarying* and *apparent lack of function'* cover a broad range of behaviours that are essentially heterogeneous (Mason, 1991). Obviously, the resulting diversity is at odds with a unifying theory. For example, a theory of stereotypy would need to cover appetitive or redirected behaviours that are bound to specific motivational systems and environmental cues (e.g. the pre- and post-feeding oral behaviours of sows; Chapter 2, this volume) as well as established stereotypies that have become largely independent of further environmental guidance (e.g. stereotypic jumping in aged bank voles; Cooper *et al.*, 1996). Furthermore, the current definition does not distinguish between low and high levels of stereotypy performance, even though the level of performance might actually radically change the significance of a stereotypy. Changing definitions is generally difficult and may not solve the problem. What we really need is a biologically meaningful classification system, based on causal, developmental and functional aspects of the behaviour. Therefore, future research should aim at differentiating between different forms and levels of performance of repetitive behaviours based on their dynamic developmental properties (increasing performance, decreasing variation, emancipation and establishment), and also examine the underlying neurophysiological mechanisms. This would help our fundamental understanding of stereotypies, and also help our use of them as welfare indicators, as these features have important implications for the significance of stereotypies for animal welfare (see below).

4.5.2.2. Studying causal factors in other species and other stereotypies

Unfortunately, not many studies have looked into the motivational problems underlying caged rodents' stereotypies. Apart from the examples discussed above (bar-mouthing and jumping in mice; digging in gerbils), we can as yet only speculate about the motivations underlying other forms of common rodent stereotypies (e.g. back-flips, cage-top twirling, route-tracing) as well as stereotypies in other rodents (e.g. hamsters, chinchillas, black rats, African striped mice, deer mice). In laboratory rodents, there is rather little variation in the housing and management within species. Therefore, epidemiological studies to identify risk factors for stereotypy development might be of limited value. However, pet rodents might be a rewarding target for epidemiological studies that could be followed up by developmental and experimental studies under controlled laboratory conditions.

4.5.2.3. Testing the coping hypothesis

As discussed above, evidence as to whether stereotypies reflect a behavioural strategy to cope with adversity is as yet ambiguous. Part of the problem may be that not many predictions have been generated that allow testing of the coping hypothesis unambiguously.

According to this hypothesis, stereotypies are based on motivational processes and learning processes. It is thought that animals learn an association between a behavioural response (their response to initial conflict or thwarting) and some positive outcome (at the physiological or psychological level). They then become highly motivated to use this strategy to obtain the positive outcome, possibly extending it to situations other than the originally eliciting ones. As mentioned before, there may be many physiological or psychological targets for such a coping response (e.g. stress, fear, boredom, pain, hunger, etc.). Therefore, a more promising approach towards testing the coping hypothesis would be to test general predictions that do not invoke a specific physiological or psychological system. Integrating theories of instrumental conditioning (e.g. Dickinson, 1994) with concepts of motivation (e.g. Toates and Jensen, 1991) might therefore greatly help to generate more powerful predictions to test the coping hypothesis.

One promising approach in this direction would be the use of consumer demand techniques (e.g. Dawkins, 1990; Houston, 1997; Mason et al., 1998). If stereotypies help animals cope with adverse conditions, they should be prepared to incur costs in order to perform their stereotypies. This seems to be true for wheel-running, but wheel-running may be highly rewarding regardless of whether or not it is performed in a stereotypical manner (Sherwin, 1998). So far, consumer demand techniques have not been used with any other stereotypy.

4.5.2.4. Testing the pathology hypothesis

Evidence in favour of the pathology hypothesis is similarly ambiguous, and the nature and aetiology of the pathology has remained elusive. For example, Dantzer's (1986) hypothesis of behavioural sensitization refers to a specific behavioural pathology that does not necessarily affect other aspects of behavioural control. By contrast, Garner (1999; Garner and Mason, 2002) and implicitly Cabib (Cabib and Bonaventura, 1997; see also Chapter 8) proposed that stereotypy is just one symptom of a more profound dysfunction of behavioural control. It is unclear as to whether Dantzer's model would predict the correlated behavioural changes found by Garner. Conversely, it is unclear how Garner's hypothesis accounts for the fact that stereotypic performance is usually restricted to one or two specific behaviour patterns, while other frequently performed behaviours do not become stereotypic in the same way (see also Box 5.4, Chapter 5, this volume).

It is also unclear whether stereotypies can occur simply by the frequent repetition of a behaviour pattern, or whether it depends on the arousal (or stress) associated with chronic thwarting and conflict. Attempts to dissociate these two factors clearly represent a worthwhile target of future research. This could be done, for example, by forcing animals to perform repeatedly some arbitrary behaviour (e.g. by using operant techniques) that would be paired with either a stressor or a neutral stimulus. If the behaviour became progressively stereotypic when paired with the stressor, this would be strong evidence for a mechanism based on stress-induced behavioural sensitization. Conversely, if pairing with a neutral stimulus is sufficient to induce stereotypy development, this would be strong evidence in favour of a mechanism similar to habit formation.

Furthermore, individual levels of stereotypy performance are often continuously distributed between zero and high levels of performance within study populations (e.g. Würbel and Stauffacher, 1997). Very low, but stable levels of stereotypy performance may be difficult to explain in terms of behavioural sensitization, or indeed brain pathology. In order to generate predictions that can be tested against each other, present hypotheses about stereotypy development need to be stated more explicitly in terms of both correlated behavioural signs and underlying neurobiological changes. Importantly, at some point hypotheses about CNS dysfunction need to be examined at the neural level. Very few groups have used neurobiological approaches to cage-induced stereotypies (see Chapter 7 for a welcome exception). I hope that more and more ethologists succeed in convincing neuroscientists that cage-induced stereotypies provide a fascinating target of research, with implications for a better understanding of fundamental biological processes, as well as important practical issues such as the validity of animal experiments (Würbel, 2001) and, of course, animal welfare.

4.5.3. Welfare implications

The welfare implications of caged rodents' stereotypies obviously depend on their exact causes and consequences. For one thing, rodents, like other

animals, readily develop stereotypies when reared in barren environments, but do not do so in the wild or when reared in an appropriate
environment. However, this does not necessarily mean that stereotypy
indicates poor welfare, nor that the absence of stereotypy indicates good
welfare (see also Mason and Latham, 2004; and Chapter 11, this volume).
First, as we have seen, not all rodent species develop stereotypies under
barren housing conditions, and even in species where stereotypies are
commonly observed, not all individuals develop them. Furthermore, in
mice stereotypic bar-mouthing was persistently prevented by reducing
the gaps between the bars (Würbel and Stauffacher, 1997), a treatment
which was unlikely to have improved their welfare. Second, individual
variation in stereotypy performance might reflect individual differences
that are unrelated to well-being (e.g. general activity levels, propensity for
routine formation), but may become apparent only under barren housing
conditions. Third, although rather unlikely, stereotypies might be an
effective means of coping with environmental restrictions, with stereotyping animals being better off in terms of their well-being than non-
stereotyping animals.

Even if stereotypies reflect impaired brain function, however, the
question still remains as to how such impairment is perceived by the
animals. Dantzer (1986) suggested that stereotypies become 'hard-wired'
with establishment, thereby losing emotional significance. However, as
long as the source of chronic thwarting that initially gave rise to stereotypy development had not been removed, stereotyping animals might still
continue to suffer (unless stereotypy was an effective coping strategy). By
contrast, Wemelsfelder (1993) considered stereotypy performance to be a
direct sign of animal suffering as a consequence of 'boredom' induced by
a lack of control over the environment. While this hypothesis has
remained vague and did not generate predictions that could be tested,
an interesting finding by Garner (1999; Garner and Mason, 2002) suggests
that stereotypy might indeed reflect some form of suffering, as indicated
by a knowledge–action dissociation in stereotyping animals. Such knowledge–action dissociation is also found in stereotypic human patients, and
patients suffering from basal ganglia pathology sometimes report feeling
frustrated by their inability to turn decisions and preferences into action
(e.g. Luria, 1965; Turner, 1997; reviewed by Garner, 1999; see also Chapter 5, this volume). Garner's conclusion that therefore stereotypic animals
may feel frustrated about the same inability is of course highly speculative, and has been criticized (e.g. Mason and Latham, 2004). However, it
provides a significant challenge to the arguably equally speculative coping hypothesis which considers stereotypies to improve subjective well-
being.

Determining the conscious emotional correlate of stereotypies and
other abnormal behaviours is likely to remain one of the most challenging
future targets in applied ethology. Excitingly, it could be the first in which
human subjects come to play guinea pigs for animals.

Acknowledgements

Many thanks to Georgia Mason and Jeff Rushen for constructive comments that greatly helped to improve this chapter.

References

Ågren, G., Zhou, Q. and Zhong, W. (1989) Ecology and social behaviour of Mongolian gerbils, *Meriones unguiculatus*, at Xilinhot, Inner Mongolia, China. *Animal Behaviour* 37, 11–27.

Akil, H., Watson, S.J., Young, E., Lewis, M.E., Khachaturian, H. and Walker, J.M. (1984) Endogenous opioids: biology and function. *Annual Reviews in the Neurosciences* 7, 223–255.

Altemus, M., Glowa, J., Galliven, E., Leong, Y. and Murphy, D. (1996) Effects of serotonergic agents on food-restriction-induced hyperactivity. *Pharmacology Biochemistry and Behavior* 53, 123–131.

Amir, S.Z., Brown, Z.W. and Amit, Z. (1981) The role of endorphins in stress: evidence and speculations. *Neuroscience and Biobehavioural Reviews* 4, 77–86.

Cabib, S. and Bonaventura, N. (1997) Parallel strain-dependent susceptibility to environmentally induced stereotypies and stress-induced behavioural sensitization in mice. *Physiology & Behavior* 61, 499–506.

Cabib, S., Puglisi-Allegra, S. and Oliverio, A. (1984) Chronic stress enhances apomorphine-induced climbing behaviour in mice: role of endogenous opioids. *Brain Research* 298, 138–140.

Cabib, S., Oliverio, A. and Puglisi-Allegra, S. (1989) Stress-induced decrease of 3-methoxytyramine in the nucleus accumbens of the mouse is prevented by naltrexone pre-treatment. *Life Sciences* 45, 1031–1037.

Callard, M.D., Bursten, S.N. and Price, E.O. (2000) Repetitive backflipping behaviour in captive roof rats (*Rattus rattus*) and the effects of cage enrichment. *Animal Welfare* 9, 139–152.

Clark, M.M. and Galef, B.G. (1977) The role of the physical rearing environment in the domestication of the Mongolian gerbil (*Meriones unguiculatus*). *Animal Behaviour* 25, 298–316.

Clark, M.M. and Galef, B.G. (1981) Environmental influence on development, behavior, and endocrine morphology of gerbils. *Physiology and Behavior* 27, 761–765.

Clubb, R. (2001) The roles of foraging niche, rearing condition and current husbandry on the development of stereotypies in carnivores. Ph.D. thesis, Oxford University, Oxford, UK.

Cooper, J.J. and Nicol, C.J. (1991) Stereotypic behaviour affects environmental preference in bank voles, *Clethrionomys glareolus*. *Animal Behaviour* 41, 971–977.

Cooper, J.J. and Nicol, C.J. (1996) Stereotypic behaviour in wild caught and laboratory bred bank voles (*Clethrionomys glareolus*). *Animal Welfare* 5, 245–257.

Cooper, J.J., Ödberg, F. and Nicol, C.J. (1996) Limitations on the effectiveness of environmental improvement in reducing stereotypic behaviour in bank voles (*Clethrionomys glareolus*). *Applied Animal Behaviour Science* 48, 237–248.

Cronin, G.M., Wiepkema, P.R. and van Ree, J.M. (1985) Endogenous opioids are involved in abnormal stereotyped behaviours of tethered sows. *Neuropeptides* 6, 527–530.

Dantzer, R. (1986) Behavioural, physiological and functional aspects of stereotypic behaviour: a review and reinterpretation. *Journal of Animal Science* 62, 1776–1786.

Dantzer, R. (1991) Stress, stereotypy and welfare. _Behavioural Processes_ 25, 95–102.

Dawkins, M.S. (1990) From an animal's point of view: motivation, fitness and animal welfare. _Brain and Behaviour Science_ 13, 1–61.

Dickinson, A. (1994) Instrumental conditioning. In: Mackintosh, N.J. (ed.) _Animal Learning and Cognition._ Academic Press, London, pp. 45–79.

Duncan, I.J.H., Rushen, J. and Lawrence, A.B. (1993) Conclusions and implications for animal welfare. In: Lawrence, A.B. and Rushen, J. (eds) _Stereotypic Animal Behaviour: Fundamentals and Applications to Welfare._ CAB International, Wallingford, UK, pp. 193–206.

Fentress, J.C. (1976) Dynamic boundaries of patterned behaviour: interaction and self-organisation. In: Bateson, P.P.G. and Hinde, R.A. (eds) _Perspectives in Ethology_, Vol. 1. Plenum Press, New York, pp. 155–224.

Fuster, J.M. (2002) Prefrontal cortex in temporal organization of action. In: Arbib, M.A. (ed.) _The Handbook of Brain Theory and Neural Networks_, 2nd edn. MIT Press, Cambridge, pp. 905–910.

Garner, J.P. (1999) The aetiology of stereotypy in caged animals. Ph.D. thesis, University of Oxford, Oxford, UK.

Garner, J.P. and Mason, G.J. (2002) Evidence for a relationship between cage stereotypies and behavioural disinhibition in laboratory rodents. _Behavioural Brain Research_ 136, 83–92.

Hebebrand, J., Exner, C., Hebebrand, K., Holtkamp, C., Casper, R., Remschmidt, H., Herpertz-Dahlmann, B. and Klingenspor, M. (2003) Hyperactivity in patients with anorexia nervosa and in semi-starved rats: evidence for a pivotal role of hypoleptinemia. _Physiology and Behavior_ 79, 25–37.

Hinde, R.A. (1962) The relevance of animal studies to human neurotic disorders. In: Richter, D., Tanner, J.M., Lord Taylor and Zangwill, O.L. (eds) _Aspects of Psychiatric Research._ Oxford University Press, Oxford, UK, pp. 240–261.

Hinde, R.A. (1970) _Animal Behaviour: A Synthesis of Ethology and Comparative Psychology_, 2nd edn. McGraw-Hill, New York, 876 pp.

Houston, A.I. (1997) Demand curves and welfare. _Animal Behaviour_ 53, 983–990.

Hurst, J.L. (1999) Introduction to rodents. In: Poole, P. (ed.) _The UFAW Handbook on the Care and Management of Laboratory Animals_, 7th edn. Blackwell Science, Oxford, UK, pp. 262–273.

Hurst, J.L., Barnard, C.J., Hare, R., Wheeldon, E.B. and West, C.D. (1996) Housing and welfare in laboratory rats: time-budgeting and pathophysiology in single-sex groups. _Animal Behaviour_ 52, 335–360.

Hurst, J.L., Barnard, C.J., Nevison, C.M. and West, C.D. (1997) Housing and welfare in laboratory rats: welfare implications of isolation and social contact among caged males. _Animal Welfare_ 6, 329–347.

Hurst, J.L., Barnard, C.J., Nevison, C.M. and West, C.D. (1998) Housing and welfare in laboratory rats: the welfare implications of social isolation and social contact among females. _Animal Welfare_ 7, 121–136.

Hurst, J.L., Barnard, C.J., Tolladay, U., Nevison, C.M. and West, C.D. (1999) Housing and welfare in laboratory rats: effects of cage stocking density and behavioural predictors of welfare. _Animal Behaviour_ 58, 563–586.

Johnson, R., Rhodes, J., Jeffrey, S., Garland, T. and Mitchell, G. (2003) Hippocampal brain-derived neurotrophic factor but not neurotrophin-3 increases more in mice selected for increased voluntary wheel running. _Neuroscience_ 121, 1–7.

Kennedy, J.S. (1985) Displacement activities and post-inhibitory rebound. _Animal Behaviour_ 34, 1375–1377.

Kennes, D. and de Rycke, P.H. (1988) The influence of the performance of stereotypies on plasma corticosterone and eosinophil levels in bank voles (*Clethrionomys glareolus*). In: Unselm, J., van Putten, G., Zeeb, K. and Ekesbo, I. (eds) *Proceedings of the International Congress of Applied Ethology in Farm Animals*. KTBL, Darmstadt, pp. 238–240.

Kennes, D., Ödberg, F.O., Bouquet, Y. and de Rycke, P.H. (1988) Changes in naloxone and haloperidol effects during the development of captivity-induced jumping stereotypy in bank voles. *European Journal of Pharmacology* 153, 19–24.

Kersten, A.M.P. (1997) Behaviour and welfare of chinchillas in commercial farming: a preliminary study. In: Hemsworth, P.H., Spinka, M. and Kostal, L. (eds) *Proceedings of the 31st International Congress of the International Society for Applied Ethology*. The International Society for Applied Ethology, Prague, p. 171.

Knight, J. (2001) Animal data jeopardized by life behind bars. *Nature* 412, 669.

Kuhnen, G. (2002) Comfortable quarters for hamsters in research institutions. In: Reinhardt, V. and Reinhardt, A. (eds) *Comfortable Quarters for Laboratory Animals*. Animal Welfare Institute, Washington, DC, pp. 33–37.

Latham, N. and Mason, G. (2004) From house mouse to mouse house: the behavioural biology of free-living *Mus musculus* and its implications in the laboratory. *Applied Animal Behaviour Science* 86, 261–289.

Lawrence, A.B. and Terlouw, E.M.C. (1993) A review of behavioural factors involved in the development and continued performance of stereotypic behaviours in pigs. *Journal of Animal Science* 71, 2815–2825.

Lewis, R.S. and Hurst, J.L. (2004) The assessment of bar-chewing as an escape behaviour in laboratory mice. *Animal Welfare* 13, 19–25.

Longoni, R., Spina, L., Mulas, A., Carboni, E., Garau, L., Melchiorri, P. and Di Chiara, G. (1991) (D-Ala2)deltrophin II: D1-dependent stereotypies and stimulation of dopamine release in the nucleus accumbens. *Journal of Neuroscience* 11, 1565–1576.

Luria, A.R. (1965) Two kinds of motor perseveration in massive injury of the frontal lobes. *Brain* 88, 1–11.

Lyon, M. and Robbins, T.W. (1975) The action of central nervous system stimulant drugs: a general theory concerning amphetamine effects. *Current Developments in Psychopharmacology* 2, 79–163.

Mason, G.J. (1991) Stereotypies: a critical review. *Animal Behaviour* 41, 1015–1037.

Mason, G.J. (1993) Forms of stereotypic behaviour. In: Lawrence, A.B. and Rushen, J. (eds) *Stereotypic Animal Behaviour: Fundamentals and Applications to Welfare*. CAB International, Wallingford, UK, pp. 41–64.

Mason, G. (1996) Early weaning enhances the later development of stereotypy in mink. In: Duncan, I., Widowski, T. and Haley, D. (eds) *Proceedings of the 30th International Congress of the International Society for Applied Ethology*. The International Society for Applied Ethology, Guelph, p. 6.

Mason, G.J. and Latham, N.R. (2004) Can't stop, won't stop: is stereotypy a reliable animal welfare indicator? *Animal Welfare* 13(Supplement), S57–S69.

Mason, G.J. and Turner, M.A. (1993) Mechanisms involved in the development and control of stereotypies. In: Bateson, P.P.G., Klopfer, P.H. and Thompson, N.S. (eds) *Behaviour and Evolution. Perspectives in Ethology*, Vol. 10. Plenum Press, New York, pp. 53–86.

Mason, G., McFarland, D. and Garner, J. (1998) A demanding task: using economic techniques to assess animal priorities. *Animal Behaviour* 55, 1071–1075.

McFarland, D. (1989) *Problems of Animal Behaviour*. Longman, Harlow, 172 pp.

Meyer-Holzapfel, M. (1968) Abnormal behaviour in zoo animals. In: Fox, M.W. (ed.) *Abnormal Behaviour in Animals.* WB Saunders, Philadelphia, pp. 476–503.

Moore, A. (2001) Of mice and mendel. *EMBO Reports* 2, 554–558.

Morrow, N., Schall, M., Grijalva, C., Geiselman, P., Garrick, T., Nuccion, S. and Novin, D. (1997) Body temperature and wheel running predict survival times in rats exposed to activity-stress. *Physiology and Behavior* 62, 815–825.

Nestler, E.J. (2004) Molecular mechanisms of drug addiction. *Neuropharmacology* 47, 24–32.

Nevison, C.M., Hurst, J.L. and Barnard, C.J. (1999a) Why do male ICR(CD-1) mice perform bar-related (stereotypic) behaviour? *Behavioural Processes* 47, 95–111.

Nevison, C.M., Hurst, J.L. and Barnard, C.J. (1999b) Strain-specific effects of cage enrichment in male laboratory mice (*Mus musculus*). *Animal Welfare* 8, 361–379.

Nicol, C.J. (1987) Behavioural responses of laying hens following a period of spatial restriction. *Animal Behaviour* 35, 1709–1719.

Ödberg, F.O. (1978) Abnormal behaviours: stereotypies. *Proceedings of the First World Congress on Ethology Applied to Zootechnics.* Industrias Graficas Espana, Madrid, pp. 475–480.

Ödberg, F.O. (1986) The jumping stereotypy in the bank vole (*Clethrionomys glareolus*). *Biology of Behaviour* 11, 130–143.

Ödberg, F.O. (1987) The influence of cage size and environmental enrichment on the development of stereotypies in bank voles (*Clethrionomys glareolus*). *Behavioural Processes* 14, 155–173.

Passingham, R. (1995) *The Frontal Lobes and Voluntary Action.* Oxford University Press, Oxford, UK, 322 pp.

Powell, S.B., Newman, H.A., Pendergast, J. and Lewis, M.H. (1999) A rodent model of spontaneous stereotypy: initial characterization of developmental, en-vironmental, and neurobiological factors. *Physiology and Behavior* 66, 355–363.

Powell, S.B., Newman, H.A., McDonald, T.A., Bugenhagen, P. and Lewis, M.H. (2000) Development of spontaneous stereotyped behavior in deer mice: effects of early and late exposure to a more complex environment. *Developmental Psychobiology* 37, 100–108.

Robbins, T.W. and Sahakian, B.J. (1981) Behavioural and neurochemical determinants of drug-induced stereotypy. In: Rode, F.C. (ed.) *Metabolic Disorders of the Nervous System.* Pitman, London, pp. 244–291.

Rushen, J. (1993) The coping hypothesis of stereotypic behaviour. *Animal Behaviour* 48, 91–96.

Rushen, J., Lawrence, A.B. and Terlouw, E.M.C. (1993) The motivational basis of stereotypies. In: Lawrence, A.B. and Rushen, J. (eds) *Stereotypic Animal Behaviour: Fundamentals and Applications to Welfare.* CAB International, Wallingford, UK, pp. 41–64.

Sachser, N. (1998) Of domestic and wild guinea pigs: studies in sociophysiology, domestication, and social evolution. *Naturwissenschaften* 85, 307–317.

Sachser, N., Lick, C. and Stanzel, K. (1994) The environment, hormones, and aggressive behaviour – a 5-year study in guinea-pigs. *Psychoneuroendocrinology* 19, 697–707.

Schönecker, B. and Heller, K.E. (2000) Indication of a genetic basis of stereotypies in laboratory-bred bank voles (*Clethrionomys glareolus*). *Applied Animal Behaviour Science* 68, 339–347.

Schwaibold, U. and Pillay, N. (2001) Stereotypic behaviour is genetically transmitted in the African striped mouse, *Rhabdomys pumilio. Applied Animal Behaviour Science* 74, 273–280.

Sherwin, C.M. (1998) Voluntary wheel running: a review and novel interpretation. *Animal Behaviour* 56, 11–27.

Sherwin, C.M. (2004) The influences of standard laboratory cages on rodents and the scientific validity of research data. *Animal Welfare* 13, S9–S15.

Toates, F. and Jensen, P. (1991) Ethological and psychological models of motivation – towards a synthesis. In: Meyer, J.A. and Wilson, S. (eds) *From Animals to Animats*. MIT Press, Cambridge, pp. 194–205.

Turner, M. (1997) Towards an executive dysfunction account of repetitive behaviour in autism. In: Russell, J. (ed.) *Autism as an Executive Disorder*. Oxford University Press, New York, pp. 57–100.

Van Praag, H., Christie, B., Sejnowski, T. and Gage, F. (1999) Running enhances neurogenesis, learning and long-term potentiation in mice. *Proceedings of the National Academy of Sciences USA* 96, 13427–13431.

Van Praag, H., Kempermann, G. and Gage, F.H. (2000) Neural consequences of environmental enrichment. *Nature Reviews Neuroscience* 1, 191–198.

Vonlanthen, E.M. (2003) Einflüsse der Laufradnutzung auf ausgewählte ethologische, moorphologische und reproduktionsbiologische Parameter beim Syrischen Goldhamster (*Mesocricetus auratus*). Dissertation (Dr. med. vet.), Universität Bern, Bern, Schweiz.

Waiblinger, E. (2003) Behavioural stereotypies in laboratory gerbils (*Meriones unguiculatus*): causes and solutions. Dissertation (Dr. phil. Nat.), Universität Zürich, Zürich, Schweiz.

Wemelsfelder, F. (1993) The concept of animal boredom and its relationship to stereotyped behaviour. In: Lawrence, A.B. and Rushen, J. (eds) *Stereotypic Animal Behaviour: Fundamentals and Applications to Welfare*. CAB International, Wallingford, UK, pp. 65–95.

Werme, M., Lindholm, S., Thoren, P., Franck, J. and Brene, S. (2002a) Running increases ethanol preference. *Behavioural Brain Research* 133, 301–308.

Werme, M., Messer, C., Olson, L., Gilden, L., Thoren, P., Nestler, E.J. and Brene, S. (2002b) ΔFosB regulates wheel running. *Journal of Neuroscience* 22, 8133–8138.

Wiedenmayer, C. (1996) Effect of cage size on the ontogeny of stereotyped behaviour in *gerbils*. *Applied Animal Behaviour Science* 47, 225–233.

Wiedenmayer, C. (1997a) Causation of the ontogenetic development of stereotypic digging in gerbils. *Animal Behaviour* 53, 461–470.

Wiedenmayer, C. (1997b) The early ontogeny of bar-gnawing in laboratory gerbils. *Animal Welfare* 6, 273–277.

Wiedenmayer, C. and Brunner, C. (1993) Is stereotyped behaviour in gerbils determined by housing conditions? In: Nichelmann, M. Wierenga, H.K. and Braun, S. (eds) *Proceedings of the International Congress on Applied Ethology 1992*. KTBL, Darmstadt, Germany, pp. 276–278.

Wiepkema, P.R., van Hellemond, K.K., Roessingh, P. and Romberg, H. (1987) - Behaviour and abomasal damage in individual veal calves. *Applied Animal Behaviour Science* 18, 257–268.

Würbel, H. (2001) Ideal homes? Housing effects on rodent brain and behaviour. *Trends in Neuroscience* 24, 207–211.

Würbel, H. and Stauffacher, M. (1994) Standard-Haltung für Labormäuse – Probleme und Lösungsansätze. *Tierlaboratorium* 17, 109–118.

Würbel, H. and Stauffacher, M. (1996) Prevention of stereotypy in laboratory mice: effects on stress-physiology and behaviour. *Physiology and Behavior* 59, 1163–1170.

Würbel, H. and Stauffacher, M. (1997) Age and weight at weaning affect corticosterone level and subsequent development of stereotypy in ICR-mice. *Animal Behaviour* 53, 891–900.

Würbel, H. and Stauffacher, M. (1998) Physical condition at weaning affects exploratory behaviour and stereotypy development in laboratory mice. *Behavioural Processes* 43, 61–69.

Würbel, H., Stauffacher, M. and von Holst, D. (1996) Stereotypies in laboratory mice – quantitative and qualitative description of the ontogeny of 'wire-gnawing' and 'jumping' in ICR and ICR nu – mice. *Ethology* 102, 371–385.

Würbel, H., Freire, R. and Nicol, C.J. (1998a) Prevention of stereotypic wire-gnawing in laboratory mice: effects on behaviour and implications for stereotypy as a coping response. *Behavioural Processes* 42, 61–72.

Würbel, H., Chapman, R. and Rutland, C. (1998b) Effect of feed and environmental enrichment on development of stereotypic wire-gnawing in laboratory mice. *Applied Animal Behaviour Science* 60, 69–81.

5

Perseveration and Stereotypy – Systems-level Insights from Clinical Psychology

J.P. GARNER

Animal Sciences Department, Purdue University, 125 South Russell Street, West Lafayette, IN 47907, USA

Editorial Introduction

In this chapter, Garner reviews how the brain organizes and effects the performance of behaviour; examines what happens when the systems involved malfunction; and presents evidence that such malfunctions play a role in stereotypies. Data from human clinical conditions, patients with selective brain damage and experimentally manipulated research animals, all show that behaviour patterns are controlled at multiple levels, with discrete 'executive' systems responsible for, say, keeping on task (e.g. sticking to a plan in the face of distraction, yet shifting to a new one when appropriate) versus sequencing the specific motor patterns needed. Garner reviews evidence that a key system of the former type is centred in the mammalian prefrontal cortex. Its malfunction causes 'dithering', distractibility and impaired abilities to shift from one goal to another or shift attention from one stimulus type to another, potentially resulting in impulsive/compulsive behaviour. In contrast, the main system of the latter type comprises the basal ganglia, notably the dorsal striatum and its outputs, with dysfunction here potentially causing hyperactivity, rapid repetitions of and/or switches between different motor patterns, and stereotypy.

This brain region is thus the main site of action for stereotypy-inducing drugs (e.g. psychostimulants like amphetamine); and is often affected by abnormal rearing conditions (e.g. those reviewed in the following chapters). In these instances, the changes specifically implicated in stereotypy involve underactivity in the neurons of the so-called 'indirect pathway', which then fail to inhibit ongoing behavioural responses resulting in forms of inappropriate repetition termed 'recurrent' or 'continuous perseveration'. How such perseveration is manifest can vary from one moment to the next (perseverative eating could be rapidly followed by perseverative drinking for example), but stereotypies seem likely to be the perseverative performance of one type of behaviour (e.g. digging or escape attempts; *cf.* Chapter 4), that is itself then repeated bout after bout. After building on evidence for this from clinical and experimental psychology, Garner then describes a series of experiments that ask, do captive animals with high levels of stereotypy also show strong tendencies to perseverate? The answer is yes, they do. In every species looked at to date, from jumping bank voles to pacing sun bears, the most stereotypic

individuals also show the most persistent, repetitive responding in tasks designed
to assess their abilities to give up a learnt response in extinction, or their tendencies
to spontaneously generate variable versus predictable responses.

   This work has important implications. Potentially, it gives us a means of
distinguishing 'true' stereotypies from other forms of repetitive abnormal behaviour. It also allows ethologists to discuss 'pathology' or 'dysfunction' much more
precisely than they have done to date, and it opens up a world of non-invasive,
diagnostic tests for investigating captive animals' behaviour. It could even encourage neuroscientists to look anew at their research animals, in suggesting that caged
mice that somersault for hours a night may perhaps not be very normal 'models' on
which to work. However, the chapter also highlights two major caveats, which in
turn represent important directions for future research. The first is that not all
correlates of captive animals' stereotypies are consistent with the specific form of
dysfunction proposed here, and nor can all aspects of stereotypy be explained by it
(e.g. form and timing): we thus need to investigate alternative or additional processes too. The second is that the correlational studies reported here cannot tell us
whether the perseveration of stereotypers is abnormal or even induced by captivity.
The following chapters suggest this to be likely, but the issue is far from resolved.

GM

5.1. Introduction

Ethologists traditionally emphasize the adaptive value of behaviour,
which perhaps explains why much of stereotypy research has – paradoxically (Garner *et al.*, 2003b) – sought functions for behaviours that are
defined as functionless (see 'coping', in Box 1.3, Chapter 1). However,
some theorists in applied ethology have suggested that captive animals'
stereotypies might involve abnormal brain function. For example, hypotheses such as the 'sensitization of neuronal pathways' (Dantzer,
1986; Kennes *et al.*, 1988; Cooper and Nicol, 1994), or an abnormal
dominance of 'central control' (Fentress, 1976) or the 'stimulus–response'
control of behaviour (Toates, 2000) have been proposed (see also Chapter
4). Authors outside of ethology have also been struck by the possible
parallels between captive animals' stereotypies and some clinical
human symptoms. For instance, the psychiatric literature has often linked
human stereotypies (e.g. autism) to those of isolation-reared primates (see
Chapters 6 and 7, and Lewis *et al.*, 1996). Similarly, the veterinary literature has linked several abnormal behaviours in animals to human psychiatric disorders (see Chapter 10, this volume).

   Despite this precedent, the connections between stereotypies in captive
animals and those in human mental disorder have rarely been investigated
empirically outside the primate social isolation literature. In part, this
reflects the (understandable) ethological bias mentioned earlier for functional explanations; and, in part, the emphasis on different levels of explanation in the theoretical versus the experimental work on stereotypy in
applied ethology. Thus, most hypotheses for a pathological basis to stereotypy in captive animals are at a systems-level of explanation (see definition

later), whereas much experimental work in applied ethology concentrates on higher levels of explanation (e.g. motivational) or lower ones, e.g. physiological (see Box 1.1, Chapter 1 and Chapters 2–4, this volume).

This chapter therefore focuses on new insights into captive animal stereotypy provided by stereotypy-like behaviours in human mental disorder, specifically at a systems-level of explanation. In contrast to motivational-level explanations that (by definition) reflect the relationships between stimuli and behaviour independent of brain design (e.g. Hinde, 1970), systems-level explanations of behaviour attempt to reflect the actual division of labour in the brain, whereby different systems have discrete functions. Systems-level explanations have two important features (see also Box 1.2, Chapter 1, this volume). First they are hierarchical, moving from the functions of systems distributed across brain areas, through the sub-system functions mediated by particular brain areas, to the components of processing occurring in individual circuits. They thus provide a framework that links motivational explanations with reductionist (i.e. molecular, neurophysiological or pharmacological) levels of explanation. Second, because systems-level explanations attempt to describe how the brain actually mediates particular functions, they explain the software glitches (i.e. subtle arbitrary features of behaviour, such as the performance of displacement behaviours or intention movements) produced as an artefact of the brain's design – in either healthy individuals (e.g. the failure of depth perception in equiluminescent images: Hubel, 1988), or those suffering from CNS damage (i.e. physical tissue loss or deterioration) or dysfunction (i.e. abnormal neurotransmission).

In this chapter, I initially take two steps back from the conventional literature on stereotypies in captive animals. The first, as outlined in Box. 5.1, is to understand the place of stereotypy in the constellation of Abnormal Repetitive Behaviours in human mental disorder. Human stereotypies are most conspicuous in autism, stereotyped movement disorder and a range of syndromes or disorders that involve mental retardation. Stereotypies (albeit called by a different name, e.g. 'tics') are also common in schizophrenia and Tourette's syndrome, and occur as rare secondary symptoms in other disorders (e.g. in some obsessive–compulsive disorder [OCD] patients) (see Frith and Done, 1990; Hymas *et al.*, 1991; Lewis *et al.*, 1996; Turner, 1997, 1999; Lewis and Bodfish, 1998). Of these disorders, mental retardation (see Chapter 7, this volume) and autism serve as the best starting point for comparing human and animal stereotypies. This is primarily because many of these patients are unable to speak, and therefore stereotypies in these disorders are defined entirely behaviourally (Frith and Done, 1990) as repetitive unvarying, purposeless patterns of movement. This is in contrast to other disorders, such as Tourette's syndrome, where a patient's reported experience is part of the definition of the behaviour.

My second step back will be to review the theory and neurobiological implementation of behavioural control. This will provide a framework for understanding Abnormal Repetitive Behaviours in brain lesion patients, following administration of particular drugs, and in mental disorder. For

Box 5.1. Abnormal Behaviour and Abnormal Repetitive Behaviour in Human Mental Disorder

J.P. Garner

Abnormal behaviours are central to human mental disorder (Davison and Neale, 1998). Abnormal Repetitive Behaviours are a subset of abnormal behaviours that are: (i) heavily repeated; (ii) invariant in either motor output, environmental interaction or goal or theme; and (iii) either apparently functionless, maladaptive or self-injurious or additionally inappropriate or 'odd' (Turner, 1997). A key problem in linking animal and human Abnormal Repetitive Behaviour is that human Abnormal Repetitive Behaviours are categorized and labelled very inconsistently in different disorders. Different behaviours may be distinguished in one disorder according to theories unique to that disorder, or on the basis of patient reports of their experiences. Thus few disorders have a purely behavioural categorization of their symptoms, which greatly complicates comparisons between disorders (Frith and Done, 1990; Lewis and Bodfish, 1998; Bodfish and Lewis, 2002) and extrapolation from humans to animals (Ödberg, 1993; Garner, 1999). For instance, in Tourette's syndrome the full range of Abnormal Repetitive Behaviour from single repeated muscle movements (often called 'tics' in other disorders: Frith and Done, 1990), to complex identical sequences of muscle movements (i.e. stereotypies), to flexible goal-directed Abnormal Repetitive Behaviours (i.e. impulsive/compulsive behaviours) are all called 'tics' (American Psychiatric Association, 1994).

There is no definitive list of, or distinctions between, different human Abnormal Repetitive Behaviours. Nevertheless, symptoms commonly grouped in this category include obsessions, compulsive behaviours, impulsive behaviours, echo-phenomena, sameness behaviours, self-injurious behaviours, stereotypies, tics, dyskinesias and akathisia (for descriptions and reviews of these symptoms see: Frith and Done, 1990; Hollander and Wong, 1995; Turner, 1997, 1999; Lewis and Bodfish, 1998; also see Chapter 6). Some movement disorders, such as tremor, chorea, athetosis and dystonia, are also included in discussions of Abnormal Repetitive Behaviour; however, these symptoms may involve very different mechanisms from the behaviours more commonly included in Abnormal Repetitive Behaviour (for a review of these mechanisms see: Albin *et al.*, 1995; Mink, 2003).

instance, as we will see, the categories of Abnormal Repetitive Behaviour distinguished behaviourally in autism are also distinguishable biologically because they involve different components of behavioural control (Turner, 1997). These biological distinctions provide a means to categorize Abnormal Repetitive Behaviours (Garner, 1999), via neuropsychological tests that assess the performance of discrete brain systems via specific tasks. From this broader perspective, testable hypotheses linking stereotypies in humans and other animals can be formulated. Having thus set the scene, I will then examine the evidence that the dysfunctions underlying stereotypies in autism and other human disorders are also involved in captive animals' stereotypies.

5.2. Executive Systems and Behavioural Control – Insights from Brain Lesion Patients

Data from patients where lesions (physical damage to the brain) have eliminated particular systems or sub-systems provide essential insights into the systems-level design of the brain. Certain brain lesions affect a patient's ability to control behaviour in very specific ways. The particular symptoms seen in such patients led to the proposal that the brain contains a set of 'executive systems' that filter, integrate and ultimately translate external and internal stimuli, plus contextual information that might not be immediately present, into expressed behaviour (see Box. 1.1, Chapter 1; plus Shallice, 1982; Norman and Shallice, 1986; D'Esposito *et al.*, 1995; Robbins, 1996). For instance, consider a trained rat placed in a maze. It must first integrate the external stimuli encoding the start box, internal stimuli reflecting its motivation for reward and its memory of the task before it can even begin. In doing so it forms a 'cognitive attentional set' representing the range of behaviours needed to complete the maze. This cognitive attentional set must be maintained once the start box is out of sight, and while the goal box is still 'around the corner', and is used to suppress unnecessary responses towards irrelevant stimuli in the maze, or other less suitable behaviours such as grooming. At each junction in the maze, the rat must pay attention to relevant stimuli, integrate them with its memory of the maze, its memory of what has just happened and select a suitable response. Finally when it reaches the end of the maze, it must switch off locomotor behaviour and switch to food consumption. Each of these problems represents a separate function of the executive systems. However, with a few exceptions (e.g. Fentress, 1976; Toates, 2000), executive processes have been overlooked in ethology. For example, in the case of the rat in the maze, motivational models simply assume that stimuli and cognitions flow smoothly into behaviour, without considering how this is actually effected.

This chapter focuses on the executive systems for three reasons. First, damage to the brain areas involved in executive processing produces characteristic behavioural signs, called perseveration (Box. 5.2), which share some similarities with stereotypy. Second, these brain areas are consistently implicated in disorders that involve stereotypies, compulsions or other Abnormal Repetitive Behaviours (Crider, 1997; Turner, 1997; Garner, 1999; Casey *et al.*, 2002). Third, a wide variety of disease processes, both between and within these disorders, appear to converge to similar systems-level dysfunctions – i.e. the unifying feature of impulsive/compulsive behaviours or (and in contrast to) stereotypies might be best understood at the executive system level (Turner, 1997, see also Würbel's plea for biologically meaningful ways of classifying stereotypies, Chapter 4; Garner, 1999; Garner and Mason, 2002). Thus a systems-level perspective allows the integration and reconciliation of different levels of explanation – especially the bewilderingly complex neuropsychopharmacology of Abnormal Repetitive Behaviours (see Chapters 6–8 and 10; and Cabib, 1993).

5.2.1. A systems model for how the brain organizes behaviour

The Supervisory Attentional System (SAS)/Contention Scheduling System (CSS) model of 'willed and automatic control of behaviour' (Shallice, 1982; Norman and Shallice, 1986) divides executive function between

Box 5.2. Forms of Perseveration

J.P. GARNER

In contrast to Abnormal Repetitive Behaviours, which occur during the patient's day-to-day life, perseveration is the inappropriate repetition of behaviour elicited in an experimental or diagnostic context (Turner, 1997), and is a diagnostic sign of dysfunction in the brain areas involved in executive processing (Luria, 1965; Sandson and Albert, 1984; Turner, 1997). Three fundamentally different forms of perseveration are recognized, each one of which involves repetition at a different level of organization.

- Continuous perseveration is the inappropriate repetition of individual movements, or motor sub-programs. For example, when asked to write down their name, a patient might write the first letter of their name over and over again.
- Recurrent perseveration is the inappropriate repetition of responses or complex motor programs. For example, when asked a series of questions, a patient might respond to every question identically with the answer they gave to the first.
- Stuck-in-set perseveration is the inappropriate repetition of goals, or abstract rules where individual responses remain flexible. For example, when shown a deck of playing cards and asked to name the suit of each card a patient will be able to do so readily. However, when asked to instead name the number of the card (2, 3, 4, . . . , Queen, King, Ace), the patient will continue to name the suit.

Stuck-in-set perseveration therefore represents the repetition of a particular form of abstract internal information that encodes the current behavioural goal and the particular stimuli or aspects of stimuli that pertain to it (called a cognitive attentional set – hence 'stuck-in-set'); while recurrent and continuous perseveration represent the repetition of components of motor programming. Stuck-in-set perseveration is seen following damage to the prefrontal cortex, while recurrent and continuous perseveration are seen following damage to the basal ganglia (Luria, 1965; Sandson and Albert, 1984; Turner, 1997).

Perseveration is associated with a phenomenon called 'knowledge–action dissociation', whereby if asked, patients know that their responses are inappropriate, and they know the correct response to make, but are unable to execute the response in behaviour (e.g. Owen *et al.*, 1991). The knowledge–action dissociation often causes patients distress and frustration at their inability to control their own behaviour (for examples see: Luria, 1965; Turner, 1997).

two basic systems, called the CSS and the SAS. It provides a powerful systems-level description of the executive systems that has been widely applied in a range of literatures. The model is derived from clinical data from patients suffering from brain lesions, and the neuropsychological and amphetamine research literature.

The CSS selects and sequences behavioural responses. Each response is linked to an optimal set of stimuli, so that the CSS activates the response whose optimal stimuli best match the stimuli presented to it (a 'contention' being the matching of a set of stimuli to a response). Responses 'compete' through localized mutual inhibition of each other's activation, and the more similar responses are, the more they inhibit each other. This simple mechanism has many important properties. First, an active response 'locks out' similar behaviours, although if pertinent stimuli (e.g. predator cues) activate a radically different behaviour (e.g. escape) the animal can switch to the new behaviour. Second, responses are considered to be self-activating, that is once switched on, a response activates itself above and beyond its match to the current stimuli. This self-activation persists. Thus, responses must inhibit themselves once completed. Third and finally, the CSS also sequences individual movements within a response using the same basic mechanisms.

The SAS, in contrast, primes and edits the selection of responses by the CSS on the basis of abstract non-physiological internal information. As an example, in a shopping centre, I must remember which shop I intend to visit, and what products I am looking for. Thus the SAS selects behaviour whenever external stimuli cannot unequivocally direct the CSS. Such situations include: (i) planning and adhering to a plan even in the face of new circumstances, distractions or temptations; (ii) error-correction, that is adjusting stimulus-directed responses when they fail to achieve their goal; (iii) selecting between two responses that equally match the set of stimuli presented to the CSS (such as choosing between two items on a menu); (iv) selecting a response when the meaning of an external stimulus is dictated by an internal uncued goal or rule (for instance in poker, a player might bet completely differently on the same hand depending on what he knows about other players and whether he intends to bluff); and (v) over-coming habitual responses (Shallice, 1982; Norman and Shallice, 1986; Shallice and Burgess, 1991, 1996). Importantly, the SAS can only influence behaviour through modulating the action of the CSS.

In terms of neuroanatomy, the CSS is located within an area of the brain called the basal ganglia (Norman and Shallice, 1986). Thus disruption to basal ganglia motor systems results in a wide range of disorders in the initiation, inhibition and control of movement (Albin *et al.*, 1995), including psychostimulant-induced stereotypies (Joyce and Iversen, 1984). Lesions of the basal ganglia motor systems also lead to two neurological signs, which can be quantified in test situations, called recurrent and continuous perseveration (see Box 5.2) (Luria, 1965; Sandson and Albert, 1984). Recurrent and continuous perseveration can be thought of as software glitches explained by the SAS/CSS systems-level model. Thus recurrent perseveration results from a failure of the CSS to inhibit a response, and continuous perseveration results in a failure to inhibit a single movement within a response, once each has been completed. This type of failure, in which responses do not inhibit themselves appropriately once completed, seems integral to stereotypies (Turner, 1997).

The SAS is based in the prefrontal cortex, and influences the CSS through connections to an area of the basal ganglia called the caudate. Human patients with prefrontal damage therefore show impaired planning (e.g. dithering and distractibility), and problems with novelty, whilst their routine stimulus-directed behaviours remain unimpaired (Shallice, 1982; Norman and Shallice, 1986; Shallice and Burgess, 1991, 1996; Turner, 1997). When tested, these patients also show 'stuck-in-set perseveration' (see Box 5.2), whereby they find it difficult to shift plans or rules in an appropriate way. A key strength of the CSS/SAS model is that it explains this paradoxical presence of both extreme distractibility *and* very repetitive behaviour in patients with lesions of the prefrontal cortex. Both these symptoms can be thought of as software glitches explained by the SAS/CSS systems-level model. Thus stuck-in-set perseveration is the failure of the SAS to shift cognitive attentional set; distractibility is its failure to maintain the activation of a response in the face of competing stimuli, while dithering is its failure to select between two equally cued responses (Frith and Grasby, 1995).

The SAS/CSS model therefore explains why different forms of perseveration are linked to different brain areas. It also helps to select neuropsychological tasks that quantify these important signs and which therefore can be used as direct measures of the function of the CSS and SAS. Box 5.3 outlines the tasks used to quantify perseveration in humans. In humans, language is often used to constrain the subject's choices in such tasks. In animals, mazes can be used to physically constrain choices, and additional statistical methods or internal controls may also be used to clean the data of biases that are not to do with perseveration (e.g. Garner and Mason, 2002). Tasks can be implemented in different ways depending upon the sensory ecology of the species – for example, the tasks can be implemented directly with touchscreens (e.g. in primates: Dias *et al.*, 1996; songbirds: Garner *et al.*, 2003a), in mazes (e.g. rats: Birrell and Brown, 2000; mice: Garner *et al.*, 2003c) or by other means within the home cage (e.g. in parrots: Garner *et al.*, 2003b). Stimuli, rewards and operant responses can also be modified to take the species' ecology into account. For instance, rats and mice readily learn to dig for food rewards in pots of sawdust. The IntraDimensional–ExtraDimensional set-shifting task (see Box 5.3) is then performed by presenting two pots in a simple maze, and altering the outer surface, odour and the digging material in the pots (Birrell and Brown, 2000). For example, the initial rule might be to choose between the two pots on the basis of the digging material, ignoring odour. Then the extradimensional shift (the stage testing for stuck-in-set preservation; see Box 5.3) would be to choose between the pots on the basis of outer texture, ignoring digging material.

5.3. The Neurobiology of Behavioural Control – Insights from Basal Ganglia Physiology and Drug-induced Stereotypies

As we shall see, the implementation of the CSS in the basal ganglia is central to linking human and animal stereotypies. The role of basal ganglia physiology in stereotypy is discussed in detail in Chapter 7. To briefly summarize, the basal ganglia select and sequence behaviour through competition between two separate pathways, called the 'direct pathway' and the 'indirect pathway'. The direct pathway activates responses, while the indirect pathway inhibits responses. The indirect pathway both inhibits current responses when another response needs to be activated and inhibits competing responses when the current response needs to be completed. Drug-induced stereotypies consistently and selectively involve suppression of the

Box 5.3. Measuring Perseveration

J.P. GARNER

Recurrent perseveration can be measured using the 'two-choice gambling task' (Frith and Done, 1983). On each trial of this task the subject chooses one of two boxes on a computer screen (four such trials are shown below), and is instructed to 'find the rule' governing which box to choose.

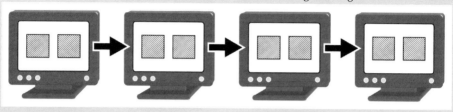

   No such rule exists: the subject is merely told that their choice was correct at random on 50% of trials. Control subjects then produce random sequences of choices (e.g. L L R L R R L R R L), while subjects showing recurrent perseveration produce repetitive sequences of choices (e.g. L L L L R R R R R L). The degree of perseveration is quantified using information analysis (e.g. Frith and Done, 1983).
   Stuck-in-set perseveration is measured using a variety of set-shifting tasks where choices between stimuli must be guided by cognitive attentional set, such as the Wisconsin card sort test, the trail-making Test part B and the IntraDimensional–ExtraDimensional set-shifting task. This latter task involves a series of problems for the subject to solve (Owen *et al.*, 1993). On each trial two stimuli are presented on a computer screen. The stimuli in a trial differ in two independent ways, called 'dimensions'. In the four trials illustrated below, both shape and pattern of the stimuli vary.

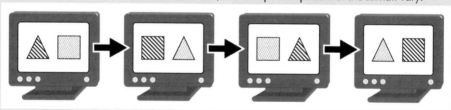

   The subject chooses a stimulus, and the computer informs them whether they chose correctly. In this example the triangle might be the correct choice. Once the subject has learnt this rule, the discrimination is 'reversed' – now the square is the correct choice on every trial. Once the subject has learnt this reversal, an 'intradimensional shift' is performed. Thus, a new set of stimuli are presented, but again the same dimension (i.e. shape) dictates the correct choice – now the cross might be the correct choice.

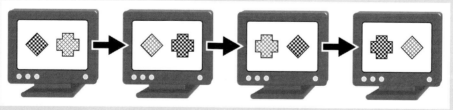

   Once this discrimination is learnt, it is reversed. By the completion of this reversal, subjects have formed a cognitive attentional set (i.e. a generalized abstract rule) – in this example to attend to shape, to solve each new problem – that is evident in the increased speed with which they solve each new problem. The key stage of the task follows, where an 'extradimensional shift' is performed. Thus a new set of stimuli are presented that vary in the dimension

Continued

Box 5.3. *Continued*

that was previously dictating choice (i.e. shape) and in a new dimension (size in the example below) – but now the new dimension dictates the correct choice, so now the smaller shape might be the correct choice.

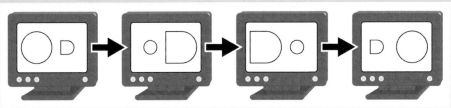

   To solve this new problem the subject must first switch off the previous set ('attend to shape'), and even healthy subjects find this difficult, taking longer to learn an extradimensional shift than an intradimensional shift. Patients showing stuck-in-set perseveration find this new problem extremely difficult to solve, hence stuck-in-set perseveration is quantified as the number of trials taken to complete this stage.

indirect pathway, and hence behavioural disinhibition. This is true for classic stereotypy-inducing dopaminergic drugs such as amphetamine acting on dopamine D2 receptors (e.g. Lyon and Robbins, 1975; Waddington *et al.*, 1990; Cabib, 1993), and also for opiate drugs such as morphine acting on enkephalin receptors (i.e. mu and delta opiate receptors) (e.g. Longoni *et al.*, 1991; Cabib, 1993; Steiner and Gerfen, 1998), and a variety of serotonergic drugs (Curzon, 1990). Importantly, as direct pathway activity is required to activate all behaviour (including stereotypy), direct pathway activity is a necessary but not sufficient prerequisite of stereotypy. Thus, drugs that selectively activate the direct pathway only induce hyperactivity, not stereotypy; and drugs that selectively inhibit the direct pathway suppress all behaviour, including stereotypy (Waddington *et al.*, 1990; Cabib, 1993). (See Chapters 7 and 8 for further details.)

   This link between the basal ganglia and stereotypy goes beyond drug effects. Isolation-reared primates, for instance (see Chapter 6, and Lewis *et al.*, 1990; Martin *et al.*, 1991) show both stereotypy and many neurophysiological changes in the basal ganglia, including altered dopamine and opiate metabolism. Stereotypies in some captive animals also involve altered dopamine function (Ödberg *et al.*, 1987), and may share common mechanisms with amphetamine sensitization (see Chapter 8, and Cabib and Bonaventura, 1997). In parallel, socially isolated rodents also show amphetamine sensitization (Robbins *et al.*, 1996), and a wide array of changes in brain physiology (including the basal ganglia) (Robbins *et al.*, 1996). Indeed the effects of chronic stress on dopamine function may play an important role in stereotypy development (see Chapter 8). Enrichment induces many changes in neurotransmitter function (e.g. Rosenzweig *et al.*, 1978), and also affects many aspects of brain growth and development (see Chapter 7). The effects of enrichment on some neurophysiological measures are magnified between high and low stereotypy individuals, and these stereotypy-specific differences are concentrated in the basal ganglia (Turner and Lewis, 2003).

5.4. The Neuropsychology of Stereotypy – Insights from Human Mental Disorder

5.4.1. Stereotypies and impulsive/compulsive behaviours – two fundamental categories of Abnormal Repetitive Behaviour

Despite the confusing differences in terminology used to describe human disorders (Box. 5.1), Abnormal Repetitive Behaviours are often differentiated into two broad categories. For simplicity, we will term these two categories 'impulsive/compulsive behaviours' and 'stereotypies'. Impulsive/compulsive behaviours are repetitive, but unlike stereotypies they usually vary in the form of the motor pattern and are goal-directed. Indeed impulsive/compulsive behaviours can be seen as goal-directed behaviours that are directed towards an inappropriately repeated goal. That is, the goal persists in directing behaviour after it has been met or despite being inappropriate to the current circumstances. Examples include compulsions in OCD, such as hand-washing; 'impulsive' behaviours such as 'focused' hair-plucking in trichotillomania; most 'complex tics' in Tourette's syndrome; the inappropriate repetition of topics in speech or thought (called 'circumscribed interests' in autism); and echophenomena (where the patient mimics the speech or gestures of another person). (For a discussion of these symptoms see: American Psychiatric Association, 1994; Hollander and Wong, 1995; Turner, 1997; Garner, 1999.)

In contrast, in human clinical work, stereotypies are described in a similar way to the ethological definition, that is, as repetitive, unvarying, goalless behaviours. Examples include hand-flapping, body-rocking and many forms of self-injurious behaviour in autism, mental retardation and other pervasive development disorders (see Chapters 6 and 7, this volume); many 'simple tics' in Tourette's syndrome; non-drug-induced dyskinesias in schizophrenia; and in autism and schizophrenia the inappropriate repetition of single words in speech (see: Owens *et al.*, 1982; Frith and Done, 1990; American Psychiatric Association, 1994; Turner, 1997; Lewis and Bodfish, 1998; Garner, 1999).

These categories are subdivided (Owens *et al.*, 1982; Hollander and Wong, 1995; Lewis and Bodfish, 1998), though these subdivisions are often hard to distinguish behaviourally or physiologically. For instance, the distinction between impulsive and compulsive symptoms is important in psychiatry (Hollander and Wong, 1995), but is at a finer level, and is less well understood, than that between stereotypies and impulsive/compulsive behaviours (Garner, 1999); as well as being harder to make for animals (see Chapters 10 and 11, this volume).

Thus, overall, the key difference between impulsive/compulsive behaviours and stereotypies is in what is repeated: with impulsive/compulsive behaviours an inappropriate goal is repeated (e.g. the seeking out and plucking of particular types of hair, or 'focused plucking' in trichotillomania), while with stereotypies, a particular motor pattern is repeated (e.g. the inappropriate repeated insertion of identical words or phrases in speech by

some schizophrenic patients: Frith and Done, 1990). As a result, stereotypies are by definition inherently unvarying, while impulsive/compul-
sive behaviours are not (though some may secondarily lose variability with
repetition; see e.g. Chapters 2–4 and 11, on channelling and habit-formation). The inappropriately repeated goal behind an impulsive/compulsive
behaviour is sometimes apparent as an obsession – an inappropriate,
intrusive, repeated thought or mental image that causes mounting anxiety.

5.4.2. Back to the CSS/SAS model – the neuropsychology of human stereotypies and impulsive/compulsive behaviours

The distinction between impulsive/compulsive behaviours and stereotypies is best developed in the 'behavioural inhibition hypothesis' of
Abnormal Repetitive Behaviour in autism (Turner, 1997). Turner proposes
that Abnormal Repetitive Behaviour involves a failure of inhibitive executive processes in either the SAS or the CSS, and that consequentially
Abnormal Repetitive Behaviour is inherently related to perseveration.
Thus, Abnormal Repetitive Behaviour represents the day-to-day expression
of a deficit in behavioural control, and perseveration represents the same
deficit elicited in a quantifiable form by an experimental task. Ethologists
have similarly suggested that stereotypies might involve a failure in behavioural control (e.g. Fentress, 1976; Toates, 2000), and psychiatrists have
suggested that Abnormal Repetitive Behaviour might represent a day-to-
day form of perseveration (e.g. Frith and Done, 1990; Crider, 1997). Turner's
'behavioural inhibition hypothesis' advances these ideas in two ways. First,
it emphasizes the failure of the SAS or CSS to inhibit behaviours, which
dovetails nicely with the ubiquitous failure of behavioural inhibition in
drug-induced stereotypies. Second, the 'behavioural inhibition hypothesis' predicts that, just as there are two fundamentally different forms of
perseveration, Abnormal Repetitive Behaviour should be divided into two
similar categories corresponding to disinhibition of the SAS or of the CSS.

Turner thus predicted that stereotypies in autistic children would
correlate with recurrent perseveration, since both were hypothesized to
involve a failure of the CSS to inhibit responses, while their 'higher level'
Abnormal Repetitive Behaviours, such as repetitively sticking to the same
topic in conversation, would correlate with stuck-in-set perseveration,
since both were hypothesized to involve a failure of the SAS to inhibit
cognitive attentional sets. As described earlier (and in more detail in Box
5.3), perseveration can be quantified using a variety of neuropsychological
tasks. Thus to test these hypotheses, Turner scored each child for the
severity of their stereotypies, for the severity of their 'higher level' Abnormal Repetitive Behaviours, and then independently scored each child for
the severity of recurrent and stuck-in-set perseveration that they showed.
All the psychiatric experiments discussed in this chapter follow this basic
design where symptom severity is scored separately from measuring perseveration – thus both measures are being implicitly treated as traits.

Autistic children show recurrent perseveration in a two-choice gambling task compared to controls (Frith, 1970; Turner, 1997). As predicted the more stereotypy performed by an autistic child in day-to-day life, the more severe their recurrent perseveration in the two-choice gambling task (i.e. the two measures were correlated) (Turner, 1997). In addition to stereotypy, repeated identical behaviours within circumscribed interests (e.g. repeating the same phrase from a book over and over again) lead to a correlation between circumscribed interests in day-to-day life and recurrent perseveration (Turner, 1997). Autistic children also show more severe stuck-in-set perseveration compared to controls on the Wisconsin Card Sort Test (Ozonoff *et al.*, 1991) and the IntraDimensional–ExtraDimensional set-shifting task (Hughes *et al.*, 1994; Turner, 1997). Again, as predicted, the more severe the stuck-in-set perseveration shown by an autistic child on the IntraDimensional–ExtraDimensional set-shifting task, the more severe their 'higher level' Abnormal Repetitive Behaviours in day-to-day life – including repeated topics in conversation, echolalia (the automatic mimicking of speech), and the repetition of themes within circumscribed interests (e.g. drawing differing pictures of differing animals repeatedly) (Turner, 1997).

Generally, scores for recurrent and stuck-in-set perseveration were uncorrelated; scores for stuck-in-set perseveration were uncorrelated with scores for stereotypy; and scores for recurrent perseveration were uncorrelated with impulsive/compulsive behaviour. This near-perfect 'double dissociation' (i.e. with the exception of circumscribed interests, the correlation of stereotypy *only* with recurrent perseveration; *and* the correlation of 'higher level' Abnormal Repetitive Behaviours *only* with stuck-in-set perseveration) confirms that the two classes of Abnormal Repetitive Behaviour are distinct entities with distinct underlying mechanisms as measured by the two distinct forms of perseveration (Turner, 1997). Furthermore, perseveration in these experiments was also associated with reports of knowledge–action dissociation, and frustration. Thus, these data also illustrate another important point. Perseveration and knowledge–action dissociations are traditionally discussed with reference to patients suffering from brain lesions. However, there is little evidence of a single consistent structural or metabolic brain abnormality in autism (Robbins, 1997b). Thus these data emphasize the point that perseveration does not automatically indicate brain damage, but can also indicate dysfunction, and that such dysfunction may involve subtle changes in brain activity.

Because Turner's 'behavioural inhibition hypothesis' makes testable predictions, it can be extended to other human disorders (Garner, 1999). Here I review four disorders (schizophrenia, OCD, trichotillomania and Tourette's syndrome) of particular relevance to animal Abnormal Repetitive Behaviour. First, in schizophrenia, stereotypies are correlated with recurrent perseveration on the two-choice gambling task (Frith and Done, 1983) and the recurrent perseveration of words in speech (Manschreck *et al.*, 1981; Frith and Done, 1990). In contrast, symptom severity scores that include impulsive/compulsive behaviours correlate with stuck-in-set perseveration (Liddle and Morris, 1991). Second, in OCD, the severity of impulsive/compulsive behaviour is correlated in several studies with abnormalities in the prefrontal cortex and caudate (amongst other brain areas) (Baxter *et al.*, 1992; Saxena *et al.*, 1998); and damage to the prefrontal cortex can also sometimes result in the development of OCD (Berthier *et al.*, 2001). OCD patients also show stuck-in-set perseveration in many studies (Lacerda *et al.*, 2003). Although few studies have related symptom severity directly to perseveration, one study did find symptom severity in OCD to be correlated with stuck-in-set perseveration (Lucey *et al.*, 1997). Third, in trichotillomania (which is characterized by impulsive/compulsive hair pulling, but also can include stereotypic hair pulling), there is some evidence of abnormal metabolism in prefrontal cortex to caudate circuits (Stein *et al.*, 2002). In trichotillomania, symptom severity (scored on a scales measuring impulsive/compulsive Abnormal Repetitive Behaviour) is again correlated with impaired performance on tasks sensitive to stuck-in-set perseveration (Rettew *et al.*, 1991; Keuthen *et al.*, 1996). Fourth, in Tourette's syndrome, the symptoms are divided into 'simple tics' and 'complex tics', the distinctions between which correspond roughly to those between stereotypies and impulsive/compulsive behaviours, respectively. The impulsive/compulsive symptoms (i.e. complex tics) in Tourette's syndrome are highly intercorrelated, but only poorly correlated with stereotypies (i.e. 'simple tics') (Robertson *et al.*, 1988). Accordingly, stuck-in-set perseveration is correlated with impulsive/compulsive behaviours, but not with stereotypies (Bornstein, 1990, 1991). Finally, in healthy individuals, sub-clinical symptom scores for OCD correlate with stuck-in-set perseveration (Zohar *et al.*, 1995).

Thus across a wide range of disorders the same basic pattern can be found. In summary, Abnormal Repetitive Behaviours in humans do seem to be divided into two broad categories – stereotypies and impulsive/compulsive behaviours – in several different disorders. These two categories are distinguishable behaviourally, neuropsychologically and biologically, and correspond to a fundamental division of labour between brain systems that sequence and inhibit individual movements (i.e. the CSS), and brain systems that sequence and inhibit abstract goals, abstract internal information and cognitive attentional set (i.e. the SAS). However, as few studies explicitly correlate perseveration, Abnormal Repetitive Behaviour, and structural or metabolic brain abnormalities, the true generality of this observation remains to be properly tested. Next, I develop the potential importance of this distinction for animals.

5.5. The Neuropsychology of Animal Stereotypy and Impulsive/ Compulsive Behaviours

Mammals and birds show the same fundamental division of labour between a system that automatically sequences behaviour in response to external stimuli, and one that sequences goals, and abstract internal information (Fentress, 1976; Toates, 2000). Throughout chordates the basal ganglia show the same basic circuitry and are involved in behavioural control (Reiner *et al.*, 1998). Thus in non-human primates (Dias *et al.*, 1996) and rodents (Birrell and Brown, 2000) damage to the prefrontal cortex results in stuck-in-set perseveration measured using the IntraDimensional–ExtraDimensional set-shifting task. Could animal Abnormal Repetitive Behaviour involve similar fundamental mechanisms to human Abnormal Repetitive Behaviour, and similarly divide into stereotypies and impulsive/compulsive behaviours? If so, animal stereotypies should correlate with recurrent perseveration, and animal impulsive/compulsive behaviours should correlate with stuck-in-set perseveration (Garner, 1999; Garner and Mason, 2002; Garner *et al.*, 2003b). As already reviewed, the involvement of the basal ganglia in drug-induced and isolation-induced stereotypies provides preliminary support for this hypothesis. Indeed, some of the physiological changes seen in isolated rats, such as increased dopamine release in the basal ganglia (e.g. Robbins *et al.*, 1996), do indeed increase perseveration (e.g. Jones *et al.*, 1991).

To test this hypothesis in captive animals, we investigated stereotypy and extinction learning in bank voles (Garner and Mason, 2002). Extinction learning (see Fig. 5.1) measures a subject's ability to suppress a previously learnt response. Increasing recurrent perseveration should lead to an inappropriate and increasing persistence of the previously learnt response, measured as the number of trials to complete extinction learning (Garner and Mason, 2002). We initially worked with extinction learning because it is impaired in isolation-reared animals (Jones *et al.*, 1991) and is relatively easy to measure. As predicted, stereotypy was highly correlated with impaired extinction learning in bank voles. Thus the vole showing the least amount of stereotypic behaviour completed extinction learning in 26 trials, while the vole showing the highest level took 244 trials (Garner and Mason, 2002) (see Table 5.1). The correlation of stereotypy with impaired extinction learning has been replicated in blue tits and marsh tits (Garner *et al.*, 2003a), and in Asiatic black bears and Malayan sun bears (Vickery and Mason, 2003, 2005) (see Fig. 5.2).

However, extinction learning is an equivocal measure of recurrent perseveration as it may be affected by several other processes (e.g. habit-formation or perhaps even stuck-in-set perseveration). We therefore looked for further evidence of CSS system-level disinhibition in the voles (Garner and Mason, 2002). Learning is apparent not only in the choices that a subject makes, but also in the timing of choices, such that correct choices are made progressively more rapidly than incorrect choices as a subject learns (Olton, 1972). Furthermore, the disinhibition of responses caused by amphetamine's action in the basal ganglia leads to inappropriately rapid responding in a variety of tasks (Robbins, 1997a). We examined the timing of responses in extinction, and found that increasing stereotypy was correlated with an increasing persistence of inappropriate rapid responses (Garner and Mason, 2002) (Table 5.1). We

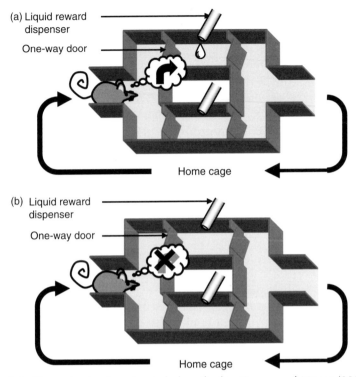

Fig. 5.1. The extinction learning task described in Garner and Mason (2002). A two-choice maze is attached to the home cage. On entering the maze, the vole can enter one of the two corridors through a one-way door. Each corridor contains a computer-controlled reward dispenser. The vole exits either corridor through another one-way door and returns to the home cage. (a) First, the vole is taught that only one corridor will contain reward (the upper corridor in the figure). Voles rapidly learn to always enter this corridor and collect reward. (b) Both reward dispensers are turned off, and extinction learning is measured by the number of trials the vole takes to return to choosing both corridors equally often. The animal's initial learning of the spatial discrimination (a) assesses many components of general learning ability that are common to extinction learning (b). Thus each animal acted as its own control, giving a refined measure that approximated recurrent perseveration more closely than simple extinction performance.

then used these two independent measures of the vole's knowledge of the lack of reward in extinction to test for a knowledge–action dissociation. Increasing stereotypy was correlated with an increasing discrepancy between knowledge measured by choices and knowledge measured by the timing of choices, suggesting a knowledge–action dissociation (Garner and Mason, 2002) (Table 5.1). We also examined behaviour in the home cage for evidence of disinhibition of the CSS. The disinhibition of behaviour caused by amphetamine leads to a general increase in the rate of switching between behaviours as stereotypy develops. We therefore examined the rate at which the voles switched between behaviours in their home cage and found a positive correlation with stereotypy (Garner and Mason, 2002) (Table 5.1). Finally, we investigated whether

Table 5.1. The intercorrelation of stereotypy and measures indicative of Contention Scheduling System disinhibition of response selection in bank voles.

	Activity	Rate of initiation of behaviours	Repetition of choices during extinction	Repetition of latencies during extinction	Knowledge–action dissociation
Stereotypy	$r = 0.857$ ($P = 0.0195$)	$r = 0.840$ ($P = 0.0035$)	$r = 0.642$ ($P = 0.0005$)	$r = 0.751$ ($P = 0.0095$)	$r = 0.790$ ($P = 0.0015$)
Activity		$r = 0.871$ ($P = 0.020$)	$r = 0.791$ ($P = 0.030$)	$r = 0.878$ ($P = 0.007$)	$r = 0.883$ ($P = 0.020$)
Rate of initiation			$r = 0.556$ ($P = 0.015$)	$r = 0.840$ ($P = 0.008$)	$r = 0.926$ ($P = 0.0005$)
Repetition of choices				$r = 0.772$ ($P = 0.0145$)	$r = 0.830$ ($P = 0.0105$)
Repetition of latencies					$r = 0.820$ ($P = 0.017$)

Redrawn from Garner and Mason (2002) with permission from Elsevier. Partial correlation coefficients are given thereby controlling for the various internal controls in each measure. The *P*-value of each is also given. This pattern of data indicates a single underlying causal factor consistent with Contention Scheduling System systems-level disinhibition of response selection. $N = 8$ ($N = 6$ for activity).

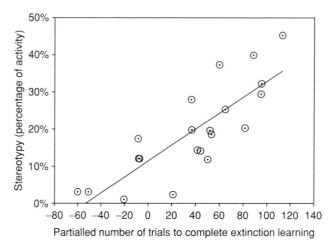

Fig. 5.2. Extinction learning is correlated with stereotypy in Asiatic black bears and Malayan sun bears. Redrawn from Vickery and Mason 2005, see also Chapter 3, this volume. The x-axis is partialled (statistically controlled) for various confounding factors, including species, sex, age and performance on the initial spatial discrimination task (hence the negative values). Recurrent perseveration will increase the number of trials taken to complete the extinction task.

this general disinhibition of all behaviour might explain why high stereotypy animals can also be hyperactive (Ödberg, 1986). Stereotypy was correlated with general activity levels in the home cage (Garner and Mason, 2002) (Table 5.1). More important than any one of these results however, was the fact that each of these measures was correlated with every other measure, indicating that a single process was leading to all of these behavioural differences (Garner and Mason, 2002). The most parsimonious – if not the only – candidate for this central causal process is a disinhibition of behaviour selection in the CSS, consistent with indirect pathway suppression. For instance, it is very hard to explain why evidence of a knowledge–action dissociation would be correlated with the rate of behaviour switching in the home cage, without both reflecting a disinhibition of response selection (see Table 5.1).

These results have been harder to replicate in full in the other species studied to date. In blue tits and marsh tits, both stereotypy and perseveration were correlated with the rate at which birds switched between behaviours – but these relationships were curvilinear, suggesting the involvement of a second unknown process (Garner et al., 2003a). Furthermore, although stereotypy and impaired extinction learning were correlated in bears, neither of these measures correlated with behavioural switching, though some correlations are seen between these three measures and activity (Vickery and Mason, 2005).

Thus, overall, the data from voles argue that recurrent perseveration is responsible for impaired extinction learning in stereotypic animals, while the patchy correlation of other measures of CSS function in bears suggests that other processes may be involved in the relationships between these latter measures and stereotypy and perseveration (Vickery and Mason, 2005). Therefore, in order to better replicate the human results, we adapted the human two-choice gambling task (Box. 5.3) for use with animals, and found that stereotypy in blue tits was indeed correlated with recurrent perseveration measured by the two-choice gambling task (Garner *et al.*, 2003a). However, there remains a minor conceptual problem with the two-choice gambling task in that on any trial, no choice is any more likely to be rewarded than the other. There is no 'right' or 'wrong' choice and so the task only approximates recurrent perseveration, which is defined as the repetition of *inappropriate* responses (Turner, 1997). In psychiatry, this has led to more complex tasks that directly measure recurrent perseveration (e.g. Turner, 1997). However, these tasks are too reliant on spoken instructions for application to animals. Instead, we modified the gambling task, so that although the animal is rewarded at random, the chance of being rewarded changes with the animal's choices. Thus in the 'bias-corrected gambling task', the probability of being rewarded for a particular choice on any given trial equals the probability with which the animal has chosen the other option over the previous 20 trials (Garner *et al.*, 2003b). As a result, even though the animal cannot predict which choice will be rewarded on any particular trial, it can maximize its chances of being rewarded by choosing both options equally often. As a result the bias-corrected gambling tasks does measure recurrent perseveration properly, and recurrent perseveration measured by this task correlates with stereotypy in orange-wing Amazon parrots (Garner *et al.*, 2003b) (see Fig. 5.3a). Furthermore, in this experiment, we found that repeated responses were made faster in higher stereotypy animals (see Fig. 5.3b), while no such relationship was seen for responses that switched from that chosen in the previous trial. This pattern can only be explained if the repetitive responses of high stereotypy animals involve a disinhibition of response selection – in other words, true recurrent perseveration.

We have recently begun to extend this work to Abnormal Repetitive Behaviour in mice. In addition to stereotypic behaviour (Würbel *et al.*, 1996), mice also perform an Abnormal Repetitive Behaviour called barbering, where one mouse plucks the fur and whiskers from its cagemates or itself in idiosyncratic repeated patterns (Sarna *et al.*, 2000; Garner *et al.*, 2004a,b). Barbering involves flexible goal-directed behaviour (Sarna *et al.*, 2000) and therefore likely represents an impulsive/compulsive behaviour rather than a stereotypy. Barbering also shows many behavioural and epidemiological similarities to human hair-plucking in trichotillomania (Garner *et al.*, 2004b). We have examined the relationships between stereotypy, barbering, recurrent perseveration and stuck-in-set perseveration in mice. Accordingly, stereotypy and recurrent perseveration in the bias-corrected gambling task are correlated, and the severity of barbering and stuck-in-set perseveration on the IntraDimensional–ExtraDimensional set-shifting task is correlated (for a preliminary report of these data see: Garner, 2002; Garner *et al.*, 2003c), while, just as

predicted, stereotypy and the severity of barbering are uncorrelated, stereotypy and stuck-in-set perseveration are uncorrelated and the severity of
barbering and recurrent perseveration are uncorrelated (unpublished
data). Thus barbering and stereotypy in mice appear to be two very

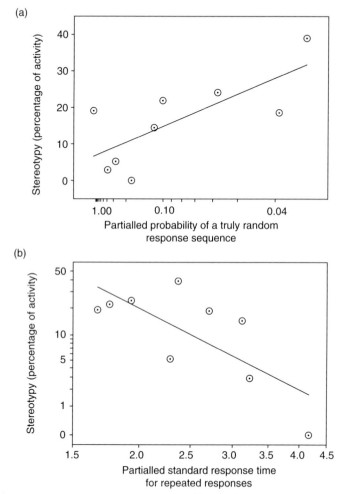

Fig. 5.3. Recurrent perseveration measured by the bias-corrected gambling task is
correlated with stereotypy in orange-wing Amazon parrots. Redrawn from Garner *et al.*
(2003b) with permission from Elsevier. (a) Recurrent perseveration is shown on the
x-axis. We calculated perseveration using third-order Markov analysis, which gives
the probability that each choice was independent of the previous three choices, for a
sequence of 102 consecutive responses. Thus increasingly random sequences have
higher probabilities of sequential independence (to the left of the x-axis), and
increasingly perseverative responses have lower probabilities (to the right of the x-axis).
The x-axis is partialled for sex. (b) The response time for repetitive choices is shown on
the x-axis.

different forms of Abnormal Repetitive Behaviour. The potentially crucial distinction between different forms of abnormal behaviours that might otherwise erroneously all be classed together is something I return to in Box. 10.2, Chapter 10.

Thus cage stereotypy really does seem to involve the same basic mechanism as drug-induced, lesion-induced and psychopathological stereotypies. Furthermore, in mice at least, stereotypy and impulsive/compulsive behaviours are distinct categories of Abnormal Repetitive Behaviour that involve the same distinct mechanisms as these two categories of Abnormal Repetitive Behaviour in human disorders.

However, there are also probably caveats to this general result (and see contributed box by Würbel, Box 5.4). For instance, Latham and Mason (Latham, 2005) investigated the relationship between recurrent perseveration in extinction learning and stereotypy in mice. They found that this relationship held only for some subsets of mice (e.g. early-weaned animals). Similarly, in bears the intercorrelation of measures related to CSS function is rather patchy, suggesting that while perseveration probably does play an important role in stereotypy, other processes (such as stress-related effects, see Chapters 8 and 11; or perhaps habit-formation, Vickery and Mason, 2005) are undoubtedly also important.

Furthermore, stepping back from these experiments and thinking more broadly about typical cage stereotypies, recurrent perseveration does not account for all the properties observed. In particular, recurrent perseveration does not explain between-individual and between-species differences in form or type of stereotypy – where motivational or phylogenetic explanations are particularly important (see Chapters 2–4). As a hypothesis, recurrent perseveration does explain why a motor pattern is repeated inflexibly, and why the inflexible repetition of motor patterns appears to be a discrete and meaningful behavioural category, something that other hypotheses mentioned in Box 5.4 do not. These latter hypotheses (repeated elicitation of responses; coping; habit-formation) only explain why a behaviour is initiated repeatedly – not why it is repeated so incessantly or why each repetition is identical (i.e. they do not really explain why a stereotypy is a stereotypy). The study of stereotypy form is therefore extremely important, partly to understand the limits of perseveration as a hypothesis, and partly to distinguish between behaviours that are flexible in form, and those that are not. For example, recurrent perseveration can explain why stereotypies are invariant within a bout, or between bouts performed in rapid succession (as recurrent perseveration can persist over hour-long timescales: Luria, 1965) – but not why they are invariant from one day to the next.

Würbel make rather similar points in Box 5.4, but I would disagree with some of his other issues. For instance, he argues that for recurrent perseveration to be a common explanation, all stereotypies must be mechanistically identical, perhaps directly produced by enhanced dopamine action. However, as discussed earlier, the power of a systems-level approach is that the proximate mechanisms of drug-induced and cage stereotypies do not need to be identical, merely that their ultimate systems-level effects (i.e. recurrent perseveration) are similar. Indeed the

Box 5.4. Stereotypies and Abnormal Perseveration – a Unifying Theory?

H. Würbel

According to Garner, cage stereotypies represent recurrent perseveration and decreased dorsal striatal inhibitory behavioural control. He also hypothesizes – while acknowledging a lack of conclusive evidence – that this decrease in inhibition is abnormal. Is this the 'unifying theory of stereotypy'? Or are there aspects it cannot explain? Garner himself recognizes puzzles in some of the data generated in animal perseveration tests. Here, I take a broader view, because, as seen in previous chapters, stereotypies have several key features that a unifying theory needs to account for.

1. First, they originate from repeated behavioural responses that seem to change in nature with development. Garner's hypothesis is potentially consistent with this. Thus, increasing performance, decreasing variance and emancipation from the originally eliciting circumstances, may all reflect progressive disruption of inhibitory behavioural control – changes arguably paralleling those seen in stereotypies induced by stimulant drugs such as dopamine (DA) agonists, where, with repeated administration, the behaviour becomes more intense and rigid (e.g. Robbins and Sahakian, 1983). However, other hypotheses e.g. 'coping' (see Box 1.3, Chapter 1 and Chapter 4), or habit-formation (see Chapters 2–4), could equally account for these developmental changes.

2. There are species differences in typical incidence. These could be due to differences in natural tendencies to perseverate, or in the effects of captivity on underlying neural substrates; thus in carnivores, for instance, wide-rangers could be naturally more perseverative, or more induced to become so by captivity (cf. Chapter 3). However, systematic comparative studies, in both natural and captive environments, are still needed to investigate this.

3. Third, different species' stereotypies differ in typical form, and individuals' stereotypies similarly vary in appearance and intensity. Species and individual variation in stereotypy form are likely to depend on motivational differences and individual choices between alternative behavioural responses (e.g. Chapter 4), not on perseveration. Furthermore, some animals show very low, yet stable, levels of stereotypy performance that it may be difficult to explain in terms of dysfunction (see Chapter 4).

4. Relatedly, caged animals typically display just one or two stereotypies. When stimulant drugs induce stereotypy, these generally affect the animals' entire behavioural output, with increasing doses narrowing down the behavioural repertoire while increasing performance of all remaining activities (Lyon and Robbins, 1975; Robbins and Sahakian, 1983). Why is it that in contrast, typically only one or two, but not other, frequently performed behaviours develop into cage stereotypies in captive animals? It is unclear how the same systems-level effects could explain this difference between drug- and environment-induced stereotypies. We need more data on how single behaviours are regulated by the basal ganglia, and under what conditions (reinforcement? stress? specific motivational states?) particular behaviours can become stereotypic (Box 8.1, Chapter 8 presents one interesting possibility).

5. Fifth, if all stereotypies were based on the same underlying mechanism, there should be similarities in their neural bases independent of causation (e.g. barren cages, drugs and brain damage). Thus there should also be similarities in the behavioural changes induced by captivity and stimulant drugs. However, above we saw one apparent difference, and furthermore, in bank voles and deer mice, drugs induce stereotypies that differ from those induced by captivity, and also fail to enhance the performance of cage-induced stereotypies (Vandebroek and Ödberg, 1997; Vandebroek *et al.*, 1998; Presti *et al.*, 2002, 2004).

6. Finally, drug-induced stereotypies are normally associated with altered DA function in the basal ganglia. However, no evidence was found for a relationship between cage-induced

Continued

> **Box 5.4.** *Continued*
>
> stereotypies and striatal DA function (density and sensitivity of DA receptors, concentration of DA or its metabolites) in stereotyping deer mice (Powell *et al.*, 1999; Presti *et al.*, 2002). This could mean that the focus on the dorsal striatal circuit alone is too narrow. For example, reduced inhibitory control by several loci of the cortex (a brain area whose development is impaired by barren housing conditions: van Praag *et al.*, 2000) have also been implicated in the development of both drug-induced stereotypies (e.g. Karler *et al.*, 1997) and cage stereo-typies (see Chapter 8).
>
> Overall, Garner's hypothesis has great appeal. However, the issues mentioned above suggest that a more complete analysis of the psychological and neural processes underlying cage stereotypies is warranted before we draw final conclusions.

differences between different kinds of stereotypy in a systems perspective reflect differences in proximate mechanisms – once again reinforcing the role of motivational mechanisms in shaping stereotypies' form and timing. Würbel also argues that frontal disinhibition may be important in stereo-typies. There is certainly a grain of truth in this suggestion, as areas of frontal cortex are involved in most levels of behavioural control (thus both super-visory attentional, and basal ganglia motor system circuitry, involve frontal areas; see also Chapters 7 and 8, this volume). However, this also suggests that a general frontal disinhibition is an unlikely primary mechanism – especially as frontal brain damage in human generally results in stuck-in-set perseveration and associated behaviours, rarely stereotypies. SAS frontal inhibition might help animals resist *existing* stereotypies, without actually being involved in the aetiology of the stereotypy itself (much as serotonergic drugs help to treat OCD, even though OCD is not caused by a serotonin imbalance: Rauch and Jenike, 1993). Nevertheless, resolving these issues to understand the causes behind perseveration, and the influ-ence of other processes on both perseveration and stereotypy is an exciting new direction for this line of research (see also Chapter 11, this volume).

5.6. Key Issues and Implications of a Role for Perseveration in Captive Animals' Stereotypies

5.6.1. The causes of perseveration

Just like Abnormal Repetitive Behaviour in human disorders, the devel-opment of Abnormal Repetitive Behaviour in captive animals involves a mixture of genetic and environmental mechanisms. For instance, marked strain differences in mice interact with several environmental factors to affect the development of barbering (Garner *et al.*, 2004a,b), and stereo-typy in mice is similarly affected by gene–environment interactions (see Chapter 8). Furthermore, even in inbred strains of mice, which are essen-tially genetically identical, barbering (Garner *et al.*, 2003c, 2004b) and stereotypy (Cabib and Bonaventura, 1997; Nevison *et al.*, 1999) vary considerably between individuals. Thus if Abnormal Repetitive Behaviour is correlated with perseveration in both genetically heterogeneous (Garner and Mason, 2002) and homogenous (Garner et al., 2003c) animals, how might environmental influences lead to physiological differences that affect CSS or SAS function? A number of different perspectives on this issue are presented in Chapters 6–8, all of which integrate in different ways with the systems-level perspective of this chapter.

5.6.2. Is the perseveration associated with Abnormal Repetitive Behaviour abnormal?

Although the current data paint a picture whereby stereotypy is indeed consistently correlated with recurrent perseveration, the interpretation of this result rests on a critical further question (Garner, 2005): is the range of perseveration seen in stereotypic animals merely normal and simply expressed as Abnormal Repetitive Behaviour in captive situations (cf. correlation of stuck-in-set perseveration and sub-clinical obsessive–compulsive scores in healthy humans: Zohar et al., 1995)? Or does it reflect abnormal (i.e. an environmentally induced pathology of) brain development? In other words, does Abnormal Repetitive Behaviour reflect a response of normal behavioural control mechanisms to an abnormal environment, or the output of abnormal behavioural control mechanisms produced by an abnormal environment (these possibilities correspond to the general categories of maladaptive versus malfunctional behaviours, beautifully described in Box 1.4, Chapter 1, this volume and Mills, 2003).

The crux of this question does not lie in whether stereotypies are an abnormal behaviour (they are, as they rarely if ever occur in the wild, and they impair health, survival or reproductive success in many species: Garner, 2005), but in *how* they are abnormal. Answering whether or not stereotypy is a symptom of a pathologically abnormal and altered brain is crucial to understanding how to manage stereotypy, and to understanding its potential effects (in the case of laboratory animals) on experimental outcomes (Garner, 2005).

There are a number of different mechanisms by which the environment could alter brain function during development, not all of which need be irreversible. Chapters 6–8, for instance, point to alterations in neurotransmitter function that could affect the CSS and SAS. In addition, environmental influences affect the structure and connectivity of the brain (see Chapter 7). For instance, the visual cortex trains itself to detect only the features of the environment present in early development (Hubel, 1988). The behavioural control systems of the CSS and SAS might similarly train themselves during early experience to process only the environmental variability and behavioural consequences experienced in the environment. Barren unchanging environments, where behaviour has few real consequences, might then induce Abnormal Repetitive Behaviour through the brain programming the CSS and SAS to be perseverative and inflexible in response to behavioural consequences.

Consequently, studying the effects of environmental influences (including those of environmental enrichment) and genetic influences on perseveration and Abnormal Repetitive Behaviour is crucial to establishing whether Abnormal Repetitive Behaviour indicates abnormal (i.e. pathological) levels of perseveration (Garner, 2005). There are two general approaches to such studies. First, one can compare the effects of treatments within individuals, examining the changes in perseveration associated with changes in Abnormal Repetitive Behaviour. For instance, the perseveration seen in drug-induced stereotypies is clearly abnormal, in that the function of the brain has been altered to a pathological state (Mills, 2003). Similarly the correlation between changes in brain metabolism and OCD symptom improvement with treatment in humans (Baxter *et al.*, 1992), implies that the original levels of brain metabolism might have been abnormal. Accordingly, the enrichment of blue tits and marsh tits reduces stereotypy, and the change in stereotypy is correlated with the change in recurrent perseveration, again indicating that the level of perseveration seen in stereotyping individuals might have been abnormal (Garner *et al.*, 2003a). However, these kinds of experiments, though persuasive, are not unequivocal. Thus reducing perseveration might be expected to reduce Abnormal Repetitive Behaviour even if the original level of perseveration was within the 'normal' or 'healthy' range (*cf.* Mills, 2003).

A second approach is really needed to resolve this issue, whereby the levels of perseveration and Abnormal Repetitive Behaviour are compared between individuals exposed to different environments. At the extreme, one might compare the range of perseveration and Abnormal Repetitive Behaviour of wild animals with that of captive conspecifics, effectively using the wild population to define the 'normal' or 'healthy' range of perseveration. However, for many species domestication often means that an appropriate wild conspecific population does not exist to make such a comparison. This is especially true for laboratory mice which are a complex hybrid of several (sub)species (Silver, 1995). Nevertheless, enrichment experiments that simultaneously compare both perseveration and Abnormal Repetitive Behaviour provide a solution. Thus if the relationship between Abnormal Repetitive Behaviour and perseveration is the same in both enriched and unenriched individuals, and the range of perseveration seen in individuals that do not perform Abnormal Repetitive Behaviour is the same in both populations, then one can define the extremes of perseveration seen in unenriched animals as abnormal (Garner, 2005). Garner and his colleagues are currently embarking exactly on these experiments in order to disentangle this issue (see also Chapter 11).

5.6.3. Implications for our use of captive animals: laboratory animal model validity

The 'good welfare is good science' approach in laboratory animal welfare research argues that changes in husbandry, housing or experimental practices which improve the laboratory animal welfare might also improve the scientific quality of the experiments in which they serve (e.g. Russell and Burch, 1959; Würbel, 2001; Garner, 2002, 2005). In particular, many of the 'high-throughput' behavioural tests in common use show poor replicability between laboratories (i.e. an experiment performed identically in two different laboratories gives different results: Crabbe *et al.*, 1999). In part, this reflects the major influence of environmental variables on such tests (e.g. Chesler *et al.*, 2002). However, many of these tests could be affected by perseveration, or the other behavioural effects of CSS dysfunction seen in stereotyping animals (Garner and Mason, 2002; Garner, 2005). For instance, the perseveration correlated with stereotypy in marsh tits affects their behaviour in a widely used food storing paradigm (Garner *et al.*, 2003a). As husbandry directly affects stereotypy, might Abnormal Repetitive Behaviour and perseveration represent an important source of uncontrolled variability – particularly between laboratories?

Furthermore, if the perseveration seen in laboratory animals that perform Abnormal Repetitive Behaviour indicates abnormal brain function, then such animals are clearly poor subjects for use in behavioural research. As the vast majority of mice perform stereotypy (Würbel *et al.*, 1996; Nevison *et al.*, 1999), what impact would Abnormal Repetitive Behaviour and perseveration have on the validity of laboratory-based behavioural research? This scenario applies to other captive animals too – for instance if stereotypy reflects an abnormal level of perseveration then could stereotypy be indirectly related to poor survival when captive-bred animals are released into the wild (Vickery and Mason, 2003)? At this point we simply do not know the answers to these questions.

5.6.4. Welfare implications – suffering and function

The association of perseveration with the experience of being unable to control behaviour (the knowledge–action dissociation) is often acutely distressing to human patients when under test (e.g. Luria, 1965; Turner, 1997). The experiences associated with Abnormal Repetitive Behaviour are, in contrast, more difficult to determine systematically for three reasons: first, patients often try and hide the inappropriateness of their perseverative behaviour by claiming that they 'just decided' to respond inappropriately; second, Abnormal Repetitive Behaviour is seen in several disorders where communication is impaired; and third, Abnormal Repetitive Behaviour may be reinforcing or anxiolytic in some patients in some disorders. Nevertheless, Abnormal Repetitive Behaviour in animals could therefore indicate the possible presence of a novel kind of suffering (and hence poor welfare) whereby the animal experiences frustration at its inability to properly control its behaviour (Garner and Mason, 2002). In this case stereotypy would indicate poor welfare according to the 'suffering' or 'feelings' definition of welfare (Duncan and Fraser, 1997). Furthermore, if Abnormal Repetitive Behaviour in captive animals does indicate abnormal brain function, then welfare is impaired *per se* according to definitions of poor welfare as compromised physiological integrity (Duncan and Fraser, 1997).

5.7. Conclusions and Future Work

Stereotypies in human mental disorder, stereotypies in isolation-reared animals and drug-induced stereotypies all involve abnormalities of the basal ganglia, and occur as part of a general syndrome of disinhibited response selection. In autistic and schizophrenic patients, stereotypies correlate with recurrent perseveration – the inappropriate repetition of responses. Recurrent perseveration is indicative of dysfunction in the basal ganglia motor system that is responsible for selecting and sequencing individual behavioural responses. These observations suggest a system-level mechanism underlying stereotypies that helps to unify these disparate examples. Spontaneously occurring stereotypies in many different species of captive animals also correlate with recurrent perseveration, and other measures consistent with differences in the basal ganglia motor system. Thus stereotypies in captive animals do seem to involve the same systems-level mechanism as stereotypies in human mental disorder. However, at this point there are several key unanswered issues: (i) why are other aspects of response selection, such as rates of behavioural switching, less reliably correlated with perseveration and stereotypy?; (ii) how do the individual differences in perseveration arise?; specifically (iii) is the perseveration seen in stereotyping animals indicative of abnormal brain function?; (iv) does the perseveration of stereotypic animals affect experimental outcomes?; and (v) do stereotyping animals suffer the same distressing experiences as stereotyping human patients? Although these unanswered questions are challenging, they are answerable, and they represent exciting directions for future work.

Acknowledgements

Many of the ideas in this chapter were first developed while I held a BBSRC PhD studentship, and have benefited over the years from discussion with many colleagues, in particular Georgia Mason, Cheryl Meehan, Joy Mench, Danny Mills, Trevor Robbins and Hanno Würbel. I was supported in part by NIMH grant No. 5 R03 MH 63907-2 while writing this chapter.

References

Albin, R.L., Young, A.B. and Penney, J.B. (1995) The functional anatomy of disorders of the basal ganglia. *Trends in Neurosciences* 18, 63–64.

American Psychiatric Association (1994) *Diagnostic and Statistical Manual of Mental Disorders*, 4th edn. American Psychiatric Association, Washington, DC.

Baxter, L.R., Schwartz, J.M., Bergman, K.S. and Szuba, M.P. (1992) Caudate glucose metabolic rate changes with both drug and behavior therapy for obsessive–compulsive disorder. *Archives of General Psychiatry* 49, 681–689.

Berthier, M.L., Kulisevsky, J., Gironell, A. and Lopez, O.L. (2001) Obsessive–compulsive disorder and traumatic brain injury: behavioral, cognitive, and neuroimaging findings. *Neuropsychiatry Neuropsychology and Behavioral Neurology* 14, 23–31.

Birrell, J.M. and Brown, V.J. (2000) Medial frontal cortex mediates perceptual attentional set shifting in the rat. *Journal of Neuroscience* 20, 4320–4324.

Bodfish, J.W. and Lewis, M.H. (2002) Self-injury and comorbid behaviors in developmental, neurological, psychiatric, and genetic disorders. In: Schroeder, S.R., Oster-Granite, M.L. and Thompson, T. (eds) *Self-injurious Behavior: Gene–Brain Behavior relationships.* American Psychological Association, Washington, DC, pp. 23–40.

Bornstein, R.A. (1990) Neuropsychological performance in children with Tourette's syndrome. *Psychiatry Research* 33, 73–81.

Bornstein, R.A. (1991) Neuropsychological correlates of obsessive characteristics in Tourette syndrome. *Journal of Neuropsychiatry and Clinical Neurosciences* 3, 157–162.

Cabib, S. (1993) Neurobiological basis of stereotypies. In: Lawrence, A.B. and Rushen, J. (eds) *Stereotypic Animal Behaviour: Fundamentals and Applications to Welfare.* CAB International, Wallingford, UK, pp. 119–145.

Cabib, S. and Bonaventura, N. (1997) Parallel strain-dependent susceptibility to environmentally-induced stereotypies and stress-induced behavioral sensitization in mice. *Physiology and Behavior* 61, 499–506.

Casey, B.J., Tottenham, N. and Fossella, J. (2002) Clinical, imaging, lesion, and genetic approaches toward a model of cognitive control. *Developmental Psychobiology* 40, 237–254.

Chesler, E.J., Wilson, S.G., Lariviere, W.R., Rodriguez-Zas, S.L. and Mogil, J.S. (2002) Identification and ranking of genetic and laboratory environment factors influencing a behavioral trait, thermal nociception, via computational analysis of a large data archive. *Neuroscience and Biobehavioral Reviews* 26, 907–923.

Cooper, J.J. and Nicol, C.J. (1994) Neighbour effects on the development of locomotor stereotypies in bank voles, *Clethrionomys glareolus. Animal Behaviour* 47, 214–216.

Crabbe, J.C., Wahlsten, D. and Dudek, B.C. (1999) Genetics of mouse behavior: interactions with laboratory environment. *Science* 284, 1670–1672.

Crider, A. (1997) Perseveration in schizophrenia. *Schizophrenia Bulletin* 23, 63–74.

Curzon, G. (1990) Stereotyped and other motor responses to 5-hydroxytryptamine receptor activation. In: Cooper, S.J. and Dourish, C.T. (eds) *Neurobiology of Stereotyped Behaviour.* Oxford University Press, Oxford, UK, pp. 142–168.

Dantzer, R. (1986) Behavioural, physiological and functional aspects of stereotypic behaviour: a review and reinterpretation. *Journal of Animal Science* 62, 1776–1786.

Davison, G.C. and Neale, J.M. (1998) *Abnormal Psychology*, 7th edn. John Wiley & Sons, New York.

D'Esposito, M., Detre, J.A., Alsop, D.C. and Shin, R.K. (1995) The neural basis of the central executive system of working memory. *Nature* 378, 279–281.

Dias, R., Robbins, T.W. and Roberts, A.C. (1996) Dissociation in prefrontal cortex of affective and attentional shifts. *Nature* 380, 69–72.

Duncan, I.J.H. and Fraser, D. (1997) Understanding animal welfare. In: Appleby, M.C. and Hughes, B.O. (eds) *Animal Welfare.* CAB International, Wallingford, UK, pp. xiii, 316.

Fentress, J.C. (1976) Dynamic boundaries of patterned behaviour: interaction and self-organisation. In: Bateson, P.P.G. and Hinde, R.A. (eds) *Perspectives in Ethology*, Vol. 1. Plenum Press, New York, pp. 155–224.

Frith, U. (1970) Studies in pattern detection in normal and autistic children: II. Reproduction and production of color sequences. *Journal of Experimental Child Psychology* 10, 120–135.

Frith, C.D. and Done, D.J. (1983) Stereotyped responding by schizophrenic-patients on a 2-choice guessing task. *Psychological Medicine* 13, 779–786.

Frith, C.D. and Done, D.J. (1990) Stereotyped behaviour in madness and in health. In:

Cooper, S.J. and Dourish, C.T. (eds) *Neurobiology of Stereotyped Behaviour*. Oxford University Press, Oxford, UK, pp. 232–259.

Frith, C.D. and Grasby, P.M. (1995) rCBF studies of prefrontal function and their relevance to psychosis. In: Boller, F. and Grafman, J. (eds) *Handbook of Neuropsychology*, Vol. 10. Elsevier Science, New York, pp. 383–403.

Garner, J.P. (1999) The aetiology of stereotypy in caged animals. DPhil thesis, University of Oxford, Oxford, UK.

Garner, J.P. (2002) Why every scientist should care about animal welfare: Abnormal Repetitive Behavior and brain function in captive animals. In: Rowan, A. and Spielmann, H. (eds) *Fourth World Congress on Alternatives and Animal use in the Life Sciences*, New Orleans, USA, p. 95.

Garner, J.P. (2005) Stereotypies and other Abnormal Repetitive Behaviors: potential impact on validity, reliability, and replicability of scientific outcomes. *ILAR Journal* 46, 106–117.

Garner, J.P. and Mason, G.J. (2002) Evidence for a relationship between cage stereotypies and behavioural disinhibition in laboratory rodents. *Behavioural Brain Research* 136, 83–92.

Garner, J.P., Mason, G. and Smith, R. (2003a) Stereotypic route-tracing in experimentally-caged songbirds correlates with general behavioural disinhibition. *Animal Behaviour* 66, 711–727.

Garner, J.P., Meehan, C.L. and Mench, J.A. (2003b) Stereotypies in caged parrots, schizophrenia and autism: evidence for a common mechanism. *Behavioural Brain Research* 145, 125–134.

Garner, J.P., Wayne, C.M., Würbel, H. and Mench, J.A. (2003c) Barbering (whisker trimming) in laboratory mice involves the same brain systems as compulsive behaviors in trichotillomania, autism and other obsessive–compulsive spectrum disorders. In: Ferrante, V. (ed.) *Proceedings of the 37th International Congress of the International Society for Applied Ethology*. Fondazione Iniziative Zooprofilattiche e Zootecniche, Abano Terme, Italy, p. 75.

Garner, J.P., Dufour, B., Gregg, L.E., Weisker, S.M. and Mench, J.A. (2004a) Social and husbandry factors affecting the prevalence and severity of barbering ('whisker trimming') in laboratory mice. *Applied Animal Behaviour Science* 89, 263–282.

Garner, J.P., Weisker, S.M., Dufour, B. and Mench, J.A. (2004b) Barbering (fur and whisker trimming) by laboratory mice as a model of human trichotillomania and obsessive–compulsive spectrum disorders. *Comparative Medicine* 54, 216–224.

Hinde, R.A. (1970) *Animal Behaviour: A Synthesis of Ethology and Comparative Psychology*, 2nd edn. McGraw-Hill, New York.

Hollander, E. and Wong, C.M. (1995) Obsessive–compulsive spectrum disorders. *Journal of Clinical Psychiatry* 56, 3–6.

Hubel, D.H. (1988) *Eye, Brain, and Vision*, Vol. 22. Scientific American Library, Distributed by W.H. Freeman, New York.

Hughes, C., Russell, J. and Robbins, T.W. (1994) Evidence for executive dysfunction in autism. *Neuropsychologia* 32, 477–492.

Hymas, N., Lees, A., Bolton, D., Epps, K. and Head, D. (1991) The neurology of obsessional slowness. *Brain* 114, 2203–2233.

Jones, G.H., Marsden, C.A. and Robbins, T.W. (1991) Behavioral rigidity and rule-learning deficits following isolation-rearing in the rat – neurochemical correlates. *Behavioural Brain Research* 43, 35–50.

Joyce, E.M. and Iversen, S.D. (1984) Dissociable effects of 6-OHDA-induced lesions of neostriatum on anorexia, locomotor-activity and stereotypy – the role of behavioral competition. *Psychopharmacology* 83, 363–366.

Karler, R., Bedingfield, J.B., Thai, D.K. and Calder, L.D. (1997) The role of the frontal cortex in the mouse in behavioral sensitization to amphetamine. *Brain Research* 757(2), 228–235.

Kennes, D., Ödberg, F.O., Bouquet, Y. and Derycke, P.H. (1988) Changes in naloxone and haloperidol effects during the development of captivity-induced jumping stereotypy in bank voles. *European Journal of Pharmacology* 153, 19–24.

Keuthen, N.J., Savage, C.R., O'Sullivan, R.L. and Brown, H.D. (1996) Neuropsychological functioning in trichotillomania. *Biological Psychiatry* 39, 747–749.

Lacerda, A.L.T., Dalgalarrondo, P., Caetano, D., Haas, G.L., Camargo, E.E. and Keshavan, M.S. (2003) Neuropsychological performance and regional cerebral blood flow in obsessive–compulsive disorder. *Progress in Neuro-Psychopharmacology and Biological Psychiatry* 27, 657–665.

Latham, N. (2005) Refining the role of stereotypic behaviour in the assessment of welfare: stress, general motor persistence and early environment in the development of abnormal behaviour. PhD thesis, Oxford University, Oxford, UK.

Lewis, M.H. and Bodfish, J.W. (1998) Repetitive behavior disorders in autism. *Mental Retardation & Developmental Disabilities Research Reviews* 4, 80–89.

Lewis, M.H., Gluck, J.P., Beauchamp, A.J. and Keresztury, M.F. (1990) Long-term effects of early social isolation in *Macaca mulatta*: changes in dopamine receptor function following apomorphine challenge. *Brain Research* 513, 67–73.

Lewis, M.H., Gluck, J.P., Bodfish, J.W. and Beauchamp, A.J. (1996) Neurobiological basis of stereotyped movement disorder. In: Sprague, R.L. and Newell, K.M. (eds) *Stereotyped Movements: Brain and Behavior Relationships.* American Psychological Association, Washington, DC, pp. 37–67.

Liddle, P.F. and Morris, D.L. (1991) Schizophrenic syndromes and frontal lobe performance. *British Journal of Psychiatry* 158, 340–345.

Longoni, R., Spina, L., Mulas, A., Carboni, E., Garau, L., Melchiorri, P. and Di Chiara, G. (1991) (d-Ala2)deltorphin II: D1-dependent stereotypies and stimulation of dopamine release in the nucleus accumbens. *Journal of Neuroscience* 11, 1565–1576.

Lucey, J.V., Burness, C.E., Costa, D.C., Gacinovic, S., Pilowsky, L.S., Ell, P.J., Marks, I.M. and Kerwin, R.W. (1997) Wisconsin card sorting test (WCST) errors and cerebral blood flow in obsessive–compulsive disorder (OCD). *British Journal of Medical Psychology* 70, 403–411.

Luria, A.R. (1965) Two kinds of motor perseveration in massive injury of the frontal lobes. *Brain* 88, 1–11.

Lyon, M. and Robbins, T. (1975) The action of central nervous system stimulant drugs: a general theory concerning amphetamine effects. *Current Developments in Psychopharmacology* 2, 79–163.

Manschreck, T.C., Maher, B.A. and Ader, D.N. (1981) Formal thought disorder, the type-token ratio, and disturbed voluntary motor movement in schizophrenia. *British Journal of Psychiatry* 139, 7–15.

Martin, L.J., Spicer, D.M., Lewis, M.H., Gluck, J.P. and Cork, L.C. (1991) Social deprivation of infant rhesus monkeys alters the chemoarchitecture of the brain: I. Subcortical regions. *Journal of Neuroscience* 11, 3344–3358.

Mills, D.S. (2003) Medical paradigms for the study of problem behaviour: a critical review. *Applied Animal Behaviour Science* 81, 265–277.

Mink, J.W. (2003) The basal ganglia and involuntary movements: impaired inhibition of competing motor patterns. *Archives in Neurology* 60, 1365–1368.

Nevison, C.M., Hurst, J.L. and Barnard, C.J. (1999) Strain-specific effects of cage enrichment in male laboratory mice (*Mus musculus*). *Animal Welfare* 8, 361–379.

Norman, D.A. and Shallice, T. (1986) Attention to action: willed and automatic control of behaviour. In: Davidson, R.J., Schwartz, G.E. and Shapiro, D. (eds) *Consciousness and Self-regulation: Advances in Research and Theory*, Vol. 4. Plenum Press, New York, pp. 1–18.

Ödberg, F.O. (1986) The jumping stereotypy in the bank vole *(Clethrionomys glareolus). Biology of Behaviour* 11, 130–143.

Ödberg, F.O. (1993) Future research directions. In: Lawrence, A.B. and Rushen, J. (eds) *Stereotypic Animal Behaviour: Fundamentals and Applications to Welfare.* CAB International, Wallingford, UK, pp. 173–191.

Ödberg, F.O., Kennes, D., Derycke, P.H. and Bouquet, Y. (1987) The effect of interference in catecholamine biosynthesis on captivity-induced jumping stereotypy in bank voles (*Clethrionomys glareolus*). *Archives Internationales de Pharmacodynamie et de Therapie* 285, 34–42.

Olton, D.S. (1972) Behavioral and neuroanatomical differentiation of response-suppression and response-shift mechanisms in the rat. *Journal of Comparative and Physiological Psychology* 78, 450–456.

Owen, A.M., Roberts, A.C., Polkey, C.E., Sahakian, B.J. and Robbins, T.W. (1991) Extra-dimensional versus intra-dimensional set shifting performance following frontal-lobe excisions, temporal-lobe excisions or amygdalo-hippocampectomy in man. *Neuropsychologia* 29, 993–1006.

Owen, A.M., Roberts, A.C., Hodges, J.R., Summers, B.A., Polkey, C.E. and Robbins, T.W. (1993) Contrasting mechanisms of impaired attentional set-shifting in patients with frontal lobe damage or Parkinson's disease. *Brain* 116, 1159–1175.

Owens, D.G.C., Johnstone, E.C. and Frith, C.D. (1982) Spontaneous involuntary disorders of movement – their prevalence, severity, and distribution in chronic-schizophrenics with and without treatment with neuroleptics. *Archives of General Psychiatry* 39, 452–461.

Ozonoff, S., Pennington, B.F. and Rogers, S.J. (1991) Executive function deficits in high-functioning autistic individuals: relationship to theory of mind. *Journal of Child Psychology and Psychiatry and Allied Disciplines* 32, 1081–1105.

Presti, M.F., Powell, S.B. and Lewis, M.H. (2002) Dissociation between spontaneously emitted and apomorphine-induced stereotypy in *Peromyscus maniculatus bairdii*. *Physiology and Behavior* 75, 347–353.

Presti, M.F., Gibney, B.C. and Lewis, M.H. (2004) Effects of introstriatal administration of selective dopaminergic ligands on spontaneous stereotypy in mice. *Physiology & Behavior* 80, 433–439.

Powell, S.B. Newman, H.A., Pendergast, J. and Lewis, M.H. (1999) A rodent model of spontaneous stereotypy: initial characterzation of developmental, environmental, and neurobiological factors. *Physiology & Behavior* 66, 355–363.

Rauch, S.L. and Jenike, M.A. (1993) Neurobiological models of obsessive–compulsive disorder. *Psychosomatics* 34, 20–32.

Reiner, A., Medina, L. and Veenman, C.L. (1998) Structural and functional evolution of the basal ganglia in vertebrates. *Brain Research Reviews* 28, 235–285.

Rettew, D.C., Cheslow, D.L., Rapoport, J.L. and Leonard, H.L. (1991) Neuropsychological test performance in trichotillomania: a further link with obsessive–compulsive disorder. *Journal of Anxiety Disorders* 5, 225–235.

Robbins, T.W. and Sahakian, B.J. (1983) Behavioral effects of psycho-motor stimulant drugs; clinical and neuropsychological implications. In: Creese, I. (ed.) *Stimulants: Neurochemical, Behavioural and Clinical Perspectives.* Rowen Press, New York, pp. 301–338.

Robbins, T.W. (1996) Dissociating executive functions of the prefrontal cortex. *Philosophical Transactions of the Royal Society of London Series B-Biological Sciences* 351, 1463–1470.

Robbins, T.W. (1997a) Arousal systems and attentional processes. *Biological Psychology* 45, 57–71.

Robbins, T.W. (1997b) Integrating the neurobiological and neuropsychological dimensions of autism. In: Russell, J. (ed.) *Autism as an Executive Disorder.* Oxford University Press, New York, pp. 21–53.

Robbins, T.W., Jones, G.H. and Wilkinson, L.S. (1996) Behavioural and neurochemical effects of early social deprivation in the rat. *Journal of Psychopharmacology* 10, 39–47.

Robertson, M.M., Trimble, M.R. and Lees, A.J. (1988) The psychopathology of the Gilles de la Tourette syndrome: a phenomenological analysis. *British Journal of Psychiatry* 152, 383–390.

Rosenzweig, M.R., Bennett, E.L., Hebert, M. and Morimoto, H. (1978) Social grouping cannot account for cerebral effects of enriched environments. *Brain Research* 153, 563–576.

Russell, W.M.S. and Burch, R.L. (1959) *The Principles of Humane Experimental Technique.* Methuen, London.

Sandson, J. and Albert, M.L. (1984) Varieties of perseveration. *Neuropsychologia* 22, 715–732.

Sarna, J.R., Dyck, R.H. and Whishaw, I.Q. (2000) The Dalila effect: C57BL6 mice barber whiskers by plucking. *Behavioural Brain Research* 108, 39–45.

Saxena, S., Brody, A.L., Schwartz, J.M. and Baxter, L.R. (1998) Neuroimaging and frontal-subcortical circuitry in obsessive–compulsive disorder. *British Journal of Psychiatry* 173, 26–37.

Shallice, T. (1982) Specific impairments of planning. *Philosophical Transactions of the Royal Society of London. Series B: Biological Sciences* 298, 199–209.

Shallice, T. and Burgess, P.W. (1991) Deficits in strategy application following frontal lobe damage in man. *Brain* 114, 727–741.

Shallice, T. and Burgess, P. (1996) The domain of supervisory processes and temporal organization of behaviour. *Philosophical Transactions of the Royal Society of London Series B-Biological Sciences* 351, 1405–1411.

Silver, L.M. (1995) *Mouse Genetics: Concepts and Applications.* Oxford University Press, Oxford, UK.

Stein, D.J., van Heerden, B., Hugo, C., van Kradenburg, J., Warwick, J., Zungu-Dirwayi, N. and Seedat, S. (2002) Functional brain imaging and pharmacotherapy in trichotillomania: single photon emission computed tomography before and after treatment with the selective serotonin reuptake inhibitor citalopram. *Progress in Neuro-Psychopharmacology and Biological Psychiatry* 26, 885–890.

Steiner, H. and Gerfen, C.R. (1998) Role of dynorphin and enkephalin in the regulation of striatal output pathways and behavior. *Experimental Brain Research* 123, 60–76.

Toates, F. (2000) Multiple factors controlling behaviour: implications for stress and welfare. In: Moberg, G.P. and Mench, J.A. (eds) *The Biology of Animal Stress: Basic Principles and Implications for Animal Welfare.* CAB International, Wallingford, UK, pp. 199–226.

Turner, M. (1997) Towards an executive dysfunction account of repetitive behaviour in autism. In: Russell, J. (ed.) *Autism as an Executive Disorder.* Oxford University Press, Oxford, UK, pp. 57–100.

Turner, M. (1999) Annotation: repetitive behaviour in autism: a review of psychological research. *Journal of Child Psychology and Psychiatry and Allied Disciplines* 40, 839–849.

Turner, C.A. and Lewis, M.H. (2003) Environmental enrichment: effects on stereotyped behavior and neurotrophin levels. *Physiology & Behavior* 80, 259–266.

Vandebroek, I. and Ödberg, F.O. (1997) Effect of apomorphine on the conflict-induced jumping stereotypy in bank voles. *Pharmacology, Biochemistry and Behavior* 57, 863–868.

Vandebroek, I., Berckmoes, V. and Ödberg, F.O. (1998) Dissociation between MK-801 and captivity-induced stereotypies in bank voles. *Psychopharmacology* (Berlin) 137, 205–214.

van Praag, H., Kempermann, G. and Gage, F.H. (2000), Neural consequences of environmental enrichment. *Nature Reviews Neuroscience* 1, 191–198.

Vickery, S.S. and Mason, G.J. (2003) Behavioural persistence in captive bears: implications for reintroduction. *Ursus* 14, 35–43.

Vickery, S.S. and Mason, G.J. (2005) Stereotypy in caged bears correlates with perseverative responding on an extinction task. *Applied Animal Behaviour Science* 91, 247–260.

Waddington, J.L., Molloy, A.G., O'Boyle, K.M. and Pugh, M.T. (1990) Aspects of stereotyped and non-stereotyped behaviour in relation to dopamine receptor subtypes. In: Cooper, S.J. and Dourish, C.T. (eds) *Neurobiology of Stereotyped Behaviour.* Oxford University Press, Oxford, UK, pp. 64–90.

Würbel, H. (2001) Ideal homes? Housing effects on rodent brain and behaviour. *Trends in Neurosciences* 24, 207–211.

Würbel, H., Stauffacher, M. and von Holst, D. (1996) Stereotypies in laboratory mice – quantitative and qualitative description of the ontogeny of 'wire-gnawing' and 'jumping' in Zur:ICR and Zur:ICR nu. *Ethology* 102, 371–385.

Zohar, A.H., LaBuda, M. and Moschel-Ravid, O. (1995) Obsessive–compulsive behaviors and cognitive functioning: a study of compulsivity, frame shifting and type A activity patterns in a normal population. *Neuropsychiatry, Neuropsychology, and Behavioral Neurology* 8, 163–167.

6

Deprived Environments: Developmental Insights from Primatology

M.A. Novak,[1,2] J.S. Meyer,[1] C. Lutz[2]
and S. Tiefenbacher[2]

[1]Department of Psychology, Tobin Hall, University of Massachusetts, Amherst, MA 01003, USA; [2]New England Primate Research Center, Harvard Medical School, One Pine Hill Road, PO Box 9102, Southborough, MA 01772-9102, USA

Editorial Introduction

Young primates have a long period of dependency on their mothers, who have important roles beyond the mere provision of milk (e.g. transporting the infants, and providing comfort). Thus it is perhaps unsurprising that being deprived of maternal care has major effects on behaviour, including promoting stereotypy. Novak and her co-authors discuss a series of past experiments on rhesus monkeys that documented these effects. Social deprivation had the severest impact when infants were raised for the first months of life without mothers or peers: these animals spent much of their time in stereotypies. Some forms seemed to relate to specific aspects of mother–infant interaction, with self-clasping appearing to reflect the loss of physical contact; rocking, the absence of maternal movement; and digit-sucking presumably reflecting diminished opportunities to suckle. However, frustrated motivations were not the whole story. As the animals moved out of infancy, these stereotypies waned, but were replaced by other abnormal behaviours such as somersaults, head bobs and sometimes self-injurious behaviour (SIB). Furthermore, they showed other changes too, including: poor abilities to extinguish learnt responses (*cf.* the perseveration described in the previous chapter); heightened fearfulness; inappropriate social interactions; and long-lasting disturbances of serotinergic function. Dopaminergic systems were also affected, although the longevity of effects seemed affected by subsequent housing conditions; amongst isolation-housed adults, however, those which had previously been isolation-reared as infants showed exaggerated responses to the catecholamine agonist amphetamine, even though tested nearly two decades later. Less extreme rearing situations (e.g. peer-rearing, or being hand-raised by human surrogates) had much less extreme effects, but could still increase stereotypy levels over those of control animals long after housing conditions were normalized.

Although some of these accounts of maternal loss come from zoos (where infant primates sometimes have to be hand-reared), most stemmed from the research of Harlow and colleagues several decades ago on the role of experience in the development of species-typical social behaviour. These experiments can be very distressing to read about, but it should be remembered that scientifically, the work was rather pioneering for its time (when species-typical behaviour was thought largely instinctive). Furthermore, even today it has some practical relevance, both to humans with the terrible discovery of extremely deprived 'orphans' in Romania in the early 1990s, and perhaps to other species too, given that early maternal separation is currently near ubiquitous in agricultural animals. Two contributed boxes develop these last two themes further.

Novak and colleagues explore two further principal issues here. First, they emphasize that early social deprivation is not the only cause of primate abnormal behaviour: moving normally-reared adolescents or adults to individual cages can also have profound effects. Furthermore, repeated exposure to stress (again, even in adulthood) is important in the aetiology of at least one abnormal behaviour – SIB; while other factors, as yet unknown, must underlie the low levels of stereotypy evident even in zoo primates that are socially raised, socially housed and have relatively large, enriched enclosures. Second, the authors describe the physiological processes that co-occur with SIB. Individuals with this behaviour appear more stress-responsive, and typically have higher levels of corticotrophin-releasing factor (CRF) plus lower levels of enkephalin opioids in their cerebrospinal fluid (CSF). Furthermore, their bouts of SIB are often precipitated by acute stressors. However, performing a bout correlates with a fall in heart rate, and a decrease in stressor- and adrenocorticotropic hormone (ACTH)-induced adrenal responses. The authors suggest that this is because SIB is often directed to sites associated with acupuncture analgesia, and so might stimulate endogenous opioid release. We may argue as to whether SIBs are stereotypies, but they are certainly the sort of persistent, repeated, abnormal behaviour that anyone interested in 'true' stereotypies cannot afford to ignore, and it should be fascinating to follow these researchers' future work on this behaviour.

GM

6.1. Introduction

One of the significant challenges of maintaining non-human primates in captivity is the development of bizarre and unusual patterns of behaviour in some individuals. These range from stereotypies (such as pacing, rocking, self-mouthing, eye-covering) to excessive self-grooming and more serious disorders such as self-inflicted wounding. This chapter reviews the possible causes of, and treatments for, primate abnormal behaviour and provides suggestions for future research.

Like the taxa surveyed in the preceding chapters, the prevailing view of how monkeys and apes acquire abnormal behaviour focuses on the captive environment. Primate abnormal behaviour is thought to be the outcome of early rearing practices (e.g. rearing infants without mothers or companions), as well as of later housing situations (e.g. moving normally reared adolescents or adults to individual cages). Thus, social aspects of the housing environment seem to be particularly crucial. Stage of development must also be added to the equation. Because growing organisms undergo radical changes in morphology, physiology and behaviour across development, the same impoverished environment impacts monkeys differently depending on whether they are infants, juveniles, adolescents or adults.

This view is based largely on the early experience research of Harlow and colleagues with infant rhesus monkeys (Harlow and Harlow, 1962, 1965), and on studies of normally-reared adolescent and adult monkeys subjected to individual cage housing (Bayne *et al.*, 1992). As we will see, two basic findings emerge from this work, both emphasizing the importance of social interaction. The first is that infants reared from birth in socially restricted environments are particularly vulnerable. The second is that even normally-reared monkeys can develop stereotypy in response to separation from companions later in their lives. This distinction between impoverished early rearing and exposure to impoverishment after normal rearing is important. The term 'social deprivation' is commonly used to characterize rearing animals from birth in an impoverished environment, whereas the term 'social separation' is used for animals that are permitted to develop social ties and then subject to removal from these companions (Gilmer and McKinney, 2003). Both conditions can yield abnormal behaviour; however, it appears that the underlying mechanisms may be somewhat different.

Environmental characteristics are not the whole story: genetic factors are likely to make monkeys more or less vulnerable to the effects of differing environments. For example, self-injurious behaviour (SIB) is much more common in rhesus monkeys housed individually than in monkeys housed in social groups, but only 10–15% of individually housed monkeys develop SIB (Novak, 2003), suggesting that some animals have genetic and/or experiential risk factors making them vulnerable to the effects of this type of housing (see Chapters 5 and 7 for similar issues in laboratory rodents).

Ultimately, the causes of abnormal behaviour will be found in some interaction of environmental factors (including prenatal, maternal and physical) and genetic factors, which play out differently across development. Therefore we must study both the consequences of captive environments and the risk factors that predispose some monkeys to develop abnormal behaviour under these conditions. However, before we develop these issues, we begin with some descriptions of primate abnormal behaviour.

6.2. The Nature of Laboratory Primate Abnormal Behaviour

Abnormal behaviour in non-human primates often takes the form of stereotypic behaviour, defined (as in earlier chapters) as repetitive, highly ritualized motor actions not serving any apparent purpose. The word 'apparent' is important because it acknowledges that research may ultimately reveal a purpose for various types of stereotypies, an issue we return to below. Those who study primates also make a distinction between stereotypy and other kinds of abnormal behaviour that can lead to serious injury (e.g. self-mutilation or head-banging). These latter activities are considered pathological because of their potential for self-harm (Bayne and Novak, 1998).

In primates, abnormal behaviour is often very idiosyncratic (Berkson, 1968; Ridley and Baker, 1982; Bayne and Novak, 1998), with stereotypic behaviour taking many different forms both across and within species (Walsh *et al.*, 1982; Bayne *et al.*, 1992). At least two classification schemes have been developed to characterize this variability. The first scheme emphasizes form, differentiating whole-body, gross motor actions from fine motor movements. Whole-body stereotypies involve repetitive movements through space and time, e.g. pacing, somersaulting and bouncing. Fine motor stereotypies consist of activities directed to the animal's own body, e.g. digit-sucking, eye-saluting, hair-pulling, and self-biting (Berkson, 1968; Bayne *et al.*, 1992). (See Figure 1 in our section of the book's website.)

However, because the intensity of primate abnormal behaviour can vary greatly, we developed a second classification scheme based on symptom severity (Bayne and Novak, 1998). Severity is assessed in terms of the frequency and intensity, and behaviour is divided into two categories: non-pathological and pathological. Non-pathological stereotypies often involve both whole-body motions and some of the fine motor activities previously described. However, any stereotypy can become pathological if its frequency of occurrence disrupts basic biological functions (e.g. parental behaviour) or if it replaces other species-typical behaviours such as grooming or play. This category also includes activities involving self-injury. SIB, such as head-banging or self-biting, is observed in a small percentage of captive non-human primates. Unlike the stereotypic patterns described previously, SIB is dangerous and can result in substantial tissue damage and increased risk of infection (Bayne and Novak, 1998). Note that our definition of pathological is based on the consequences, rather than causes, of stereotypies (see Box 1.3, Box 10.2, and Chapter 11 for further discussion and alternative definitions).

In laboratory primates, most or all of these forms of stereotypy, however they are classified, seem to occur in response to a diverse array of acutely stressful events (see Box 8.3). Some also appear to have stress-reducing consequences, an issue we return to later in this chapter.

6.3. The Effects of Early Impoverished Rearing on Abnormal Behaviour and Physiology of Primates

For nearly two decades (1958–1975), Harry Harlow conducted pioneering research on the role of early social experience in the development of species-typical social behaviour in rhesus monkeys. His work came at a time when species-typical behaviour was thought instinctive, requiring little in the way of experience for its expression. It was designed to examine two general questions, one related to developmental plasticity (are certain social experiences necessary for normative development?), the other to reversibility (could the deleterious effects of impoverished early rearing environments be reversed?). Harlow established a strong connection between early social experience and adequate social development. He also showed that certain kinds of socially restricted environments were associated with the development of bizarre and atypical behaviour, including stereotypic behaviour. These behaviour patterns could only be reversed partially, and through highly specialized interventions.

Harlow's basic rearing paradigm consisted of four phases. Infants were separated from their mothers shortly after birth, and then reared in incubators in a nursery for a month where they were hand-fed by humans. They were subsequently exposed to a particular rearing environment (described below) for a period of time ranging in length from 3 to 12 months (typically 6 months). After treatment, monkeys were typically housed in social groups with other like-reared monkeys or in some cases housed individually because of aggressiveness (see isolation-rearing). Thus, even post-treatment environments should not be thought of as fully normalized. These monkeys were compared directly or indirectly with normally reared animals (i.e. animals reared with their mothers and peers). Such comparisons were made both during the period of rearing itself (i.e. during infancy) and during later stages of development.

6.3.1. Isolation-rearing

In this treatment, monkeys were removed from their mothers shortly after birth and reared in a nursery for 30 days as described above. After this period, they were placed alone in small cages where they could not see, hear or physically interact with members of their species for the first 6 months of life. In a variant called 'partial isolation', infants could see and hear other monkeys but not physically interact with them. However, there were few outcome differences between total and partial isolates (Cross and Harlow, 1965), and for this review, they are considered together.

6.3.1.1. Behavioural effects of isolation-rearing

Isolate-reared monkeys developed a constellation of characteristics that became known as the 'isolation syndrome' (Capitanio, 1986; Cross and Harlow, 1965; Harlow and Harlow, 1962, 1965; Sackett, 1965). This syndrome featured abnormal behaviour along with heightened withdrawal from novelty, inadequate motor coordination and deficits in social interaction, e.g. socially inappropriate aggression (Mason, 1968). When isolated monkeys were tested directly with normally-reared age mates shortly after their removal from isolation, such social outcomes remained the same regardless of the length of isolation-rearing (6 months versus 12 months). These deficits were not simply results of barren, sensory-poor physical environments: monkeys reared in total isolation in sensory-rich environments containing toys and manipulanda, and exposed to pictures and videotapes, did not show reduced abnormal behaviour nor improved social behaviour (Sackett et al., 1982).

Isolate-reared monkeys showed whole-body stereotypies including rocking and bouncing as well as numerous self-directed stereotypies: self-clasping, digit-sucking and eye-covering. Furthermore, the amount of stereotypic behaviour was very high, ranging from 35–60% of an observation session, and dominating the behavioural repertoire. It was thus pathological in nature. Some of the isolate-reared monkeys' stereotypic behaviour appeared to involve the redirection of normal behaviour to their own body (e.g. self-clasping instead of clasping a mother). This hypothesis was tested by giving infants access to a warm, terry-cloth mother during the period of isolation. Rather than clasp themselves, infants clasped their surrogate mothers and used them as a base of operations when exploring novel stimuli (Harlow and Suomi, 1970; Harlow and Zimmerman, 1959; see Figure 2 on our section of the book's website). A further decrease in the development of stereotypies, particularly rocking, was achieved by adding motion to the surrogate (Mason and Berkson, 1975). However, the addition of an inanimate surrogate mother did not improve later social behaviour, nor eliminate all forms of abnormal behaviour.

It proved very difficult to normalize the post-treatment environment of isolate-reared monkeys. The direct testing of isolates with normal age mates had to be discontinued, because normally-reared monkeys typically attacked isolate-reared monkeys when they encountered them. Attempts were then made to house isolates together in social groups, but a marked increase in aggressiveness at the time of puberty necessitated placing most of the isolates in individual cages, where they remained except for brief social interactions. All the subsequent effects of isolation-rearing are described below. Still unclear is the extent to which the effects of early isolation-rearing were compounded by later individual cage-housing or stressful social interactions.

Stereotypic behaviour changed with age. By the time the isolate-reared monkeys reached 3 years of age, digit-sucking and self-clasping decreased but were replaced with other kinds of stereotypies such as somersaults, head bobs, unusual limb manipulations (e.g. leg behind neck), and in some cases, SIB (Fittinghoff et al., 1974; Mitchell et al., 1966; Mitchell, 1968; Sackett, 1967). Males were more likely to develop SIB than females (Sackett, 1974).

Isolation-rearing was also associated with deficits in learning and cognition. In early studies, isolate-reared monkeys did not differ from controls on a standard learning test battery (Harlow et al., 1969). However, in subsequent studies testing them as adults, isolate-reared monkeys showed deficits in both simple and complex forms of learning. For example, in a simple lever-pressing task to obtain food, isolate-reared animals learned as readily as control animals, but in extinction continued to press at a high rate for hundreds of unrewarded responses (Gluck and Sackett, 1976) (see also Chapter 5). Unlike controls, isolate-reared monkeys also failed to show blocking in a conditioned blocking paradigm (Beauchamp *et al.*, 1991). In this paradigm, a conditioning stimulus such as a tone is paired with an unconditioned stimulus. Eventually, all monkeys learn this conditioned response. A second stimulus (a light) was added to the tone. Monkeys were then tested for their response to the light alone. Controls showed the typical blocking response, with little or no response to the light, whereas isolates showed a conditioned response to the light, demonstrating increased sensitivity to environmental events. Isolate-reared monkeys also had greater difficulty than controls in solving 'oddity problems' (Gluck *et al.*, 1973), i.e. tasks in which monkeys are presented with three objects, two of which are identical, and must select the odd object in order to receive reward.

6.3.1.2. Physiological effects of isolation-rearing

Major changes in central nervous system (CNS) function accompanied the behavioural effects. Kraemer and colleagues conducted extensive neurochemical studies of isolate-reared monkeys, focusing on the major monoamine neurotransmitters, serotonin (5-HT), and two catecholamines – norepinephrine (NE; also known as noradrenaline) and dopamine (DA). One series of longitudinal studies examined alterations in monoamine functioning and their possible relationship to abnormal behaviour. At 8–15 months of age (thence already removed from the isolation condition), isolate-reared monkeys showed significantly higher cerebrospinal fluid (CSF) levels of 5-hydroxyindoleacetic acid (5-HIAA), the major metabolite of 5-HT in the brain, than socially-reared controls (Kraemer *et al.*, 1989). They also showed lower levels of CSF NE, suggesting an imbalance in central monoaminergic functioning. At 3 years of age, abnormal behaviour was significantly reduced in these animals by treatment with the 5-HT$_{1A}$ receptor partial agonist buspirone but not with BW A616U, an inhibitor of monoamine oxidase A (MAO-A) (Kraemer and Clarke, 1990). Furthermore, at 5 years of age, the central serotonergic system of the isolation-reared monkeys failed to respond to a number of drugs that enhanced or suppressed serotonergic activity in socially reared animals (Kraemer *et al.*, 1997). Although these findings are not easily interpreted, together they point to long-lasting disturbances of serotonergic function in isolate-reared monkeys apparently related to the expression of abnormal behaviour.

Behavioural responses to specific catecholaminergic challenges in isolate-reared monkeys were also examined. One study compared four different rearing conditions: total social isolation for the first 11 months of life, mother-rearing for 6 months followed by 1 month of isolation, mother-rearing for 6 months with no isolation, and peer-rearing (see Section 6.3.2) (Kraemer *et al.*, 1984). Following the isolation period, all animals were housed socially with like-reared age-mates until 30–36 months of age when they were challenged with varying doses of the catecholamine agonist, d-amphetamine. Amphetamine is well-known to provoke stereotyped behaviours (see e.g. Box 7.2), and this is what occurred in the non-isolated control groups. However, the two isolate groups unexpectedly failed to show amphetamine-induced stereotypy, but rather displayed high levels of agonistic behaviour. Repeated treatment with 1 or 2 mg/kg of amphetamine led to postural collapse and even convulsions in the 11-month isolated group, effects not observed in the controls. Although the 11-month isolates could not be tested neurochemically, the 1-month isolated group showed an enhanced CSF NE response compared to the combined control groups. In a later study by Lewis *et al.* (1990), monkeys reared in total isolation for 9 months were compared with socially reared (mother and peers) animals in terms of their behavioural responses to a challenge with the DA agonist apomorphine. Monkeys in both groups had all been individually housed following their initial rearing, until the study began when the animals were at least 14 years old. In this case, both groups displayed a dose-dependent increase in apomorphine-induced stereotypies, but the isolates showed significantly more whole-body stereotypy than the controls at a dose of 0.3 mg/kg. Together, these studies suggest that early isolation-rearing leads to a long-lasting enhancement of catecholaminergic function, most evident following a pharmacologic challenge. Such an hypothesis is consistent with other studies reporting significant neuroanatomical and physiological changes in the basal ganglia of social isolates (Martin *et al.*, 1991), as well as a reduction in abnormal behaviour following treatment of isolates with the DA antagonist chlorpromazine (McKinney *et al.*, 1973). However, the manifestation of the isolation-rearing-induced catecholaminergic changes seems to depend on whether the animals are subsequently housed individually or in social groups.

Isolation-rearing is presumably highly stressful, and thus could affect the hypothalamic-pituitary-adrenocortical (HPA) axis, a key component of the physiological stress-response system. Meyer and Bowman (1972) determined baseline plasma cortisol levels and the cortisol responses to adrenocorticotropic hormone (ACTH) administration and chair-restraint stress in 9-month total isolates, 9-month partial isolates, and feral-reared animals at approximately 4 years of age. No differences were found. A subsequent study by Sackett *et al.* (1973) in which isolates and peer-reared monkeys were studied at 19 months of age found elevated baseline cortisol levels in the isolates. These results raise the possibility that isolation-rearing may cause an increase in baseline HPA activity that persists for some time after emergence from isolation but eventually wanes.

The studies discussed above suggest that isolation-rearing influences the HPA stress response system as well as all three major monoamine systems. Although these changes may contribute to the abnormal behaviours also observed, several limitations must be noted. First, few monkeys were subjected to isolation-rearing, and they exhibited wide individual differences in abnormal behaviour. This has made it difficult to determine the involvement of particular abnormal behaviours (e.g. stereotypies) in disrupting normal species-typical behaviour patterns such as maternal behaviour, and also to correlate the physiological and behavioural effects observed. Thus existing data do not permit clear mechanistic explanations to be formulated; for example, although it may be tempting to conclude that the monoaminergic abnormalities discussed above underlie the abnormal behaviours exhibited by social isolates, this relationship is almost entirely correlational.

6.3.2. Peer-rearing

Could exposure to peers eliminate the development of abnormal and stereotypic behaviour, and produce normal social development? To find out, a peer-rearing paradigm was investigated in which monkeys were separated from their mothers at birth, reared in a nursery for the first 30 days, and then placed with other like-reared infants for the next 6–12 months. Monkeys continued to remain in these groups after the first year of life.

6.3.2.1. Behavioural effects of peer-rearing

In marked contrast to early isolation-rearing, peer-reared infants displayed nearly normal social behaviour and substantially lower levels of stereotypic behaviour (Chamove, 1973). Peer-reared monkeys displayed stereotypic behaviour about 4–15% of the time, mostly digit-sucking. However, the development of appropriate social behaviour was somewhat delayed compared to normals (Chamove *et al.*, 1973). This delay has been attributed to the excessive clinging that occurs when infants are reared in this manner. Other impairments were noted. For example, in a study of males, peer-reared juveniles were less likely to show affiliative contact (e.g. grooming), less able to have their stress levels alleviated by a companion, and more likely to show stereotypic behaviour than mother-peer-reared males (Winslow *et al.*, 2003). Another key characteristic of peer-rearing was heightened fearfulness: minor events, insignificant to most normally-reared monkeys, such as a caretaker walking into a colony room, elicited fearful vocalizations and prolonged clinging to one another, a pattern frequently persisting into adulthood (see image on website). In addition, peer-reared monkeys showed more severe reactions to social separation than normally reared monkeys (Higley *et al.*, 1991).

Subsequent retrospective studies revealed the continued vulnerability of peer-reared monkeys (Lutz *et al.*, 2003a) and apes (Nash *et al.*, 1999) to display abnormal behaviour into adulthood, particularly if individually housed. In some cases, peer-reared monkeys also showed decreased parental competence (Schapiro *et al.*, 1994). Survey data showed that chimpanzees performed a similar range of abnormal behaviours to monkeys, including whole-body stereotypies and self-directed stereotypies such as rocking and digit sucking (Nash *et al.*, 1999). Although the presence of abnormal behaviour *per se* was not associated with reproductive impairments or inadequate maternal behaviour (Fritz *et al.*, 1992), nursery-peer-reared chimpanzees were less maternally competent than normally reared chimpanzees (King and Mellen, 1994). Note that recent studies have also shown the long-term effects of peer-rearing in humans (see Box 6.1 on Romanian orphans, although the conditions involved here might be more akin to the 'partial isolation' of Section 6.3.1).

6.3.2.2. Physiological effects of peer-rearing

Although peer-rearing resulted in more 'normal' behaviour than isolation-rearing, this condition was none the less associated with disruptions in neurotransmitter (particularly monoamine) activity, as well as in the stress response system. Higley *et al.* (1992) examined the effects of rearing condition on the metabolites of 5-HT, DA, and NE in the CSF. Rhesus monkeys were reared with their mothers or with peers for the first 6 months of life, and CSF samples were obtained at the end of the rearing manipulation, and again at 18 months. Compared with the mother-reared group, peer-reared monkeys showed increased turnover (activity) of the noradrenergic system at both ages as indicated by higher CSF levels of the NE metabolite 3-methoxy-4-hydroxyphenylglycol (MHPG) (unlike the earlier work of Kraemer and colleagues, NE itself was not investigated in this study). Some involvement of the serotonergic system was also noted, with peer-reared females showing lower levels of 5-hydroxyindole-acetic acid (5-HIAA) but peer-reared males showing higher levels of 5-HIAA compared to their mother-reared counterparts. No differences were detected for homovanillic acid (HVA), the major metabolite of DA. A subsequent report by Clarke *et al.* (1996) confirmed the presence of higher levels of the NE-metabolite MHPG in the CSF of peer-reared monkeys. However, these authors also found that peer-reared monkeys in the latter half of the first year of life showed significantly lower CSF concentrations of both HVA and dihydroxyphenylacetic acid (DOPAC), another DA metabolite. Thus, monoamine studies suggest that peer-rearing of rhesus monkeys led to a transient decrease in central dopaminergic activity, a longer-lasting increase in central noradrenergic activity, and gender-specific changes in serotonergic activity. The relationships between these neurotransmitter alterations and the behavioural effects of peer-rearing are as yet unknown.

Box 6.1. Deprivation Stereotypies in Human Children: the Case of the Romanian Orphans

G. Mason

Following the fall of Ceaucescu in 1989, tens of thousands of Romanian children were discovered in so-called 'orphanages'. Typically abandoned by impoverished families, these children lived in bare rooms, often spending 20 h a day confined to separate cribs. They were largely silent, and adult contact was minimal. They were thus highly socially-deprived, and exposed to nutritional, physical and sensory deprivation too (e.g. Fisher *et al.*, 1997; Gunnar, 1999; Beckett *et al.*, 2002). Stereotypies were prevalent, infants rocking back and forth on hands and knees, or standing, holding their cribs' railings, shifting from foot to foot (e.g. Carlson and Earls, 1997; Fisher *et al.*, 1997).

1992–1993 saw thousands of these children adopted internationally, several hundred then being studied by developmental psychologists (e.g. references above plus O'Connor *et al.*, 2000; Chugani *et al.*, 2001; Rutter *et al.*, 2004). Adoptees showed incredible improvements in health, behaviour and cognition, yet often also profound long-term deficits including general cognitive impairments, 'executive problems' like impulsivity (*cf.* Chapter 5, this volume), and autistic-like symptoms including circumscribed interests and – our focus here – stereotypies: one of the commonest problems reported by adopting parents. Rocking was most prevalent, but they also performed hand stereotypies (often while staring at their fingers), and self-injurious behaviours (SIB) like head-banging and eye-poking (Benoit *et al.*, 1996; Fisher *et al.*, 1997; Beckett *et al.*, 2002; MacLean, 2003). Half or more of the children stereotyped, even a year post-adoption (Fisher *et al.*, 1997; Beckett *et al.*, 2002; MacLean, 2003), with SIB evident in around a quarter (Beckett *et al.*, 2002). They often rocked before falling asleep, when bored, or when anxious; while SIB was more specifically linked with acute stress (Beckett *et al.*, 2002; *cf.* Novak *et al.*, in this chapter and Box 8.3, Chapter 8). Both abnormal behaviours were more prevalent the longer the child had been institutionalized and the older she/he was at adoption (Beckett *et al.*, 2002). Within a few years in their new homes, half or more then ceased to stereotype, the remainder showing reductions (Fisher *et al.*, 1997). Stereotypy-persistence was most marked the longer the child had been institutionalized, and the less time she/he had with the new family (Fisher *et al.*, 1997; Beckett *et al.*, 2002). Persistent stereotypers also had lower IQs than other adoptees, even when length of time in the orphanage was controlled for, and typically (but not always) other 'quasi-autistic' symptoms (Beckett *et al.*, 2002).

In these age-matched cross-sectional studies, length of time in the institution was positively correlated with age at adoption (since most were institutionalized as small babies), and negatively correlated with amount of time with the new family. Thus from these papers alone, one cannot disentangle the relative roles of the ages over which the children were deprived versus the lengths of their periods of deprivation and/or of exposure to normal homes. It is also unknown how environmental variation, both within orphanages and post-adoption, affected later stereotypy; nor how stereotypy relates to other variables such as the reduced cranium size common in these children (Rutter *et al.*, 2004), their altered forebrain activity and executive dysfunction (Chugani *et al.*, 2001), and their elevated evening cortisol levels (*cf.* Gunnar, 1999). Thus what caused some children but not others to develop abnormal behaviours, and why some stereotypies then persisted post-adoption, is not fully understood. However, these infants highlight both the potentially profound long-term effects of early deprivation (*cf.* this chapter), and the impressive recovery sometimes possible in good environments.

Because heightened fearfulness is a key characteristic of peer-rearing, a number of studies have focused on the HPA axis. Unfortunately, no consistent pattern has emerged from this effort. In an initial study, peer-reared monkeys had higher concentrations of cortisol than mother-reared controls under baseline conditions (Higley *et al.*, 1992). However, opposite results were obtained by Clarke (1993). Furthermore, mother-reared animals responded to stress with larger increases in ACTH and cortisol than peer-reared monkeys (Clarke, 1993). In a subsequent study designed to examine infants longitudinally under several different conditions, mother-reared infants displayed higher concentrations of cortisol than peer-reared monkeys during the first 2 months of life but showed no difference in their response to 30-minute separation periods (Shannon *et al.*, 1998). In yet another study, neither baseline levels of cortisol or stress levels varied by rearing condition (Winslow *et al.*, 2003). This variability across studies may be related to procedural variations in the rearing paradigm and the methodology used to obtain blood samples for cortisol assay.

Several general conclusions can be drawn from the work on peer-rearing. Unlike isolation-rearing, peer-rearing did not produce major deficits in social behaviour nor excessively abnormal behaviour. However, peer-reared monkeys were none the less more fearful and showed higher levels of stereotypic behaviour than normally reared monkeys. Recent studies also revealed that affiliative processes were altered in peer-reared monkeys (e.g. social buffering effects). These behavioural differences have been associated with alterations in monoamine, neuroendocrine, and immune function. However, as in the case of isolate-reared animals, it is difficult at this time to make causal connections between any of these physiological systems and the behaviour patterns seen in peer-reared monkeys.

6.3.3. Surrogate with limited peer-rearing

Monkeys reared with peers developed a strong, clinging attachment to one another, similar to that seen in young infants with their mothers, which may have interfered with the development of play behaviour. The surrogate-peer-rearing condition was instituted to overcome the problem of infants serving in the dual role as a mother figure and as playmate. Surrogate-peer-reared monkeys were reared with continuous exposure to an inanimate 'terry cloth'-covered mother and were given brief daily exposure to similarly reared peers. Monkeys were housed in individual cages with their surrogate mother and then hand-carried daily to a playpen environment. Depending on the study, the exposure to peers ranged from 30 min to 2 h a day (Hansen, 1966; Meyer *et al.*, 1975). The brief exposure to peers was designed to mimic naturalistic early mother–infant interaction, in which infants spend most of their time with their mothers and only interact with other infants for brief periods. The brief exposure was also expected to facilitate play behaviour with peers and reduce the risk of developing a primary attachment to peers. Data were typically collected both during the treatment and afterwards when the surrogate-peer-reared monkeys were housed with each other continuously. The behaviour of surrogate-peer-reared animals was compared with the behaviour of animals reared with mothers and peers.

6.3.3.1. Behavioural effects of surrogate plus limited peer-rearing

In contrast to peer-rearing, the surrogate-peer-rearing regimen resulted in the development of normal social behaviour (Hansen, 1966; Ruppenthal *et al.*, 1991). Minor differences in vocalization such as 'geckering' and cooing between surrogate-peer-reared and normally reared monkeys disappeared after the first few months of life. Some forms of stereotypic behaviour were observed (mostly digit-sucking and some rocking against the surrogate surface) occurring about 5–10% of the time, but these patterns declined across age so that surrogate-peer-reared monkeys behaved like normally-reared monkeys at 1 year of age (Hansen, 1966). Surrogate-peer-reared animals continued to develop socially, showing adequate skills in grooming, reproduction and parental care with little expression of abnormal behaviour (which occurred about 5% of the time, providing that the animals continued to be socially housed) (Novak *et al.*, 1992; Sackett *et al.*, 2002).

6.3.3.2. Physiological effects of surrogate plus limited peer-rearing

There are only a few studies of the effect of surrogate-peer-rearing on CNS function. To date, the emphasis has been on the HPA axis. Converging evidence suggests that surrogate-peer-reared monkeys have significantly lower concentrations of circulating cortisol than mother-peer-reared monkeys. This difference was detected during the first month of life and again in juveniles ranging in age from 1–3 years (Davenport *et al.*, 2003). This latter finding is of interest because the surrogate-peer-reared juveniles in this case were housed in a large mixed rearing group containing mother-peer-reared and peer-reared animals. Thus, the differences in cortisol observed in infancy when the rearing groups were not mixed persisted even when the surrogate-peer-reared monkeys were exposed to other rearing conditions. Surrogate-peer-reared monkeys also responded significantly less than mother-peer-reared monkeys to the stress of brief social separation (Shannon *et al.*, 1998).

The information derived from this rearing condition suggests that infants can acquire all species-typical social behaviours under conditions in which their only early exposure to conspecifics is with naïve infants. Apart from the HPA system findings, we know little about how surrogate-peer-rearing may influence physiological functioning. The finding of significantly lower cortisol levels in these monkeys is interesting with respect to similar findings in rodents subjected to early handling manipulations. The surrogate-peer-rearing condition is the only one in which infant monkeys received extensive human handling during the daily transfers from their individual cage to the playroom environment. The possible significance of this variable remains to be determined.

6.3.4. Reversibility of the effects of early social deprivation

As a part of his research, Harlow was interested in the extent to which negative consequences of adverse early rearing conditions could be reversed with treatment. Were the deficits permanent due to persisting abnormalities in the CNS, or could previously isolated monkeys reduce their stereotyped behaviours and 'acquire' more appropriate social behaviours with the right experiences (i.e. therapy) after the isolation period was over?

His therapeutic interventions involved social stimulation and most failed. Isolate-reared monkeys did not show less stereotypy and more social behaviour when exposed to socially sophisticated animals of the same age (Harlow *et al.*, 1964; Rowland, 1964), nor did they benefit from extensive adaptation to social environments (Clark, 1969). It was only when the isolates were exposed to younger monkeys that substantial improvements were noted (Novak and Harlow, 1975; Suomi and Harlow, 1972; Suomi *et al.*, 1974) (Fig. 6.1). This intervention paired socially unsophisticated isolates with monkeys matched for developmental stage rather than chronological age. During the several months of exposure to younger monkeys, isolates began first to tolerate social contact and then to reciprocate with play. Associated with these social changes were substantial decreases in stereotypic behaviour. Before this intervention, the isolates spent nearly 50% of their time engaged in self-clasping, rocking, and huddling, a value that was reduced to less than 10% after 18 weeks of treatment. In a follow-up study (Novak, 1979), isolates exhibited more complex forms of social behaviour such as grooming. However, other patterns of social behaviour failed to develop (e.g. double-foot-clasp mount in males), and behaviour was not always context appropriate. Some forms of stereotypic behaviour persisted (e.g. self-clasping at low levels) and other forms of stereotypic behaviour developed (e.g. back-flipping and eye-covering). These results suggested that with exposure to younger monkeys, isolate-reared monkeys could acquire some social skills later in development. However, there were limitations to this treatment, and stereotypy was not abolished. Although physiological assessments of 'rehabilitated' isolate-reared monkeys would be of great interest, no such assessments have been performed.

Interestingly, studies of breeding female isolates also suggested that they might be able to acquire social responses through experience. Although isolate-reared females showed highly abnormal maternal behaviour (e.g. abusiveness or indifference) toward their first infant (Arling and

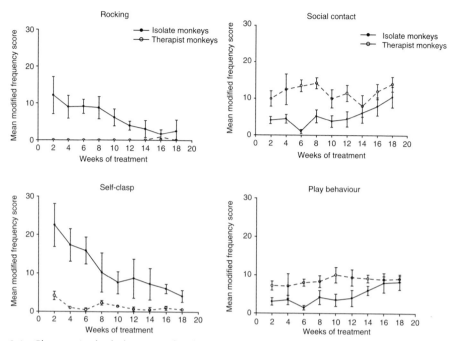

Fig. 6.1. Changes in the behaviour of isolate-reared monkeys over the course of 18 weeks of social experience with young 'therapist' monkeys. The isolates were 16 months of age at the beginning of treatment and the therapist monkeys were 7 months of age. Treatment resulted in substantial reductions in abnormal behaviour (self-clasp and rocking) and increases in social behaviour (social contact and play). (From Novak and Harlow, 1975.)

Harlow, 1967), substantial improvements were noted with exposure to their second infant (Ruppenthal *et al.*, 1976).

Overall, Harlow and colleagues' attempts at therapy are consistent with the neurophysiological studies reported earlier, which showed some lasting effects of early isolation. As we will see in Section 6.5, they are also consistent with findings in zoos that hand-reared primates are more stereotypic in adulthood despite an intervening period of normal housing.

6.3.5. Summary of the effects of early social deprivation

The work by Harlow and others established that impoverished early social environments could both affect the development of normal behaviour and produce pathological behaviour. Only the most severe forms of impoverishment (i.e. isolation-rearing) resulted in social incompetence, profoundly pathological behaviour, and major disruptions in neurotransmitter systems. In contrast, monkeys reared with naïve infants showed high levels of social behaviour, markedly reduced levels of abnormal behaviour, and some changes in CNS function (see Table 6.1). Even some of the deleterious behavioural effects of isolation-rearing could be reversed by younger

Table 6.1. Behavioural and physiological effects of rearing monkeys from birth either alone for the first 6 months of life (isolation), continuously with peers (peer-only) or with an inanimate surrogate and brief daily peer experience followed by group housing at 1 year of age (surrogate-peer).

Infant rearing condition	Stereotypic behaviour as % of the repertoire	Emotional behaviour	Social behaviour	HPA axis	Monoamine metabolites
Isolation	35–60	Extreme withdrawal in infancy; hyper-aggressive as adults	Show little, if any, appropriate social behaviour	Elevated levels of cortisol in juveniles	Dysregulation of serotonergic system; enhancement of catecholaminergic system
Peer-only	4–15	Heightened fearfulness and clinging throughout life	Show most species-typical patterns of behaviour	No consistent pattern of effects	Transient decreases in dopaminergic activity; increased noradrenergic activity Alterations in the serotonergic system
Surrogate-peer	5	Mild differences in vocalization during infancy; possibly more aggressive as adults	Display virtually all species-typical patterns of behaviour	Lower levels of cortisol in infants and juveniles	No data available

monkey 'therapy.' However, early experience effects could be long-lasting, and Harlow's original work inspired subsequent primate work on this important topic. These primate data also provide a useful context for understanding the potentially long-term effects of early social deprivation in non-primate species (see Box 6.2). In the next section, we look at how some of these early treatments affect monkeys that have already had some social experience.

6.4. The Effects of Individual Cage-housing on Abnormal Behaviour and Physiology of Normally Reared Primates

Although the early rearing environment can be an important predictor of behavioural abnormality in rhesus monkeys and other primates, normally (i.e. socially) reared monkeys can also develop abnormal behaviour if they are removed from conspecifics and placed in individual cages at some point in their life ('social separation'). Individual cage-housing of laboratory primates typically occurs in two contexts: when necessitated by research protocol or because of hyper-aggressiveness in particular animals. Individual cage-housing of normally-reared monkeys has been shown to produce both stereotypic behaviour and more serious kinds of abnormal behaviour (Lutz *et al.*, 2003b). The kinds and extents of abnormal behaviour depend, in part, on the age at which separation occurs. Removal from the primary attachment figure, usually the mother, during the first year of life is thought to be different than removal from others after the first year of life. Consequently, we will differentiate social separation occurring during the first year of life from that occurring at a later point in time.

6.4.1. Separation from attachment figures in infancy

6.4.1.1. Brief separations

Considerable early research demonstrated the powerful effects of mother–infant separation. Infants separated even briefly from their mothers during the first 6 months of life responded initially with heightened activity termed 'protest'. Within 24–48 h, the protest waned and was replaced with depressive-like symptoms (Hinde *et al.*, 1966; Kaufman and Rosenblum, 1967). Most of the early studies involved brief 1–2 week separations from the mother (for a review and theoretical perspective, see Mineka and Suomi, 1978). Abnormal behaviour in infants was confined primarily to the separation period and consisted mostly of pacing and vocalizations in the first 24 h followed by huddling and also elevated cortisol concentrations in the depressive phase (Levine and Weiner, 1988; Smotherman *et al.*, 1979). When reunited with their mother, infants spent more time

Box 6.2. Maternal Deprivation and Stereotypy in Animals other than Primates

G. Mason

Social deprivation profoundly affects primate abnormal behaviour (*cf.* this chapter, and recent complementary reviews by Sackett *et al.*, 1999; Sanchez *et al.*, 2001; Gilmer and McKinney, 2003). But what about other taxa? Most animals kept by humans are removed from their mothers earlier than would happen naturally: for instance, horses, farmed mink and laboratory mice are separated from their mothers before natural dispersal age; piglets are removed while still dependent on milk; and more extreme still, dairy calves and hatchery-raised poultry have respectively minimal maternal contact (separated on day 1) and none at all (e.g. Mason, 1995; Roden and Wechsler, 1998; Latham and Mason, 2004; and *cf.* Chapters 2 and 4). So could standard husbandry practices be predisposing animals to abnormal behaviour?

In many instances, this early separation promotes the rapid emergence of stereotypies, whose 'source behaviours' (see Box 1.1, Chapter 1) appear to be frustrated suckling or escape attempts. Dairy calves thus show intense non-nutritive sucking, which seems to have similar physiological effects (e.g. on insulin) to normal teat-sucking following milk let-down (de Passillé *et al.*, 1993); early weaned kittens 'wool-suck' (e.g. Morris, 1987); and piglets rub their snouts on the floor and 'belly-nose' the flanks of their fellow-piglets – behaviours typically more frequent the younger the piglets were weaned (e.g. Bøe, 1997; Worobec *et al.*, 1999; Widowski *et al.*, 2003). In mice, gerbils and black rats, back-flips and bar-mouthing similarly develop rapidly when young are moved from the natal cage (see Chapter 4; Callard *et al.*, 2000; Waiblinger and König, 2004). In mice (at least) these seem to begin as escape attempts; and in both mice and gerbils, individuals with the youngest developmental ages at separation go on to develop the most frequent stereotypies. Pacing the enclosure can even occur in young pygmy hippopotamuses after removal from the mother – and very intensely, albeit transiently (Stroman and Slaughter, 1972; see also the transient pacing of briefly maternally-separated primate infants, this chapter).

However, such stereotypies can persist long after frustration should have waned. Thus mouse stereotypies do not decline once natural dispersal age passes, but instead persist or even increase (e.g. Latham, 2005), and more anecdotally, the same seems true for oral stereotypies in a subset of early-weaned cats, pigs and cattle (Fry *et al.*, 1981; Morris, 1987; T. Widowski, personal communication, Guelph, 2005). Indeed sometimes maternal deprivation has effects that are latent until young adulthood. Thus mink separated from their mothers around natural dispersal age (11 weeks) are no less active over the following 2 months than animals separated at 7 weeks, but when pacing and similar appear 3–4 months later, the late-separated animals start to differ, developing stereotypies that are both less frequent and more variable (Mason, 1992, 1996; see also Mason, 1995 and Jeppesen *et al.*, 2000). Likewise, poultry chicks reared with their mothers show no less feather-pecking at 2 months than chicks reared with peers alone (Roden and Wechsler, 1998), but once they are 3–7-months old pullets, they then emerge as less likely to be feather-peckers, and spend less time in the behaviour (Perré *et al.*, 2002).

Such effects are not deterministic – many factors other than maternal deprivation affect both the development and the continued performance of the stereotypies discussed here, including individual differences and the physical environment post-weaning (e.g. Mason, 1996; Bøe, 1997; Jeppesen *et al.*, 2000; Widowski *et al.*, 2003). However, these data do suggest that early maternal deprivation can have long-term effects in taxa other than primates. In species as diverse as mice, pigs and poultry, maternal deprivation also affects immediate stress levels (see Chapter 4), later anxiety and stress responsiveness (e.g. Adriani and Laviola, 2002; Perré *et al.*, 2002) and brain dopaminergic (Fry *et al.*, 1981; Sharman *et al.*, 1982; Adriani and Laviola, 2002) and serotoninergic (Sumner *et al.*, 2002) systems (*cf.* similar findings for primates) – although such data are somewhat patchy. Exactly how such effects are mediated thus looks a fascinating, practically important area for future research.

in proximity to her and showed lower levels of activity than non-separated controls (Spencer-Booth and Hinde, 1971). The reunion differences in activity, produced by one or two brief separations during the 30th week of life, were still present at 30 months of age (Spencer-Booth and Hinde, 1971); and monkeys that had experienced separation also showed heightened reactions to fearful stimuli (Young *et al.*, 1973). However, there were no lasting effects on stereotypic behaviour, which was confined primarily to the separation period.

6.4.1.2. Early permanent separations from the mother

Some research procedures require that infants be weaned from the mother before the normal weaning period and placed into individual cage-housing. We know little about the effects of such manipulations on abnormal behaviour in primates (though see Box 6.2). However, early weaning was associated with changes in brain morphology (reduced corpus callosum size), impairments in cognitive function (Sanchez *et al.*, 1998) and with reduced rates of reconciliation (Ljungberg and Westlund, 2000), suggesting that permanent removal from the mother very early in life can have substantial consequences for the offspring.

6.4.2. Separation from companions post-infancy

For juvenile, adolescent and adult monkeys, separation from companions is an important risk factor for the development of abnormal behaviour (Bayne *et al.*, 1992; Lutz *et al.*, 2003b). Rhesus monkeys placed into individual cages can develop depressive-like behaviour (Suomi *et al.*, 1975) and typically display both whole-body stereotypies (e.g. pacing and somersaulting) and self-directed stereotypies (e.g. eye-saluting and digit-sucking). Furthermore, about 10–12% of these monkeys exhibit pathological behaviour in the form of biting and wounding themselves (Novak, 2003). These effects can be observed in adult monkeys of different ages, although juveniles are thought to be at greater risk for developing these symptoms (Lutz *et al.*, 2003b).

The finding that separation from companions is a substantial risk factor for stereotypic and pathological behaviour has come from converging sources of information. One approach involved direct comparisons of individually- and socially-housed rhesus monkeys. Bayne *et al.* (1992) compared adult monkeys housed individually for more than a year with animals housed socially for 4 years. Individually-housed monkeys showed higher levels of repetitive locomotion, stereotypic behaviour (e.g. saluting), and self-directed behaviour (e.g. self-clasp, hair-pull, and self-bite) than monkeys housed in social groups.

A second approach was to compare the response of primates in social groups to removal and placement in individual cage-housing. In a study of chimpanzees housed socially before being moved into individual cages, stereotyped behaviours such as rocking, pacing, flipping and spinning, increased during the 5 weeks of individual housing in comparison to the week prior to single caging. However, self-directed behaviours such as self-injurious behaviour, self-orality and eye-saluting did not change from pre- to post-separation (Brent *et al.*, 1989). Thus in this study, individual cage-housing only increased whole-body stereotypies, a not-unexpected finding given that locomotor activity is constrained in this environment, so that active monkeys can only express activity through stereotypic patterns.

In the third approach, routine assessments of individually-housed monkeys were combined with demographic information, colony location and health records to discern important relationships. In a survey we conducted of 362 singly housed adult rhesus monkeys at the New England Primate Research Center, 321 animals (89%) displayed at least one abnormal behaviour with a mean of 2.3 different behaviours and a range of 1–8 behaviours (Lutz *et al.*, 2003b). Pacing was the most common stereotypic behaviour, which occurred in 78% of the population, a finding also observed in pigtailed macaques (Bellanca and Crockett, 2002). A substantially larger percentage of monkeys exclusively displayed whole-body stereotypies (48%) compared with those that exclusively exhibited self-directed stereotypies (4%). The two kinds of stereotypies co-occurred in 33% of the population. About 11% of the individually-housed monkeys developed self-injurious behaviour and had a veterinary record for self-inflicted wounding. The incidence of SIB was somewhat lower in pigtailed macaques where only 6% of the animals had a veterinary record of wounding (Bellanca and Crockett, 2002). The role played by the length of time housed individually – a significant risk factor for SIB – is presented and discussed in Section 6.6. We have thus far attributed the development of abnormal behaviour in individually-housed monkeys to the loss of social companions. However, it is also possible that the emergence of such behaviour may be related to the nature of the individual cage, both in terms of its small size and its lack of complexity (a standard individual cage is shown on the book's website). If true, then increasing the cage size or adding enrichment to the cage environment should reduce abnormal behaviour in individually-housed monkeys.

6.4.2.1. Cage size

No clear picture has emerged from studies in which cage size was manipulated in individually housed monkeys. The failure to obtain consistent findings across studies may stem from two factors – the actual sizes of the cages examined and the confounding of a change of location (novelty) with a change in cage size. Very large changes in cage size, such as moving monkeys to large runway cages (Paulk *et al.*, 1977) or to very large outdoor pens (Draper and Bernstein, 1963), substantially reduced stereotypic behaviour, but smaller variations in cage size yielded negligible effects (Line *et al.*, 1991a; Crockett *et al.*, 1995). However, in the above studies, a change in cage size was confounded by a change in cage location (novelty). When this confounding variable was eliminated, rhesus monkeys did not show a reduction in abnormal behaviour when moved to large runway cages (Kaufman *et al.*, 2004). Thus, stereotypic behaviour in individually housed animals appears to be largely unaffected by changes in cage size *per se*.

6.4.2.2. Enrichment

Increasing the complexity of an individual cage by adding toys, furnishings and foraging devices reduces abnormal behaviour, effectiveness varying with the device, the species tested, and the kind of abnormal behaviour (*cf.* our contributed box in Chapter 9). Modest reductions in SIB were observed in rhesus monkeys exposed to a 'feeder box that triggered music' (Line *et al.*, 1990) – a radio and a food dispenser contained within a box attached to the home cage. The decline in SIB was observed only during the 20-week period in which the box was present. When it was removed, SIB increased. Self-directed stereotypies and SIB also declined during brief daily exposure to an enriched playpen environment, but again, the change occurred only in the playpen, not carrying over to the home cage environment (Bryant *et al.*, 1988). Exposure to a foraging/grooming board led to a different outcome. Although whole-body stereotypies (e.g. pacing) decreased, self-directed stereotypies and self-injurious behaviour were unaffected (Bayne *et al.*, 1991). Again, however, the improvement in abnormal behaviour was not maintained when the foraging/grooming board was removed. A similar pattern was noted when monkeys with a history of SIB were provided with a food puzzle feeder. The presence of the feeder led to reductions in pacing but did not have any beneficial effect on SIB or self-directed stereotypies (Novak *et al.*, 1998). Thus although environmental enrichment appeared to aid in the reduction of stereotypies, more pathological behaviours such as SIB were generally resistant to this treatment (Line *et al.*, 1991b, Novak *et al.*, 1998; Schapiro and Bloomsmith, 1994, 1995). The above studies suggest a link between a reduction in environmental complexity and the development of stereotypy. However, other factors such as the loss of companions undoubtedly play an equally, if not more important, role. Furthermore, enrichment was introduced after monkeys had been individually-housed for different periods of time and after abnormal behaviour had developed. In this situation, enrichment was often therapeutic in reducing general types of stereotypies but not pathological behaviour. These data do not, however, address the issue of prevention (i.e. whether animals would have been protected from acquiring abnormal behaviour if they had been exposed to an enriched environment at the time of their placement into individual cages).

6.4.2.3. Social interaction

If the development of abnormal behaviour is tied to the loss of companions, then providing companionship might be very effective at reducing abnormal behaviour in individually housed animals. Emerging evidence suggests that providing companions can indeed lead to marked reductions in abnormal behaviour. Comparisons of individually-housed rhesus monkeys before and after introduction to a compatible partner revealed significant decreases in self-directed stereotypies such as hair-pulling and digit-sucking (Eaton *et al.*, 1994). Pair-housing also led to decreased abnormal behaviour in female long-tailed macaques, although this strategy was considerably less successful with males (Crockett *et al.*, 1994). Group-housing instituted after 5 years of individual cage-housing led to marked reductions of abnormal behaviour in baboons (Kessel and Brent, 2001). Furthermore, in contrast to the effects of enrichment, pair-housing in rhesus monkeys successfully reduced SIB (Reinhardt, 1999).

6.4.3. Summary of social separation in normally reared primates

Separation from species members can have different effects depending on the age of the monkey and the source of attachment. Infants separated from their mothers initially show heightened arousal (vocalizations and pacing) followed by withdrawal and depression (huddling behaviour). Separation in juveniles does not typically produce heightened arousal but it can lead to depressive behaviour and it is associated with an increase in stereotypic behaviour. Juveniles separated from their social group appear to have increased vulnerability to develop SIB compared to adolescents and adults undergoing a similar separation. Considerable evidence suggests that enriching the environment of individually-housed monkeys with toys and foraging devices is effective in eliminating some forms of abnormal behaviour. However, social housing is more effective than enrichment in ameliorating severe forms of abnormal behaviour. Thus, it is likely that social loss is more important in the development of pathological behaviour than the small size or physical barrenness of the individual cage.

6.5. Abnormal Behaviour in Socially-Reared and Socially-Housed Primates: From Laboratories to Zoos

The focus so far has been on laboratory animals either reared under socially deprived conditions, or socially reared but then separated from companions. However, stereotypic behaviour can also develop in socially reared animals that remain socially housed. The best evidence comes from observations or surveys of primates housed in zoological gardens (e.g. Bollen *et al.*, submitted), where the animals are maintained in social groups and the environment is generally considered to be superior to laboratory housing environments. (Swaisgood and Shepherd-

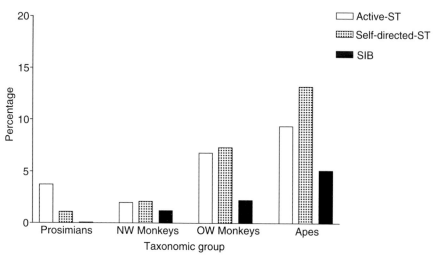

Fig. 6.2. Incidence of three types of abnormal behaviour in zoo-housed primates. ST is an abbreviation for stereotypy (with 'active' meaning whole body) and SIB, an abbreviation for self-injurious behaviour. NW signifies New World and OW denotes Old World species. (From Bollen *et al.*, submitted for publication; see also our Box 3.3, in Chapter 3.)

son in Chapter 9 deal further with such behaviours when as in zoos.) Studies of zoo-maintained animals also permit a direct comparison of several different primate taxa (see Figure 6.2, and Box 3.3, Chapter 3).

Zoo primate rearing is not always naturalistic: infant zoo primates occasionally have to be hand-reared due to maternal rejection. Hand-rearing is considered a strategy of last resort because it can produce animals with abnormal behaviour. For example, Marriner and Drickamer (1994) found that animals which had been hand-reared performed significantly more stereotypy than mother-reared animals even though hand- and mother-reared animals were born in the same year and exposed to the same changes and improvement in zoo design. More recently, Martin (2002) similarly found mother-reared chimpanzees to be the least stereotypic when housed in zoo enclosures, compared with human-reared animals.

6.6. The Causes and Correlates of Self-injurious Behaviour

In previous sections, we discussed the complex roles of early rearing experience, later environmental conditions, and other factors in the development and maintenance of stereotypies and SIB. Nevertheless, the relationship between stereotypy and SIB is not yet fully understood. We know that SIB is significantly associated with stereotypies directed to the self (e.g. self-clasping, hair-pulling – see image on the book's website), but it is also the case that such stereotypies can occur in the absence of more severe behavioural pathology (e.g. SIB). However, despite many gaps in our understanding of how stereotypy and SIB fit together, significant progress has been made in characterizing the risk factors and physiological correlates of SIB, particularly in captive rhesus monkeys. Earlier, we reviewed the occurrence of SIB and showed that it occurs with high probability in isolate-reared monkeys. However, we have also shown that SIB can develop spontaneously in a small percentage of monkeys that were normally reared but subsequently separated from companions and housed in individual cages. Our discussion of SIB will be restricted to this latter category, focusing on research from our laboratory. Only a small percentage of macaques (10–15%) appear to be vulnerable to this disorder (Bayne *et al.*, 1995; Lutz *et al.*, 2003b). However, for the animals that develop SIB, there is no generally effective treatment for this problem as long as they remain individually housed. Thus, SIB poses a significant challenge to the management of non-human primates in captivity. Complicating the picture is that monkeys with SIB come to the attention of caretakers and veterinarians only *after* the disorder is well established. Thus, both retrospective analysis of colony records and behavioural observations of individual animals are necessary to study the disorder.

Based on our research and the work of others, we have formulated an integrated developmental-neurochemical hypothesis in which SIB arises from adverse life events, is maintained by dysregulations of several neurochemical and physiological systems, and serves to reduce anxiety (Tiefenbacher *et al.*, 2005). Below we review the evidence for this hypothesis.

6.6.1. Adverse life events

Retrospective analysis of the colony records at the New England Primate Research Center supports a linkage between adverse, stressful experiences and the development of SIB. As was the case with other kinds of abnormal behaviour (see above), social separation was a significant risk factor for SIB (Novak, 2003). Monkeys with SIB were separated from companions and housed alone at a much earlier age than monkeys without SIB (mean age of 14.45 versus 25.12 months respectively; Novak, 2003). Two other factors also increased the risk of developing SIB. Monkeys with SIB had a longer tenure in individual cages and experienced more veterinary procedures (e.g. venipuncture) than animals without the disorder (Lutz *et al.*, 2003b).

We hypothesize that a combination of social deprivation and repeated exposure to veterinary procedures during the juvenile period is highly stressful to rhesus monkeys and may subsequently elicit SIB in vulnerable individuals. Since in primates, bouts of stereotypy and SIB are often elicited by acutely stressful events (see our Box 8.3, Chapter 8), such effects may be mediated by endowing individuals with a heightened sensitivity to a variety of environmental events.

6.6.2. Physiological correlates

SIB is associated with heightened reactions to the environment, and elevated CSF levels of CRF (see our Box 8.3, Chapter 8, for more details), but also with some seemingly paradoxical changes in the stress response system, particularly the HPA axis. Monkeys with a history of self-inflicted wounding showed an *attenuated* plasma cortisol response to the mild stress of blood sampling compared to non-wounding controls (Tiefenbacher *et al.*, 2000). Moreover, stress-induced cortisol levels were inversely related to the rate of self-directed biting, the main form of expression of SIB in macaques. In addition, self-biting was preferentially directed to body sites associated with acupuncture analgesia (Marinus *et al.* 2000).

Subsequent studies investigated the mechanisms underlying this blunted stress response. Monkeys with the disorder showed an attenuated cortisol response to an ACTH challenge, suggesting reduced adrenocortical sensitivity (Tiefenbacher *et al.*, 2003d). In addition, the cortisol rise at 30 min post-ACTH was correlated with the recency of self-wounding, such that animals with the most recent wounds showed the most blunted cortisol response to ACTH. Finally, we found altered glucocorticoid negative feedback in the SIB monkeys, as demonstrated by an attenuated urinary cortisol response to a low dose of dexamethasone (Tiefenbacher *et al.*, 2003d). Together these findings suggest that SIB in this cohort of rhesus monkeys is associated with complex and persistent changes in HPA activity that are related to both the outcome (i.e. wounding) and the expression (i.e. self-directed biting) of the behavioural pathology. The present evidence is only correlational in nature, and it remains to be determined whether HPA system dysregulation is a cause or a consequence of SIB. However, the persistence of SIB in some *human* populations is thought to be mediated through a reduction in anxiety. Indeed, individuals often verbally report significant tension relief after cutting, burning or other kinds of self-mutilation. Our recent studies indicate that relief of anxiety may similarly underlie the expression of SIB in rhesus monkeys. For example, episodes of SIB were found to be preceded by a rise in heart-rate, which was followed by an immediate return to baseline once the episode was over (Novak, 2003). This finding suggests that self-directed biting serves as a coping mechanism to alleviate acute episodes of arousal or stress. (See Box 1.3, Chapter 1, for more on abnormal behaviour and coping.)

We do not yet know how such 'coping' effects are mediated. However, growing evidence implicates the opioid system in the expression of SIB. Symptom reduction following treatment with an opioid receptor antagonist such as naltrexone in humans (Sandman *et al.*, 2000) raises the possibility that such individuals may injure themselves in order to stimulate endogenous opioid release (see Box 1.3, also Chapter 10, for a critique of opioid antagonist effects). Furthermore, more direct evidence relating endogenous opioid activity to SIB in monkeys comes from two recent studies. In both cases, circulating β-endorphin-like immunoreactivity

(IR) was positively correlated with the expression of SIB (Crockett *et al.*, 2003; Tiefenbacher *et al.*, 2003b). β-endorphin-like IR was also positively correlated with age of first individual caging (Tiefenbacher *et al.*, 2003b), suggesting that the observed alterations in endogenous opioid activity may precede the onset of abnormal behaviour rather than being a consequence. In a subsequent experiment, met-enkephalin-like IR in the CSF was also lower in monkeys with a veterinary record of self-inflicted wounding and was positively associated with the percentage of bites directed towards acupuncture sites (Tiefenbacher *et al.*, 2003c). Together with the β-endorphin results, these results support the hypothesis that monkeys with SIB tend to have low baseline opioid activity, which might promote self-directed biting (and occasional wounding) in order to stimulate opioid peptide release. It is well known that opioid receptor activation can be highly rewarding to both animals and humans. Thus, monkeys might persist in self-biting because it is reinforcing. In addition, early clinical work by Pickar and others (1982) reported that high doses of the opioid antagonist naloxone increased tension and anxiety, suggesting that endogenous opioids decrease such feelings. Based on this finding, the elevated anxiety thought to contribute to self-biting and SIB in rhesus monkeys may result not only from elevated CRF levels, as mentioned above, but also from a deficit in opioid activity.

Although our working hypothesis has focused on the interacting influences of stressful events, heightened anxiety, and a dysregulation of the HPA and opioid systems, other systems may also be involved. Levels of the neurotransmitter, serotonin and the gonadal hormone testosterone have been linked to aggressiveness in male rhesus monkeys and implicated in SIB. Weld *et al.* (1998) found that self-biting behaviour in male rhesus monkeys was decreased following the administration of l-tryptophan, the metabolic precursor to serotonin (see Chapter 10 for more on 'psychodietetrics'). These results suggested that self-biting was related to a deficit in serotonergic activity that could be ameliorated by appropriate pharmacologic intervention. However, this hypothesis must be tempered in light of other experimental findings. Monkeys with SIB did not differ from controls in their levels of baseline 5-HIAA levels in CSF (Tiefenbacher *et al.*, 2000; Weld *et al.*, 1998). Moreover, there were no group differences in prolactin or cortisol responses to a challenge with the 5-HT releasing agent fenfluramine (Tiefenbacher *et al.*, 2003a). Consequently, if there is a serotonergic deficit in monkeys that exhibit SIB, the nature of this deficit remains to be identified.

6.6.3. Summary

Our findings show that a small percentage of socially-reared monkeys develop self-injurious behaviour when separated from companions and placed into individual cages. Animals with this disorder tend to share several characteristics. First, they were separated at an early age (less than 2 years) and were exposed to more stressful experiences than animals without SIB. Second, they exhibit a dysregulation of the HPA axis that is manifested, in part, by a blunted response to mild stress. Third, they show reduced levels of the endogenous opioid peptides met-enkephalin and β-endorphin, and fourth, they appear to suffer from elevated arousal and/or anxiety. The hypothesis that the appearance of SIB in monkeys may involve a complex pattern of heightened anxiety, dysfunctional reward mechanisms, and abnormal stress responses fits well with much of the human literature on SIB. This hypothesis holds promise for informing future research in determining the etiology and treatment of SIB.

6.7. Conclusions: Understanding Abnormal Behaviour in Primates: Common Themes

Socially restricted environments play a substantial role in the development of abnormal behaviour in monkeys. However, this relationship is strongly influenced by development. Monkeys have a relatively long infant period and spend several more years as juveniles and adolescents before reaching adulthood. Thus, the same socially impoverished environment will have substantially different effects depending on the age of exposure. For example, as we have seen, isolation-rearing from birth leads to a very different syndrome than is produced by separation from the mother, which in turn is somewhat different than the loss of companions in adolescence or adulthood.

Social restriction in infancy clearly has particularly marked effects on the development of abnormal behaviour. In fact, the single best predictor of stereotypic and pathological behaviour is rearing infant monkeys in isolation. This rearing condition is also associated with pronounced social deficits. However, monkeys that are reared with naïve peers, or with surrogates and naïve peers, develop most species-typical social behaviours and show fewer abnormal behaviours. This latter finding suggests that social development in rhesus monkeys may be relatively buffered from altered early experiences, as long as some social interaction occurs. However, as we have also seen, the effects of early social deprivation may also be prolonged, and in some severe instances very hard to reverse.

Even if monkeys are reared socially, they may still acquire abnormal behaviour if subsequently separated from companions and housed in individual cages. Risk factors include the initial age of exposure, the duration of individual cage-housing, and also the number of stressful events experienced. In some cases and under some circumstances, abnormal behaviour can be ameliorated with social interaction, or even with non-social enrichment devices. However, the path to abnormal behaviour is not always through exposure to social separation. Both stereotypic and pathological behaviour can occur in primates maintained in social groups in enriched environments (e.g. zoological gardens). The significantly reduced rates of stereotypic behaviour in zoo environments also further suggests that the quality of the physical environment may play some role in the development of stereotypic behaviour in primates. The finding of differences in both the range and intensity of stereotypic behaviour across taxonomic groups also suggests that genetic variables may play some role in the etiology of behavioural abnormalities.

Physiological correlates of abnormal behaviour have been found to vary as a function of the type of environment and the developmental stage of the organism. For example, isolation-rearing is associated with alterations of the serotonergic system, whereas peer-rearing is not. A number of studies have linked various neuroendocrine and neuropeptide systems to abnormal behaviour in normally reared monkeys. Self-injurious behaviour is of particular interest, especially for ideas about 'coping', and is related to HPA axis dysregulation, elevated central levels of CRF, and alterations in both central and peripheral opioid activity.

We have thus established a picture of stereotypic behaviour in primates that plays out on a larger stage that includes early experiences, social competence, and physiological effects. However, the actual interrelationship between these variables remains largely unknown, except possibly for our analysis of SIB. Below we discuss future research directions.

6.8. Future Directions

Although we have made much progress in understanding abnormal behaviour in primates, there is more work to be done. For instance, at present, we know relatively little about the actual relationship between social deficits and stereotypy. We might predict that the animals with the highest rates of abnormal behaviour are the most socially impaired (see Chapter 5 for some behavioural control problems that might affect social interaction). Certainly, isolation-rearing produces both the most severe social deficits and the highest level of abnormal behaviour. However, we can also ask this question about individuals within specific rearing groups. For example, do the most socially impaired peer-reared monkeys also show the highest level of abnormal behaviour, or the most severe kinds? And if there are individual differences, are there genetic factors that underscore them?

The physiological bases of deprivation-induced and other stereotypic behaviours also remain essentially unknown. For example, relationships between specific neurotransmitter alterations and the behavioural affects of early social deprivation have yet to be determined. Chapters 5, 7 and 8 all suggest specific hypotheses that could be tested. More extensive analyses of the biological underpinnings of both stereotypy and SIB are needed to determine whether these categories represent different manifestations of the same underlying mechanism. In addition, longitudinal prospective studies would be helpful in ascertaining whether an increase in the frequency or intensity of stereotypy is a necessary precursor to the development of SIB.

In terms of experiential effects, some further questions still remain unanswered. Although there has been speculation about the role of sensitive periods in the development of species normative social behaviour, it is also likely that there may be sensitive periods for the development of abnormal behaviour (see also Chapter 7 for further discussion of this issue). Certainly, some forms of abnormal behaviour are linked to specific developmental periods. For example, monkeys reared apart from their mothers tend to develop a pattern of digit-sucking which is high during the first year of life but wanes as the animal approaches puberty. Very few animals develop a pattern of digit-sucking after the first year of life. Similarly, monkeys that are separated from their social group as juveniles are more vulnerable to developing SIB than are adolescents or adults. Furthermore, other developmental factors affect SIB; risk factors include the initial age at exposure, the direction of individual cage-housing (see Chapter 7 for somewhat similar findings for physical enrichment), and also the number of stressful events experienced (see Chapter 8 for how uncontrollable stress can affect CNS functioning). Do these examples involve sensitive periods? Are there sensitive periods for other stereotypies? How are such effects mediated? Future research is needed to answer these questions.

We also do not yet fully understand the effect of the physical environment in eliciting stereotypies in primates. Clearly, placing a monkey into a barren individual cage leads to stereotypic behaviour. But is it the loss of companionship or the loss of environmental complexity that produces this outcome? Undoubtedly, both contribute in some manner but the relative importance of each contribution remains unclear. Determining how these factors interact could help us determine whether changes to the physical environment might compensate for social loss; and zoo studies might be one route forward here.

Finally, the putative function of much abnormal behaviour is unclear. It is possible that active whole-body stereotypies, such as somersaulting and bouncing, are replacements for locomotor activity that is constrained by small cages. On the other hand, the significance of stereotypies such as eye-covering and poking remains largely unknown. Complicating the picture is the possibility that stereotypies may have different etiologies in different individuals. For example, eye-covering and -poking may be associated with decreases in visual acuity or visual distortions, may be a form of visual stimulation, or may be copied from other individuals. As for SIBs, for which there is arguably the most convincing and intriguing evidence, questions still remain as to the exact effect such behaviour may have on endogenous opioid systems.

References

Adriani, W. and Laviola, G. (2002) Spontaneous novelty-seeking and amphetamine-induced conditioning and sensitization in adult mice: evidence of dissociation as a function of age at weaning. *Neuropyschopharmacology* 27, 225–236.

Arling, G.L. and Harlow, H.F. (1967) Effects of social deprivation on maternal behaviour of rhesus monkeys. *Journal of Comparative and Physiological Psychology* 64, 371–377.

Bayne, K. and Novak, M. (1998) Behavioural disorders. In: Bennett, B., Abee, C. and Henrickson, R. (eds) *Nonhuman Primates in Biomedical Research: Diseases.* Academic Press, New York, pp. 485–500.

Bayne, K., Mainzer, H., Dexter, S., Campbell, G., Yamada, F. and Suomi, S. (1991) The reduction of abnormal behaviours in individually housed rhesus monkeys (*Macaca mulatta*) with a foraging/grooming board. *American Journal of Primatology* 23, 23–35.

Bayne, K., Dexter, S. and Suomi, S. (1992) A preliminary survey of the incidence of abnormal behaviour in rhesus monkeys (*Macaca mulatta*) relative to housing condition. *Laboratory Animal* 21, 38–46.

Bayne, K., Haines, M., Dexter, S., Woodman, D. and Evans, C. (1995) Nonhuman primate wounding prevalence: a retrospective analysis. *Laboratory Animal* 24, 40–44.

Beauchamp, A.J., Gluck, J.P., Fouty, H.E. and Lewis, M.H. (1991) Associative processes in differentially reared monkeys (*Macaca mulatta*): blocking. *Developmental Psychobiology* 24, 175–189.

Beckett, C., Bredenkamp, D., Castle, J., Groothues, C., O'Connor, T.G., Rutter, M. and the English and Romanian Adoptees (ERA) Study Team (2002) Behavior patterns associated with institutional deprivation: a study of children adopted from Romania. *Developmental and Behavioral Pediatrics* 23, 297–303.

Bellanca, R. and Crockett, C. (2002) Factors predicting increased incidence of abnormal behaviour in male pigtailed macaques. *American Journal of Primatology* 58, 57–69.

Benoit, T.C., Jocelyn, L.J., Moddemann, D.M. and Embree, J. (1996) Romanian children – the Manitoba experience. *Archives of Pediatric and Adolescent Medicine* 150, 1278–1282.

Berkson, G. (1968) Development of abnormal stereotyped behaviours. *Developmental Psychobiology* 1, 118–132.

Bøe, K. (1997) The effect of age at weaning and post-weaning environment on the behaviour of pigs. *Acta Agriculturae Scandinavica Section A* 43, 173–180.

Bollen, K.S., Well, A. and Novak, M.A. (submitted for publication). A survey of abnormal behaviour in captive zoo primates.

Brent, L., Lee, D. and Eichberg, J. (1989) The effects of single caging on chimpanzee behaviour. *Laboratory Animal Science* 39, 345–346.

Bryant, C., Rupniak, N. and Iversen, S. (1988) Effects of different environmental enrichment devices on cage stereotypies and autoaggression in captive cynomolgus monkeys. *Journal of Medical Primatology* 17, 257–269.

Callard, M.D., Bursten, S.N. and Price, E.O. (2000) Repetitive behaviour in captive roof rats (*Rattus rattus*) and the effects of cage enrichment. *Animal Welfare* 9, 139–152.

Capitanio, J.P. (1986) Behavioural pathology. In: Mitchell, G. and Erwin, J. (eds) *Comparative Primate Biology*, Vol. 2, Part A: *Behaviour, Conservation, and Ecology.* Alan R. Liss, New York, pp. 411–454.

Carlson, M. and Earls, F. (1997) Psychological and neuroendocrinological sequelae of early social deprivation in institutionalized children in Romania. *Annals of the New York Academy of Science* 807, 419–428.

Chamove, A.S. (1973) Rearing infant rhesus together. *Behaviour* 47, 48–66.

Chamove, A.S., Rosenblum, L.A. and Harlow, H.F. (1973) Monkeys (*Macaca mulatta*) raised only with peers. A pilot study. *Animal Behaviour* 21, 316–325.

Chugani, H.T., Behen, M.E., Muzik, O., Juhasz, C., Nagy, F. and Chugani, D.C. (2001) Local brain functional activity following early deprivation: a study of post-institutionalized Romanian orphans. *NeuroImage* 14, 1290–1301.

Clark, D.L. (1969) Immediate and delayed effects of early, intermediate, and late social isolation in the rhesus monkey. *Dissertation Abstracts International* B29, 4862.

Clarke, A.S. (1993) Social rearing effects on HPA axis activity over early development and in response to stress in rhesus monkeys. *Developmental Psychobiology* 26, 433–446.

Clarke, A.S., Hedeker, D.R., Ebert, M.H., Schmidt, D.E., McKinney, W.T. and Kraemer, G.W. (1996) Rearing experience and biogenic amine activity in infant rhesus monkeys. *Biological Psychiatry* 40, 338–352.

Crockett, C., Bowers, C., Bowden, D. and Sackett, G. (1994) Sex differences in compatibility of pair-housed adult long-tailed macaques. *American Journal of Primatology* 32, 73–94.

Crockett, C., Bowers, C., Shimoji, M., Leu, M., Bowden, D. and Sackett, G. (1995) Behavioural responses of longtailed macaques to different cage sizes and common laboratory experiences. *Journal of Comparative Psychology* 109, 368–383.

Crockett, C.M., Sackett, G.P., Sandman, C.A. and Chicz-DeMet, A. (2003) Beta endorphin levels in longtailed and pigtailed macaques vary by species, sex, and abnormal behaviour rating: a pilot study. *American Journal of Primatology* 60, 109–110.

Cross, H. and Harlow, H. (1965) Prolonged and progressive effects of partial isolation on the behaviour of macaque monkeys. *Journal of Experimental Research in Personality* 1, 39–49.

Davenport, M.D., Novak, M.A., Meyer, J.S., Tiefenbacher, S.T., Higley, J.D., Lindell, S.G., Champoux, M., Shannon, C. and Suomi, S.J. (2003) Continuity and change in emotional reactivity in rhesus monkeys throughout the pre-pubertal period. *Motivation and Emotion* 27, 57–76.

Draper, W. and Bernstein, I. (1963) Stereotyped behaviour and cage size. *Perceptual and Motor Skills* 16, 231–234.

Eaton, G., Kelley, S., Axthelm, M., Iliff-Sizemore, S. and Shigi, S. (1994) Psychological well-being in paired adult female rhesus (*Macaca mulatta*). *American Journal of Primatology* 33, 89–99.

Fisher, L., Ames, E.W., Chisholm, K. and Savoie, L. (1997) Problems reported by parents of Romanian orphans adopted to British Columbia. *International Journal of Behavioral Development* 20, 67–82.

Fittinghoff, N., Lindburg, D., Gomber, J. and Mitchell, G. (1974) Consistency and variability in the behaviour of mature, isolation-reared, male rhesus macaques. *Primates* 15, 111–139.

Fritz, J., Nash, L., Alford, P. and Bowen, J. (1992) Abnormal behaviours, with a special focus on rocking, and reproductive competence in a large sample of captive chimpanzees (*Pan troglodytes*). *American Journal of Primatology* 27, 161–176.

Fry, J.P., Sharman, D.F. and Stephens, D.B. (1981) Cerebral dopamine, apomorphine and oral activity in the neonatal pig. *Journal of Veterinary Pharmacology and Therapeutics* 4, 193–207.

Gilmer, W. and McKinney, W.T. (2003) Early experience and depressive disorders: human and non-human primate studies. *Journal of Affective Disorders* 75, 97–113.

Gluck, J.P. and Sackett, G.P. (1976) Extinction deficits in socially isolated rhesus monkeys (*Macaca mulatta*). *Developmental Psychology* 12, 173–174.

Gluck, J.P., Harlow, H.F. and Schiltz, K.A. (1973) Differential effect of early enrichment and deprivation on learning in the rhesus monkey. *Journal of Comparative and Physiological Psychology* 84, 598–604.

Gunnar, M. (1999) Overview of research on post-institutionalized children. In: Paper Presented at the Early Experience and Glucocorticoid Network Meeting, No. 2, 21 November 1999. Available at: http://

www.education.umn.edu/icd/EarlyExp/ Public%2011-21-99.htm (accessed December 2004).

Hansen, E.W. (1966) The development of maternal and infant behaviour in the rhesus monkey. *Behaviour* 27, 107–149.

Harlow, H.F. and Harlow, M.K. (1962) The effect of rearing conditions on behaviour. *Bulletin of the Menninger Clinic* 26, 213–224.

Harlow, H.F. and Harlow, M.K. (1965) The effect of rearing conditions on behaviour. *International Journal of Psychiatry* 1, 43–51.

Harlow, H.F. and Suomi, S.J. (1970) Nature of love: simplified. *American Psychologist* 25, 161–168.

Harlow, H.F. and Zimmermann, R.R. (1959) Affectional responses in the infant monkey. *Science* 130, 421–432.

Harlow, H.F., Rowland, G.L. and Griffin, G.A. (1964) The effect of total social deprivation on the development of monkey behaviour. *Psychiatric Research Report of the American Psychiatric Association* 19, 116–135.

Harlow, H.F., Schiltz, K.A. and Harlow, M.K. (1969) Effects of social isolation on the learning performance of rhesus monkeys. In: Carpenter, C. (ed.) *Proceedings of the Second International Congress of Primatology*, Vol. 1. S. Karger, Basel, pp. 178–185.

Higley, J.D., Suomi, S.J. and Linnoila, M. (1991) CSF monoamine metabolite concentrations vary according to age, rearing, and sex, and are influenced by the stressor of social separation in rhesus monkeys. *Psychopharmacology* 103, 551–556.

Higley, J.D., Suomi, S.J. and Linnoila, M. (1992) A longitudinal assessment of CSF monoamine metabolite and plasma cortisol concentrations in young rhesus monkeys. *Biological Psychiatry* 32, 127–145.

Hinde, R.A., Spencer-Booth, Y. and Bruce, M. (1966) Effects of 6-day maternal deprivation on rhesus monkey infants. *Nature* 210, 1021–1033.

Jeppesen, L.L., Heller, K.E. and Dalsgaard, T. (2000) Effects of early weaning and housing conditions on the development of stereotypies in farmed mink. *Applied Animal Behaviour Science* 68, 85–92.

Kaufman, I.C. and Rosenblum, L.A. (1967) The reaction to separation in infant monkeys: anaclitic depression and conservation-withdrawal. *Psychosomatic Medicine* 29, 648–675.

Kaufman, B.M., Pouliot, A.L. Tiefenbacher, S. and Novak, M.A. (2004) Short and long-term effects of a substantial change in cage size on individually housed, adult male rhesus monkeys (*Macaca mulatta*). *Applied Animal Behaviour Science* 88, 319–330.

Kessel, A. and Brent, L. (2001) The rehabilitation of captive baboons. *Journal of Medical Primatology* 30, 71–80.

King, N.E. and Mellen, J.D. (1994) The effects of early experience on adult copulatory behaviour in zoo-born chimpanzees (*Pan troglodytes*). *Zoo Biology* 13, 51–59.

Kraemer, G.W. and Clarke, A.S. (1990) The behavioural neurobiology of self-injurious behaviour in rhesus monkeys. *Progress in Neuropsychopharmacology and Biological Psychiatry* 14 (Supplement), 141–168.

Kraemer, G.W., Ebert, M.H., Lake, C.R. and McKinney, W.T. (1984) Hypersensitivity to d-amphetamine several years after early social deprivation in rhesus monkeys. *Psychopharmacology* 82, 266–271.

Kraemer, G.W., Ebert, M.H., Schmidt, D.E. and McKinney, W.T. (1989) A longitudinal study of the effect of different social rearing conditions on cerebrospinal fluid norepinephrine and biogenic amine metabolites in rhesus monkeys. *Neuropsychopharmacology* 2, 175–189.

Kraemer, G.W., Schmidt, D.E. and Ebert, M.H. (1997) The behavioural neurobiology of self-injurious behaviour in rhesus monkeys. Current concepts and relations to impulsive behaviour in humans. *Annals of the New York Academy of Sciences* 836, 12–38.

Latham, N. (2005) Refining the role of stereotypic behaviour in the assessment of welfare: stress, general motor persistence and early environment in the development of abnormal behaviour. Ph.D. thesis, Oxford University, Oxford, UK.

Latham, N. and Mason, G. (2004) From house mouse to mouse house: the behavioural biology of free-living *Mus musculus* and its implications in the laboratory. *Applied Animal Behaviour Science* 86, 261–289.

Levine, S. and Weiner, S.G. (1988) Psychoendocrine aspects of mother–infant relationships in nonhuman primates. *Psychoneuroendocrinology* 13, 143–154.

Lewis, M.H., Gluck, J.P., Beauchamp, A.J., Keresztury, M.F. and Mailman, R.B. (1990) Long-term effects of early social isolation in *Macaca mulatta*: changes in dopamine receptor function following apomorphine challenge. *Brain Research* 513, 67–73.

Line, S.W., Clarke, A.S., Markowitz, H. and Ellman, G. (1990) Responses of female rhesus macaques to an environmental enrichment apparatus. *Laboratory Animals* 24, 3–220.

Line, S., Markowitz, H., Morgan, K. and Strong, S. (1991a) Effects of cage size and environmental enrichment on behavioural and physiological responses of rhesus macaques to the stress of daily events. In: Novak, M.A. and Petto, A.J. (eds) *Through the Looking Glass: Issues of Psychological Well-being in Captive Nonhuman Primates*. American Psychological Association, Washington, DC, pp. 160–179.

Line, S., Morgan, K. and Markowitz, H. (1991b) Simple toys do not alter the behaviour of aged rhesus monkeys. *Zoo Biology* 10, 473–484.

Ljungberg, T. and Westlund, K. (2000) Impaired reconciliation in rhesus macaques with a history of early weaning and disturbed socialization. *Primates* 41, 79–88.

Lutz, C., Marinus, L., Chase, W., Meyer, J. and Novak, M. (2003a) Self-injurious behaviour in male rhesus macaques does not reflect externally directed aggression. *Physiology and Behaviour* 78, 33–39.

Lutz, C., Well, A. and Novak, M. (2003b) Stereotypic and self-injurious behaviour in rhesus macaques: a survey and retrospective analysis of environment and early experience. *American Journal of Primatology* 60, 1–15.

MacLean, K. (2003) The impact of institutionalization on child development. *Development and Psychopathology* 15, 853–884.

Marinus, L.M., Chase, W.K. and Novak, M.A. (2000) Self-biting behavior in rhesus macaques (*Macaca mulatta*) is preferentially directed to body sites associated with acupuncture analgesia. *American Journal of Primatology* 51, 71–72.

Marriner, L. and Drickamer, L. (1994) Factors influencing stereotyped behaviour of primates in a zoo. *Zoo Biology* 13, 267–275.

Martin, J.E. (2002) Early life experiences: activity levels and abnormal behaviours in resocialised chimpanzees. *Animal Welfare* 11, 419–436.

Martin, L.J., Spicer, D.M., Lewis, M.H., Gluck, J.P. and Cork, L.C. (1991) Social deprivation of infant rhesus monkeys alters the chemoarchitecture of the brain: I. Subcortical regions. *Journal of Neuroscience* 11, 3344–3358.

Mason, W.A. (1968) Early social deprivation in the nonhuman primates: implications for human behaviour. In: Glass, D.C. (ed.) *Environmental Influences*. Rockefeller University Press, New York, pp: 70–101, 273–275.

Mason, G.J. (1992) Individual variation in the stereotypies of caged mink. Ph.D. thesis, Cambridge University, Cambridge, UK.

Mason G.J. (1995) Tail-biting in mink (*Mustela vison*) is influenced by age at removal from the mother. *Animal Welfare* 3, 305–311.

Mason, G. (1996) Early weaning enhances the later development of stereotypy in mink. In: Duncan, I., Widowski, T. and

Haley, D. (eds) *Proceedings of the 30th International Congress of the International Society for Applied Ethology.* The International Society for Applied Ethology, Guelph, p. 6.

Mason, W.A. and Berkson, G. (1975) Effects of maternal mobility on the development of rocking and other behaviours in rhesus monkeys: a study with artificial mothers. *Developmental Psychobiology* 8, 197–211.

McKinney, W.T., Jr., Young, L.D., Suomi, S.J. and Davis, J.M. (1973) Chlorpromazine treatment of disturbed monkeys. *Archives of General Psychiatry* 29, 490–494.

Meyer, J.S. and Bowman, R.E. (1972) Rearing experience, stress and adrenocorticosteroids in the rhesus monkey. *Physiology and Behavior* 8, 339–343.

Meyer, J.S., Novak, M.A., Bowman, R.E. and Harlow, H.F. (1975) Behavioural and hormonal effects of attachment object separation in surrogate–peer-reared and mother-reared infant rhesus monkeys. *Developmental Psychobiology* 8, 425–435.

Mineka, S. and Suomi, S.J. (1978) Social separation in monkeys. *Psychological Bulletin* 85, 1376–1400.

Mitchell, G.D. (1968) Persistent behaviour pathology in rhesus monkeys following early social isolation. *Folia Primatologica* 8, 132–147.

Mitchell, G.D., Raymond, E.J., Ruppenthal, G.C. and Harlow, H.F. (1966) Long-term effects of total social isolation upon behaviour of rhesus monkeys. *Psychological Report* 18, 567–580.

Morris, D. (1987) *Catlore.* Jonathan Cape, London.

Nash, L.T., Fritz, J., Alford, P.A. and Brent, L. (1999) Variables influencing the origins of diverse abnormal behaviours in a large sample of captive chimpanzees (*Pan troglodytes*). *American Journal of Primatology* 48, 15–29.

Novak, M.A. (1979) Social recovery of monkeys isolated for the first year of life. II. Long-term assessment. *Developmental Psychology* 15, 50–61.

Novak, M.A. (2003) Self-injurious behaviour in rhesus monkeys: new insights into its etiology, physiology, and treatment. *American Journal of Primatology* 59, 3–19.

Novak, M.A. and Harlow, H.F. (1975) Social recovery of monkeys isolated for the first year of life. I. Rehabilitation and therapy. *Developmental Psychology* 11, 453–465.

Novak, M.A., O'Neill, P. and Suomi, S.J. (1992) Adjustments and adaptations to indoor and outdoor environments: continuity and change in young adult rhesus monkeys. *American Journal of Primatology* 28, 125–138.

Novak, M.A., Kinsey, J.H., Jorgensen, M.J. and Hazen, T.J. (1998) Effects of puzzle feeders on pathological behaviour in individually housed rhesus monkeys. *American Journal of Primatology* 46, 213–227.

O'Connor, T.G., Rutter, M., Beckett, C., Keaveny, L., Kreppner, J.M. and the English and Romanian Adoptees (ERA) Study Team (2000) The effects of global severe privation on cognitive competence: extension and longitudinal follow-up. *Child Development* 71, 376–390.

de Passillé, A.M., Christopherson, R. and Rushen, J. (1993) Nonnutritive sucking by the calf and postprandial secretion of insulin, CCK, and gastrin. *Physiology and Behaviour* 54, 1069–1073

Paulk, H., Dienske, H. and Ribbens, L. (1977) Abnormal behaviour in relation to cage size in rhesus monkeys. *Journal of Abnormal Psychology* 86, 87–92.

Perré, Y., Wauters, A.M. and Richard-Yris, M.A. (2002) Influence of mothering on emotional reactivity of domestic pullets. *Applied Animal Behaviour Science* 75, 133–146.

Pickar, D., Cohen, M.R., Naber, D. and Cohen, R.M. (1982) Clinical studies of the endogenous opioid system. *Biological Psychiatry* 11, 1243–1276.

Reinhardt, V. (1999) Pair-housing overcomes self-biting behaviour in macaques. *Laboratory Primate Newsletter* 38, 4–5.

Ridley, R. and Baker, H. (1982) Stereotypy in monkeys and humans. *Psychological Medicine* 12, 61–72.

Roden, C. and Wechsler, B. (1998) A comparison of the behaviour of domestic chicks reared with or without a hen in enriched pens. *Applied Animal Behaviour Science* 55, 317–326.

Rowland, G.L. (1964) The effects of total social isolation upon learning and social behaviour of rhesus monkeys. *Dissertation Abstracts International* 25, 1364–1365.

Ruppenthal, G.C., Arling, G.L., Harlow, H.F., Sackett, G.P. and Suomi, S.J. (1976) A 10-year perspective of motherless-mother monkey behaviour. *Journal of Abnormal Psychology* 85, 341–349.

Ruppenthal, G.C., Walker, C.G. and Sackett, G.P. (1991) Rearing infant monkeys (*Macaca nemestrina*) in pairs produces deficient social development compared with rearing in single cages. *American Journal of Primatology* 25, 103–113.

Rutter, M., O'Connor, T.G. and the English and Romanian Adoptees (ERA) Study Team (2004) Are there biological programming effects for psychological development? Findings from a study of Romanian adoptees. *Developmental Psychology* 40, 81–94.

Sackett G.P. (1965) Effects of rearing conditions upon the behaviour of rhesus monkeys (*Macaca mulatto*). *Child Development* 36, 855–868.

Sackett, G.P. (1967) Some persistent effects of different rearing conditions on pre-adult social behaviour of monkeys. *Journal of Comparative and Physiological Psychology* 64, 363–365.

Sackett G.P. (1974) Sex differences in rhesus monkeys following varied rearing experiences. In: Friedman, C., Richart, R.M. and vande Wiele, R.L. (eds) *Sex Differences in Behaviour*. John Wiley & Sons, New York, pp. 99–122.

Sackett, G.P., Bowman, R.E., Meyer, J.S., Tripp, R.L. and Grady, S.S. (1973) Adrenocortical and behavioral reactions by differentially raised rhesus monkeys. *Physiological Psychology* 1, 209–212.

Sackett, G.P., Tripp, R. and Grady, S. (1982) Rhesus monkeys reared in isolation with added social, nonsocial, and electrical brain stimulation. *Annali Dell Instituto Superiore di Sanita* 18, 203–213.

Sackett, G.P., Novak, M.F.S.X. and Kroeker, R. (1999) Early experience effects on adaptive behaviour. *Mental Retardation and Developmental Disabilities Research Review* 5, 30–40.

Sackett, G.P., Ruppenthal, G.C. and Davis, A.E. (2002) Survival, growth, health, and reproduction following nursery rearing compared with mother rearing in pig-tailed monkeys (*Macaca nemestrina*). *American Journal of Primatology* 56, 165–183.

Sanchez, M.M., Hearn, E.F., Do, D., Rilling, J.K. and Herndon, J.G. (1998) Differential rearing affects corpus callosum size and cognitive function of rhesus monkeys. *Brain Research* 812, 38–49.

Sanchez, M.M., Ladd, C.O. and Plotsky, P.M. (2001) Early adverse experience as a developmental risk factor for later psychopathology: evidence from rodent and primate models. *Development and Psychopathology* 13, 419–449.

Sandman, C.A., Hetrick, W., Taylor, D.V., Marion, S.D., Touchette, P., Barron, J.L., Martinezzi, V., Steinberg, R.M. and Crinella, F.M. (2000) Long-term effects of naltrexone on self-injurious behaviour. *American Journal of Mental Retardation* 105, 103–117.

Schapiro, S. and Bloomsmith, M. (1994) Behavioural effects of enrichment on pair-housed juvenile rhesus monkeys. *American Journal of Primatology* 32, 159–170.

Schapiro, S. and Bloomsmith, M. (1995) Behavioural effects of enrichment on singly-housed yearling rhesus monkeys: an analysis including three enrichment conditions and a control group. *American Journal of Primatology* 35, 89–101.

Schapiro, S.J., Lee-Parritz, D.E., Taylor, L.L., Watson, L., Bloomsmith, M.A. and Petto, A. (1994) Behavioural management of specific pathogen-free rhesus macaques: group formation, reproduction, and parental competence. *Laboratory Animal Science* 44, 229–234.

Shannon, C., Champoux, M. and Suomi, S.J. (1998) Rearing condition and plasma cortisol in rhesus monkey infants. *American Journal of Primatology* 46, 311–321.

Sharman, D.F., Mann, S.P., Fry, J.P., Banns, H. and Stephens, D.B. (1982) Cerebral dopamine metabolism and stereotyped behaviour in early-weaned piglets. *Neuroscience* 7, 1937–1944.

Smotherman, W.P., Hunt, L.E., McGinnis, L.M. and Levine, S. (1979) Mother–infant separation in group-living rhesus macaques: a hormonal analysis. *Developmental Psychobiology* 12, 211–217.

Spencer-Booth, Y. and Hinde, R.A. (1971) Effect of brief separations from mothers during infancy on behaviour of rhesus monkeys 6–24 months later. *Journal of Child Psychology and Psychiatry* 12, 157–172.

Stroman, H.R. and Slaughter, L.M (1972) The care and breeding of the pygmy hippopotamus, *Chereopsis liberiensis*, in captivity. *International Zoo Yearbook* 12, 126–131.

Sumner, B.E.H., Lawrence, A.B., Jarvis, S., Calvert, S.K., Stevenson, J., Farnworth, M.J., Croy, I., Douglas, A.T., Russell, J.A. and Seckl, J.R. (2002) Enduring behavioural and neural effects of early versus late weaning in pigs. *Stress* 5, (Supplement), p. 105.

Suomi, S.J. and Harlow, H.F. (1972) Social rehabilitation of isolate-reared monkeys. *Developmental Psychology* 6, 487–496.

Suomi, S.J., Harlow, H.F. and Novak, M.A. (1974) Reversal of social deficits produced by isolation rearing in monkeys. *Journal of Human Evolution* 3, 527–534.

Suomi, S.J., Eisele, C.D., Grady, S.A. and Harlow, H.F. (1975) Depressive behaviour in adult monkeys following separation from family environment. *Journal of Abnormal Psychology* 84, 576–578.

Tiefenbacher, S., Novak, M.A., Jorgensen, M.J. and Meyer, J.S. (2000) Physiological correlates of self-injurious behaviour in captive, socially-reared rhesus monkeys. *Psychoneuroendocrinology* 25, 799–817.

Tiefenbacher, S., Davenport, M.D., Novak, M.A., Pouliot, A.L. and Meyer, J.S.

(2003a) Fenfluramine challenge, self-injurious behaviour, and aggression in rhesus monkeys. *Physiology and Behaviour* 80, 327–331.

Tiefenbacher, S., Marinus, L., Novak, M.A. and Meyer, J.S. (2003b) Endogenous opioid activity in a nonhuman primate model of self-injurious behaviour. *Society for Neuroscience*, New Orleans, Louisiana.

Tiefenbacher, S., Marinus, L.M., Davenport, M.D., Pouliot, A.L., Kaufman, B.M., Fahey, M.A., Novak, M.A. and Meyer, J.S. (2003c) Evidence for endogenous opioid involvement in the expression of self-injurious behaviour in rhesus monkeys. *American Journal of Primatology*, 60, 103.

Tiefenbacher, S., Novak, M.A., Marinus, L.M., Chase, W.K., Miller, J.A. and Meyer, J.S. (2003d) Altered hypothalamic–pituitary–adrenocortical function in rhesus monkeys (*Macaca mulatta*) with self-injurious behaviour. *Psychoneuroendocrinology* 29, 501–515.

Tiefenbacher, S., Novak, M.A., Lutz, C.C. and Meyer, J.S. (2005) The physiology and neurochemistry of self-injurious behaviour: a nonhuman primate model. *Frontiers in Bioscience* 10, 1–11.

Waiblinger, E. and König, B. (2004) Refinement of gerbil housing and husbandry in the lab. *Alternatives to Laboratory Animals* 32 (Supplement 1A), 163–169.

Walsh, S., Bramblett, C. and Alford, P. (1982) A vocabulary of abnormal behaviours in restrictively reared chimpanzees. *American Journal of Primatology* 3, 315–319.

Weld, K.P., Mench, J.A., Woodward, R.A., Bolesta, M.S., Suomi, S.J. and Higley, J.D. (1998) Effect of tryptophan treatment on self-biting and central nervous system serotonin metabolism in rhesus monkeys (*Macaca mulatta*). *Neuropsychopharmacology* 19, 314–321.

Widowski, T.M., Cottrell, T., Dewey, C.E. and Friendship, R.M. (2003) Observation of piglet-directed behaviour: patterns and skin lesions in eleven swine herds. *Journal of Swine Health and Production* 11, 181–185.

Winslow, J.T., Noble, P.L., Lyons, C.K., Sterk, S.M. and Insel, T.R. (2003) Rearing effects on cerebrospinal fluid oxytocin concentration and social buffering in rhesus monkeys. *Neuropsychopharmacology* 28, 910–918.

Worobec, E.K.; Duncan, I.J.H. and Widowski, T.M. (1999) The effects of weaning at 7, 14 and 28 days on piglet behaviour. *Applied Animal Behaviour Science* 62, 173–182.

Young, L.D., Suomi, S.J., Harlow, H.F. and McKinney, W.T. (1973) Early stress and later response to separation. *American Journal of Psychiatry* 130, 400–405.

7 The Neurobiology of Stereotypy I: Environmental Complexity

M.H. Lewis,[1] M.F. Presti,[1] J.B. Lewis[2] and C.A. Turner[3]

[1]Department of Psychiatry and Neuroscience, University of Florida and McKnight Brain Institute of the University of Florida, 100 S. Newell Drive, Gainesville, FL 32610-0256, USA; [2]New College of Florida, 5700 North Tamiami Trail, Sarasota, Florida 34243-2197, USA; [3]Mental Health Research Institute, Department of Neuroscience, 205 Zina Pitcher Place, University of Michigan, Ann Arbor, MI 48109-0720, USA

Editorial Introduction

The contribution of Lewis and colleagues plays three important roles in this book. First, it builds on the previous chapter in emphasizing the lasting effects that early experience can have on stereotypy; second, it reviews the diverse changes in brain functioning induced by the barren environments typically associated with abnormal behaviour; and third, it fills in the details behind the 'systems view' of Chapter 5. This group's own research is unique here, both in its focus on environment-induced stereotypies and its parallel investigations of brain and behaviour.

While Novak illustrated the potentially lasting effects of social deprivation, this chapter focuses on the physical environment: the predictability of 'impoverished' conditions versus the space, complexity and novelty of those captive environments we label 'enriched'. Evidence is reviewed that the prolonged exposure to impoverished environments can potentially induce stereotypies that become hard to abolish with enrichment; while in contrast the early exposure to enriched environments can protect animals against the development of stereotypies if moved to barren environments (a contributed box also contrasting the behaviour of wild-caught and captive-born animals). Lewis and colleagues' own work on deer mice illustrates such effects beautifully. For example, young deer mice housed in enriched cages develop little stereotypy even if then moved to barren cages 2 months later. The stereotypy of barren-raised mice, in contrast, is affected by a change in physical complexity, declining with enrichment – unless, that is, the animals have lived in these barren conditions until well into their mature years, in which case enrichment is no longer effective. Much remains unknown about these interesting effects, however. As the authors review, barren or enriched rearing environments have many potential effects on brain development, but which – if any – have 'sensitive periods' is far from known, and indeed recent evidence suggests that many aspects stay plastic long into adulthood. Furthermore, the ages at which animals are exposed to differential rearing, their length of time in these conditions, and their ages when tested in new environments are typically all inter-correlated: which factors are key has not yet been teased apart. Finally, how the effects of physically barren environments compare with those of social deprivation (Chapter 6) or chronic stress (Chapter 8) is not resolved: are they qualitatively similar processes (e.g. all involving degrees of stress), or are they distinct?

One way to address such questions is to examine the CNS changes seen in stereotypy-inducing environments, and Lewis and colleagues have conducted a series of experiments on this issue, rearing deer mice in barren or enriched environments and investigating the high and low stereotypers that emerge. Enriched animals with negligible stereotypy prove a distinctive group, with higher levels of neuronal metabolic activity, dendritic branching and the neurotrophin BNDF in the motor cortex and striatum. In addition, although low and high stereotypers from barren cages were indistinguishable using the measures above, more focused research into the neuropeptides of the striatum revealed significant differences between these animals, the most stereotypic mice having the low enkephalin and relatively high dynorphin characteristic of under-activity in the indirect pathway. These data thus build on Chapter 5, this volume, in implicating specific changes in basal ganglia functioning, particularly in the circuits between motor cortex and dorsal striatum, in cage stereotypy. The chapter also provides an excellent and thorough overview of the neuroanatomy and physiology of this region, and its involvement in the stereotypies induced by drugs and various clinical conditions.

If we assume that the differences between enriched and barren environments represent those between normal and pathological development, then enriched environments that protect deer mice against stereotypy are thus doing so by normalizing CNS development. However, whether this means that all the stereotypies of captive animals stem from abnormal CNS development is still unknown. Nor does this mean that the motor loop is always the critical brain system involved – a topic that Cabib develops in the subsequent chapter.

GM

7.1. Introduction

In this chapter, we propose a fundamental parallel between the predictable nature of impoverished environments and the predictable nature of the stereotypies they engender (Sections 7.2 and 7.3). We argue that such effects are mediated by profound changes in the central nervous system (CNS), an idea we support with evidence from our own work on deer mice (*Peromyscus maniculatus*), data from a wide range of studies on the neurobiological effects of environmental enrichment, and research into the stereotypies resulting from pharmacological manipulations and certain clinical conditions (Sections 7.4–7.6). During our review, we also touch on a number of areas that are as yet only partially understood, some of which we return to in our final discussion (Section 7.7): What are the key features of the environment in the aetiology or prevention of stereotypy? How does the length of exposure, or its timing relative to an animal's stage of development, influence the stereotypy-inducing effects of barren environments? And at what point in development does enrichment cease being able to confer its benefits? Conversely, does exposure to enrichment early in development confer lasting neuroprotection to animals transferred to typical (i.e. impoverished) captive environments? And finally, do all stereotypies share fundamental similarities in mechanism at the molecular level, or are there instead diverse means by which the CNS can be affected (e.g. via the direct and indirect pathways of the striatum), all sharing similar behavioural outcomes?

7.2. Complexity and Regularity

Stereotyped behaviour is characterized by strikingly periodic, predictable or regular dynamics. This marked regularity contrasts with the complexity, variability and relative unpredictability that characterizes adaptive, functional behaviour. In the 'dynamical disease' model of Glass and Mackey (1988), medical illness (e.g. cardiovascular disease) can be characterized by strikingly periodic or predictable dynamics in the relevant physiological systems, whereas healthy systems are complex and variable. Indeed, within the field of dynamical diseases, the most repetitive and predictable dynamics are associated with the most severe pathological conditions (e.g. Cheyne-Stokes breathing), a phenomenon referred to as the 'law of stereotypy' (Goldberger, 1997). Abnormal Repetitive Behaviour may be viewed from this perspective as representing a loss of complexity associated with CNS insult.

The loss of behavioural complexity or variability potentially indexed by abnormal repetition may arise from a number of sources. Insult to the CNS very early in development (e.g. autism), and exposure to high and/or chronic doses of certain pharmacological agents (e.g. amphetamine) are two such factors. The marked regularity of behaviour may also be accompanied, or driven by, a shift from complexity to regularity in early experience and/or current environment. Restricted, impoverished environments arguably lack variability, complexity, unpredictability and novelty, and these environments have often been associated with stereotypy. In the rest of this chapter, we therefore review the evidence for the effects of environmental complexity on brain structure and function, and discuss which of these effects mediate the amelioration or prevention of Abnormal Repetitive Behaviour. Finally, we use these and related findings to advance a neurobiological model that accounts for the expression of spontaneous, persistent stereotypy.

The relationship of stereotypy to the lack of complexity or variability in experience and environment is the focus of several previous chapters (especially Chapters 2–4 and 6, this volume). Suffice to say, then, that stereotypies in both humans and animals have been documented following abnormal rearing and housing conditions like early maternal deprivation, social isolation or restraint. Animals exposed to the regularity arguably associated with such rearing conditions reliably develop behavioural repertoires that are also devoid of complexity or variability. For instance, the studies of Harlow and colleagues showed that rhesus monkeys subjected to social isolation or maternal deprivation develop stereotyped behaviours

(Chapter 6, this volume). In humans, case history reports have also documented this (e.g. body-rocking) in children raised under conditions of severe social deprivation (Davis, 1940, 1946; Freedman, 1968) – a devastating effect seen on a much larger scale in the orphanages established in Romania under the Ceaucescu regime of the 1980s, in which many children exhibited body-rocking and other stereotypies (Carlson and Earls, 1997; personal observations; see Box 6.1, Chapter 6 for more details). Similarly, confinement and/or the prevention of species-typical behaviours is associated with stereotypy development in laboratory, zoo and farm animals (Chapters 2–4, this volume); indeed, stereotypies are the most common category of abnormal behaviour observed in confined animals (Würbel, 2001; also Chapter 1, this volume). In our own work, for example, deer mice develop specific forms of high rate and persistent stereotyped behaviour as a consequence of being housed in standard laboratory cages (Powell *et al.*, 1999; see video images on the book's website). These are typically hindlimb jumping or backward somersaulting, and can occur at very high rates. These behaviours appear spontaneously in such housing conditions – they do not require a drug challenge, nor specific eliciting stimuli for their expression (our observational data suggesting that their occurrence is not systematically associated with any particular environmental event).

7.3. Environmental Complexity and Stereotypy

If environmental restriction or regularity induces stereotypy, environmental complexity should ameliorate or prevent it. Increasing environmental complexity is indeed a frequently and successfully employed strategy for reducing animal stereotypy. Surprisingly, however, few studies have systematically assessed the effects of environmental complexity on stereotypies (e.g. Meehan *et al.*, 2004; Chapter 9, this volume), or measured complexity *per se*. It should also be noted that despite the popular use of the term 'environmental enrichment', attempts to make captive environments more complex hardly result in environments that could be considered enriched relative to the animal's natural habitat (again see Chapter 9). None the less, even modest changes (e.g. adding straw bedding to sows' stalls; see Fraser, 1975 and other citations in Chapter 2) can have appreciable effects on stereotypic behaviour.

Not surprisingly, controlled experimental manipulations of environmental complexity have tended to use rodents. For instance, Ödberg (1987) demonstrated that bank voles exposed to an enriched environment exhibited significantly less stereotypy than voles housed in barren cages, an effect independent of cage size, and Würbel (Chapter 4, this volume) gives further examples. Our own work has focused on such a model. Our first reports showed that deer mice reared in larger, more complex environments exhibit significantly less of the stereotypy described in Section 7.2 than do deer mice reared in standard laboratory cages (Powell *et al.*, 1999, 2000). When housed in these conditions, fewer enriched animals developed stereotyped behaviours, and they developed them later and displayed them at a lower rate than the standard-caged mice.

Building on this work, our current version of environmental complexity (Turner *et al.*, 2002, 2003; Turner and Lewis, 2003) consists of placing up to six same-sex weanlings in specially customized large dog kennels (121.9 × 81.3 × 88.9 cm), consisting of two extra levels constructed from galvanized wire mesh and connected by ramps to create three interconnected levels (see figure on this volume's website). Bedding, a running wheel, a shelter (an inverted baking pan or similar opaque, concave object), and various other objects (Habitrail tubes, plastic toys, mesh structures for climbing) are also added. Novel objects are introduced on a weekly basis, and birdseed is distributed three times each week to encourage foraging-type behaviours. Furthermore, to allow us to assess the behaviour of standard and enriched-housed animals when not in such dissimilar environments, we have adapted automated activity monitors to quantify the outcomes of exposure to these different conditions. Because the stereotyped behaviours observed in our deer mice typically involve vertical activity, they can be quantified by using photobeam arrays which, when interrupted, record a count. Our activity monitors have been adapted to record accurately the high rates of stereotyped jumping or backward somersaulting that can be observed (e.g. more than 2000 jumps in a 1-h period, and in some instances as many as 4000; e.g. Turner and Lewis, 2003); and we also routinely videotape these sessions to ensure accuracy of the automated counters, as well as to identify the topography of stereotypy. These methods have shown that in general, after 60 post-weaning days exposure to the more complex environment, most deer mice (approximately 70–80%) exhibit low rates of stereotypy when tested in the standardized activity monitors. The remaining animals exhibit stereotypies at roughly the same frequency observed in standard cage mice. Conversely, after 60 post-weaning days of standard housing, only *ca.* 20% of the mice tested exhibit little or no stereotypy. Table 7.1 provides one example from a recent study (Turner and Lewis, 2003). In this study, animals ($N = 105$) were monitored for two 1-h periods over two consecutive days. Animals exhibiting fewer than 500 counts per hour were considered 'low stereotypy', mice exhibiting more than 1000 counts per hour, 'high stereotypy'.

Note that why individual mice react differently to the either standard or enriched cages is as yet unknown, although as we shall see later, brain changes co-occur with at least some of this within-group variation in behavioural response (see Section 7.5). Nevertheless, overall there is a

Table 7.1. Percentage of deer mice categorized as high or low stereotypy in a standard test apparatus.

Housing type	Low stereotypy (%)	High stereotypy (%)
Large enriched cage	67	18
Standard cage	22	60

clear treatment effect, with housing typically affecting how stereotypic the mice are even when removed from these cages and tested in a standardized environment. Such results raise a number of questions. How are such effects mediated? How might the timing of exposure to these housing environments alter their effects? And what are the key features of stereotypy-reducing environments? We try and address these issues in the sub-sections that follow.

7.3.1. The timing and duration of exposure: sensitive periods for the environmental induction and reduction of stereotypy?

Just as is the case for social deprivation (Chapter 6, this volume), rather little information is available as to whether sensitive periods exist for the induction of stereotypy by environmental restriction, or for its prevention/reversal by environmental complexity. There is also little information on how much time is needed for each such environment to have its effects, nor on whether environmental complexity effects always last in the longer term, e.g. if the environment is altered to the more typical, standard laboratory housing.

A large literature on early experience does suggest, however, that exposure to a restricted environment should have more deleterious consequences on younger than older organisms. The loss of CNS plasticity that has been reported with age (e.g. Reis *et al.*, 2005) also leads to the hypothesis that the effects of a restricted environment should be more readily reversible by complex environments in younger animals. Consistent with this, of two studies conducted on bank voles, one study (Cooper *et al.*, 1996) showed that stereotypies were harder to disrupt through environmental enrichment in older voles (14 months of age) than in young voles (2 months of age). However, the other study, looking at younger animals, found no such effects: Ödberg (1987) found that voles raised with enrichment through 60 days of age exhibited significantly less stereotypy than voles housed in barren cages; but also that animals exposed to enrichment conditions *after* day 60 of life similarly displayed decreased stereotypies, thus demonstrating the power of enrichment provided later in development. This last finding is consistent with, for example, the work of Meehan *et al.* (2004) who reported that increased environmental complexity was effective in reducing stereotypy in Amazon parrots despite their being maintained in standard caging up to 48 weeks of age (see Chapter 9, this volume, for further examples). Thus the stereotypies of older animals (or those exposed to barren conditions for longer) are sometimes, but not always, harder to alleviate (see Chapter 10, this volume for similar issues in companion animals). Turning to the effects of taking animals from enriched conditions and placing them in barren cages, Ödberg (1987) showed that voles transferred from enriched cages to standard ones at 60 days of age did not develop higher rates of stereotypy when tested 30 days later. Additionally, Mason, in Box 7.1,

Box 7.1. Are Wild-born Animals 'Protected' from Stereotypy When Placed in Captivity?

G. Mason

If stereotypies arose solely from frustrated normal behaviours (Chapters 2–4, this volume), then one would expect them to be more evident in animals captured from the wild than those bred in captivity. After all, if farmed mink are raised with swimming water, but this is removed in adulthood, they pace more than mink who have never had this enrichment (Vinke, 2004); and laboratory mice raised in small enriched cages but transferred to standard ones stereotype more than control animals (Latham, 2005). However, in almost every reported instance, wild-caught animals show less stereotypy than captive-bred conspecifics.

Thus in laboratories, wild-born chimpanzees were less stereotypic than laboratory-born animals (e.g. Davenport and Menzel, 1963). Stereotypies (plus also 'faeces-smearing') were absent altogether in the wild-caught individuals of two macaque species, in marked contrast to their zoo- and laboratory-born peers (Mason and Green, 1962, cited by Davenport and Menzel, 1963; Berkson, 1968; Wesseling *et al.*, 1988, cited by Philbin, 1998; Mallapur *et al.*, 2005). Likewise black rats, bank voles and African striped mice caught from the wild and caged as adults also typically show no stereotypies (Sørenson and Randrup, 1986; Callard *et al.*, 2000; Schoenecker *et al.*, 2000; N. Pillay, personal communication, Johannesburg, 2004), although these emerge rapidly in most of their cage-born offspring. In carnivores the picture is less clearcut, but wild-caught individuals still never seem more stereotypic than captive-borns: two studies of polar bears and other ursids detected no differences (van Keulen-Kromhout, 1978; Ames, 2001), but a more recent study's preliminary analyses suggest that wild-caught male polar bears pace less than zoo-born males (D. Shepherdson, personal communication, Oregon, 2005); while on an experimental fur farm, pacing was found in only *ca.* 25% wild-caught beech martens, but in 100% of their offspring (Hansen, 1992, who reports too that abnormal fur-chewing was absent in wild-caught but common in farm-born martens). In surveys, zoo-keepers also score zoo-born black rhinoceroses as more stereotypic than wild-caught animals (K. Carlstead, personal communication, Hawaii, 2004). My one avian case contains the only (partial) counter-example: caged wild-caught blue jays spent far more time route-tracing than did hand-reared birds; however, they were much less prone to stereotypic spot-pecking (Keiper, 1969).

Overall, these data are consistent with the idea that being raised in naturalistic environment protects animals against later stereotypy, while being raised in spatially restrictive and/or barren captive environments predisposes animals towards the behaviour (see Lewis *et al.*, this chapter). However, captive-rearing often involves some social deprivation too – most laboratory rodents and primates are removed from their mothers earlier than would happen naturally, for instance (Chapter 6, this volume) – and so the developmental importance of physical complexity *per se* cannot be determined from the cases given here.

presents evidence that wild-caught animals are less stereotypic when captive-housed than are animals born and raised in such conditions. These findings together argue for a neuroprotective effect of early enrichment against later stereotypy development. Furthermore, consistent with this neuroprotection hypothesis, a recent longitudinal study of early environmental enrichment in children showed that children receiving a structured exercise, nutrition and educational programme between the ages of 3 and 5 were significantly less likely later to exhibit symptoms of schizotypy (response patterns that mimic schizophrenia) at the ages of 17 and 23 (Raine *et al.*, 2003).

Using the environmental complexity paradigm previously described, we investigated these issues further, to explore: (i) whether exposure to a more complex environment later in development successfully ameliorates the stereotypy observed in deer mice previously housed in standard cages, and if so, whether this holds over a range of ages; and also (ii) whether early exposure to a more complex environment prevents the subsequent development of stereotypy if animals are transferred to restrictive environments later in development. We found, first, that both early (at weaning) and later (60 days after weaning) exposure to environmental complexity resulted in low rates of stereotyped behaviours (Powell *et al.*, 2000) – as consistent with the work of Ödberg (1987) on similarly aged voles. Thus both the early and later exposure to enrichment reduced the frequency of stereotyped behaviour, and resulted in fewer mice categorized as stereotypic. We recently replicated this outcome in a subsequent study (Hadley *et al.*, 2006), where a more complex environment provided after an initial period of 60 days of standard caging post-weaning, again reduced stereotyped behaviour to about the same level as exposure to the more complex environment initiated at weaning (mean stereotypic responses per hour were 886.2 ($\pm$ 157.4) for the late enrichment animals and 640.3 ($\pm$ 168.1) for the early enrichment group). However, in this further work we also found that older animals (11–14 months old) did not, in contrast, appear to significantly benefit from being exposed to environmental complexity, in terms of their stereotypy reduction. Like the Cooper *et al.* (1996) example given above, this suggests some limit to CNS plasticity.

Turning to our second question, our findings also indicate that early exposure to a larger, more complex environment conferred significant protection when our animals were moved to standard cages (Powell *et al.*, 2000; Hadley *et al.*, 2006). For example, when observed in their home cages, a significantly higher proportion of deer mice stereotyped after 60 days post-weaning in standard housing and after 120 days of such housing, than if housed for 60 days of standard housing after 60 days of enriched housing (see Table 7.2). Furthermore, their stereotypy rates in a standard test environment showed that, as before, 60 days of housing in a more complex environment following weaning substantially reduced or prevented stereotypy. However, exposing these animals to standard cage housing for a subsequent 60-day period resulted in significantly less stereotypy (907.2 ($\pm$ 226.2) stereotypy counts per hour) than that seen in animals maintained in standard cages throughout the 120 days (1820.1 ($\pm$ 282.1) stereotypy counts per hour) (Hadley *et al.*, 2006).

Such data might thus suggest that the timing of enrichment is more important than its duration. Indeed with respect to the duration of environmental enrichment, our unpublished data indicate that 30 days of environmental complexity immediately following weaning is as effective as 60 days of such experience in preventing stereotypy development.

Table 7.2. The proportion of deer mice displaying stereotypies in standard cages, as affected by prior experience (from Experiment 1, Powell *et al.*, 2000).

Housing	% standard-housed subjects with stereotypies at the end of Phase 1	% standard-housed subjects with stereotypies at the end of Phase 2
Early enrichment: enriched housing for 60 days post-weaning (Phase 1), standard for 60–120 days post-weaning (Phase 2)	N/A	13 (2/15)
Late enrichment: standard housing for 60 days post-weaning (Phase 1), enriched for 60–120 days post-weaning (Phase 2)	69 (11/16)	N/A
Control: standard housing throughout	77 (10/13)	67 (6/9)

However, to our knowledge no other work has yet been done to systematically examine how duration of exposure to more complex environments affects the prevention/amelioration of stereotypy. The ability of an organism's stereotypies to benefit from environmental complexity would thus seem affected by the developmental age of the individual, and perhaps also the duration of exposure to such an environment. But, what can studies of the effects of complex environments on brain functioning add to this picture?

7.3.2. Complex environments and CNS changes: effects of the timing and duration of exposure

Several studies have looked at how the timing of enrichment relative to an animal's stage of development, and also the length of exposure to complexity (two issues that are often confounded) influence its impact on the brain.

One early study using rats suggested that just 2 h of increased environmental complexity daily for a period of 30 or 54 days post-weaning resulted in similar alterations in neurotransmitters and brain weight as continuous 24-h exposure to enrichment over the same periods of development (Rosenzweig *et al.*, 1968). A later study also indicated that as little as 4 days in a more complex environment was sufficient to induce alterations in dendritic morphology in weanling rats (Wallace *et al.*, 1992). Likewise, Rampon *et al.* (2000a) documented significant alterations in gene expression with as little as 2 days of environmental complexity for 6 h per day, even in adult animals. Finally, mice exposed to different durations (2.5, 15 or 25 months) of environmental complexity appear to benefit equally as indexed by improved cognitive functioning (Kobayashi *et al.*, 2002). Focusing instead on timing, exposing rats to a more complex environment for 30 days, starting at either 50 days of age (juveniles) or 105 days of age (adults), also resulted in similar changes in the brain (Rosenzweig and Bennett, 1996). Enrichment effects in adult rats were also demonstrated by Kempermann *et al.* (2002) who demonstrated that hippocampal neurogenesis in mice living in more complex environments from age 10 to 20 months was fivefold higher than in standard cage controls.

Perhaps surprisingly, in the light of some of the stereotypy examples given above, exposure to a more complex environment thus appears to have significant effects on brain structure and function across a variety of ages and durations of exposure (an issue we revisit in Section 7.4 when discussing neuronal plasticity). These results indicate that – despite ideas about age and loss of plasticity – environmental manipulations providing increased complexity for almost any duration, at almost any age, can result in improved learning and memory.

Much less information is available regarding the persistence of enrichment effects on the brain after the organism is returned to a restricted environment. Camel *et al.* (1986) demonstrated that an environmental complexity-induced increase in dendritic arborization and synapse number persisted for at least 30 days after the end of an experimental housing period. It might be argued that exposure to a more complex environment earlier rather than later in development should confer greater persistence of effects or neuroprotection, but such information does not as yet seem available.

7.3.3. Key features of stereotypy-attenuating environments

Despite many studies documenting impressive effects of increased environmental complexity on brain and behaviour, relatively little experimental work has addressed the question of what specific key factors are responsible for these outcomes. In practice, environmental complexity paradigms usually share several commonalities (although they may differ across species in their details), and provide multiple factors together including increased space, increased number of conspecifics and more novelty or unpredictability (Chapters 3 and 9, this volume, further discuss the many confounded properties of typical enrichments. For a comprehensive review of methods of rodent environmental enrichment, preferences and variability across research, also see Olsson and Dahlborn, 2002). Thus, for example, environmental complexity often includes both social and inanimate factors, the many effects of which are often confounded (Schrijver *et al.*, 2002), and the differential impact of social versus inanimate features of the environment on stereotypy has not been systematically examined (though see Chapter 6, for some valuable data from primates). In addition, physical complexity is often provided along with more space, although here, some ethological studies teasing their relative effects apart are available. For instance, Ödberg (1987) demonstrated that when bank voles reared in enriched environments exhibited less stereotypy, this effect was independent of cage size; and see Chapters 2 and 6, this volume, for further studies.

In many environmental complexity protocols, a variety of objects are placed in enriched cages and those objects are rotated frequently, in some cases daily, providing novel configurations of objects and their placement to animals. However, again the importance of such variability and novelty, independent of other features of the environment, has not been assessed systematically. We are not aware of any studies examining the effects of novelty *per se* on spontaneous stereotypy. In our own studies, as we have reviewed, we place a variety of objects (e.g. inverted baking pan, Habitrail tubes, plastic toys, mesh structures for climbing) in enriched cages, and we rotate those objects weekly providing novel configurations of objects and their placement to the mice. We have not, however, assessed systematically the importance of this last manipulation, independent of other features of the environment. Interestingly, studies concerned with drug-induced stereotypy have demonstrated decreased stereotyped behaviour within novel environments and increased stereotyped behaviour within familiar environments (Sahakian and Robbins, 1975; Russell and Pihl, 1978; Einon and Sahakian, 1979).

As well as being predictable, impoverished or restricted environments often lack the specific stimuli, cues and physical dimensions necessary for the individual to engage in species-typical behaviour. As reviewed in Chapters 2–4, this volume, the inability to engage in species-typical motor behaviours is likely the very root of many stereotypies. This inference is supported, for example, by the analysis of stereotypic digging in Mongolian gerbils (by Wiedenmayer, 1997), discussed in Chapter 4, this volume. In these experiments, access to loose sand did not prevent the development of stereotypic digging in the animals' home cage. Conversely, access to burrows of a particular type prevented the development of stereotypic digging. In neuroscience studies, one mainstay of enriched environments for rodents is a wheel that allows running (although see Box 4.2, Chapter 4, this volume, for a discussion of whether or not wheel-running should itself be considered a stereotypy). Access to voluntary exercise is considered essential for enrichment-related hippocampal neurogenesis and neurotrophin expression (van Praag *et al.*, 1999; Cotman and Berchtold, 2002; Ehninger and Kempermann, 2002). Furthermore, in humans, movement restriction may be operative in the stereotypies observed in typically developing children. Thelen (1980), for example, found that infants characterized as 'high stereotypy' (based on a median split) spent much more time in infant seats, playpens and high chairs than did their 'low stereotypy' counterparts. However, despite these data, no study has as yet assessed the importance of running wheels for the prevention/amelioration of stereotypy.

Finally, no systematic attention has been directed towards the issue of individual differences in response to environmental complexity. Presumably, as our own stereotypy data suggest (see e.g. Table 7.1), not all animals benefit or benefit equally from larger, more complex environments (an issue we return to later, in Section 7.5). However, whether this stems from differential access to key resources, or from some other reason, remains unknown.

The unanswered research questions here are important because as well as being differentially effective, the various attributes of enrichment might work in different ways. It does, for instance, appear that the isolation-housing of naturally social species, regardless of the inanimate environment, enhances emotional reactivity (e.g. freezing) to novel stimuli, while a more complex environment, independent of the number of animals, accelerates animals' habituation to novelty and improves spatial learning and memory (e.g. Schrijver *et al.*, 2002).

7.3.4. Summary

To summarize Section 7.3, it appears that increased environmental complexity can robustly attenuate the development of stereotypy in a number of species. Such effects are likely to be observed with varying durations of exposure to enriched housing, and at varying developmental periods. However, older animals sometimes appear to benefit least; and early environmental complexity can also have effects that remain when animals are transferred to more barren conditions. This raises unanswered questions about possible sensitive periods. Furthermore, relatively little is yet known about what specific features of environmental enrichment are responsible for such effects, although (as we also see in Chapters 2–4, this volume), certain environmental features appear differentially effective for specific species and topographies of stereotypy.

7.4. Environmental Complexity and CNS Function

The efficacy of environmental complexity and enrichment in attenuating stereotypy leads to the question of neurobiological mechanisms. How does the experience of a more complex environment alter brain function so as to prevent or attenuate the development of abnormal behaviour? This is a challenging question, as environmental enrichment has been reported to be associated with a myriad of CNS effects (and furthermore, as discussed in the preceding paragraphs, enrichment paradigms often involve multiple attributes). Some of these CNS effects are reviewed in the following section, followed by a discussion of the specific complexity-induced brain changes that seem associated with the prevention or reversal of stereotypy.

Examining how exposure to more complex environments affects brain function has a long history. As early as the 1940s, Hebb exposed rats to a complex environment (his home) as an experimental manipulation and demonstrated that this resulted in increased maze learning and increased brain weights (Hebb, 1949). Although the specific methodology and paradigms employed since then have varied, work on the effects of environmental complexity on the CNS has continued. Starting in the early 1960s, Rosenzweig and colleagues examined environmental complexity effects on neuroanatomy (e.g. brain weight, cortical thickness, dendritic structure) and learning and memory (e.g. Rosenzweig *et al.*, 1968). The seminal contributions of Greenough and colleagues started in the 1970s and involved examination of the impact of environmental complexity on a variety of parameters of brain function (e.g. dendritic branching, spine density, synaptogenesis, angiogenesis, gliogenesis; e.g. Greenough *et al.*, 1985; Comery *et al.*, 1995). Recent work dating from the 1990s has further evaluated the effects of environmental complexity on a variety of molecular and cellular endpoints, including gene expression, apoptosis or programmed cell death and neurogenesis. Relevant to the topic of Abnormal Repetitive Behaviour, a number of these studies have involved animal models of CNS insult (a topic we explore further below).

Enrichment-related alterations in CNS structure translate to functional behavioural outcomes. For example, exposure to a more complex environment improves spatial memory acquisition and retention in a number of maze tasks (Wong and Jamieson, 1968; Schrijver *et al.*, 2002; Frick and Fernandez, 2003; Frick *et al.*, 2003). Enrichment can also improve retention in non-spatial tasks, object recognition tests and two-way active avoidance tasks (Escorihuela *et al.*, 1995). Additionally, enriched mice show improved memory for contextual fear conditioning (Duffy *et al.*, 2001). Beyond learning and memory, enrichment in rodents has also been shown to decrease voluntary alcohol consumption (Rockman and Borowski, 1986), decrease aggressive behaviour (Haemisch *et al.*, 1994) and increase foraging (Cheal, 1987). Additionally, enrichment decreases stress reactivity, as indexed by performance on an elevated plus maze (Fernandez-Teruel *et al.*, 1997), defecation in a novel environment (Fernandez-Teruel *et al.*, 1992), defensive responses to a predator (Klein *et al.*, 1994) and open field exploration (Widman and Rosellini, 1990). Francis *et al.* (2002) and Bredy *et al.* (2003) have even demonstrated that the stress hyper-reactivity resulting from maternal separation is completely attenuated by environmental enrichment. However, how any of these various enrichment-induced changes in responsiveness, cognition and behaviour relate to prevention or attenuation of stereotypy has not been established. It may, for example, be that alterations in stress responses (Chapter 8, this volume) are important, but these links have simply not as yet been investigated.

7.4.1. Neuronal plasticity

Environmental complexity is thought to exert its behavioural effects via selective, long-term alterations in synaptic efficacy. These alterations are achieved through differential transcription of neurotransmission-related target genes. For example, enriched rearing conditions reportedly produce significant elevations in the density or affinity of a variety of neurotrans-mitter receptors (e.g. Bredy *et al.*, 2003). Such changes in receptor and signalling molecules are paralleled by enhanced second-messenger cas-cades (Rampon *et al.*, 2000a). Thus enriched rearing conditions produce significant elevations in the density or binding affinity of α-amino-3-hydroxy-5-methyl-4-isoxazoleproprionic acid (AMPA) and *n*-methyl-D-aspartate (NMDA) glutamate receptors (Bredy *et al.*, 2003), D1 dopamine receptors (Anderson *et al.*, 2000), 5-HT$_{1A}$ serotonin receptors (Rasmuson *et al.*, 1998), kappa opioid receptors (Smith *et al.*, 2003) and both gluco-corticoid and mineralocorticoid 'stress' hormone receptors (Dahlqvist *et al.*, 1999, 2003). The expression of genes coding for such proteins is initially modulated by the activity of immediate early gene (IEG) products (e.g. Fos and Jun proteins), and the expression of these genomic regula-tory complexes is enhanced by environmental enrichment. As but one example, activity-regulated cytoskeletal gene (arc) is an IEG implicated in enrichment-induced neuronal plasticity. Specifically, given its site of translation in the post-synaptic density (Steward *et al.*, 1998), this IEG is thought to play a key role in dendritic growth processes associated with synaptic reorganization. This protein is significantly upregulated throughout much of the cortex and hippocampus of rats exposed to an enriched environment relative to standard-housed controls (Pinaud *et al.*, 2001). To sustain synaptic plasticity, neurons utilize endogenous signal-ling molecules that also promote cell survival, division, growth, differen-tiation and morphological plasticity. These neurotrophic factors (e.g. nerve growth factor (NGF), brain-derived neurotrophic factor (BDNF)) can also be robustly increased by environmental enrichment (for review see Pham *et al.*, 2002).

Until recently, it was widely accepted that the mature mammalian nervous system was incapable of neuronal proliferation. However, as we saw in Section 7.3, recent findings convincingly indicate that neurogen-esis actually continues throughout adulthood, and furthermore, such proliferative processes are significantly enhanced as a function of envir-onmental enrichment. For instance, as indicated in this earlier section, even long-term exposure to an enriched environment later in life (from 10 to 20 months in mice) produces a robust, fivefold enhancement of net neurogenesis relative to standard-caged animals (Kempermann *et al.*, 2002). As another example, exposure to a complex environment also significantly increases neurogenesis in the visual cortex (for review, see Kaplan, 2001), and hippocampus (Young *et al.*, 1999), even in adults. Access to voluntary wheel-running alone (as we have discussed, a com-mon feature of enriched environments for rodents) also increased neuro-genesis in adult rats (van Praag *et al.*, 1999) – and the mechanism by which physical activity increases neurogenesis appeared to be through enhanced neurotrophin expression (Cotman and Berchtold, 2002).

So to summarize, a large literature provides evidence of the effect of environmental enrichment on IEGs, neurotransmission-related genes, neurotrophic factors and neurogenesis. Such effects are consistent with the notion that the environment can exert a profound impact on the development and maintenance of neuronal plasticity that, in turn, will have important functional outcomes for the organism. Yet again, there has been little investigation of how such processes might be implicated in stereotypy-reduction.

7.4.2. Ameliorative or protective effects of complex environments in relation to insult and pathology

Finally, exposing animals to more complex environments has also been shown to attenuate or reverse the sequelae of such CNS insults as seizures, ischaemia, infarct, cortical lesion and traumatic brain injury (Kolb and Gibb, 1991; Johansson, 1996; Johansson and Ohlsson, 1996; Hamm et al., 1996; Young et al., 1999), and to be protective against genetic perturbations. For instance, improvements in motor deficits in mouse models of cerebellar degeneration (e.g. *Lurcher* mutants) (Caston et al., 1999) and non-spatial memory deficits in a glutamate receptor subtype (NMDA) knockout mouse model (Rampon et al., 2000b) have been reported. In this targeted knockout of CA1 NMDA receptor NR1 subunit, impaired spatial and non-spatial memory, object recognition, olfactory discrimination and contextual fear were attenuated by environmental enrichment. Furthermore, in the R6/2 transgenic mouse model of Huntington's disease, even limited environmental enrichment attenuated the loss of peristriatal cerebral volume, and slowed the decline of sensory-motor integration as measured by 'rotarod' performance (Hockly et al., 2002). Environmental enrichment has also been associated with resistance to age-related neurobiological impairments. In animal models, both the age-related decline in cognitive performance and the cellular markers that parallel this decline are reversed via exposure to increased environmental complexity. For example, enriched rearing conditions attenuate spontaneous cell death or apoptosis (Young et al., 1999), and the compromised spatial memory, synaptophysin expression, glutamic acid decarboxylase (GAD) activity and hippocampal synaptic density associated with ageing (Saito et al., 1994; Frick and Fernandez, 2003; Frick et al., 2003).

Collectively, these findings provide clear evidence for a protective effect of environmental enrichment against challenge-related impairments in behaviour and neurobiological function. However, once again, how such effects might play a role in stereotypy *per se* is unknown.

7.5. Environmental Complexity, CNS Function and Stereotypy

A critical question for our research group has therefore been which of the myriad of brain changes associated with increased environmental complexity operate in preventing repetitive behaviour. In three related studies (Turner et al., 2002, 2003; Turner and Lewis, 2003), we examined the enrichment-related changes in neuronal structure and function, and compared them with our animals' degree of stereotypic behaviour. In all the studies, deer mice were, as described in previous sections, reared in enriched or standard cages for 60 days post-weaning, before being tested in our automated photocell detectors, and classified as 'low stereotypy' or 'high stereotypy'. This testing paradigm thus yielded four distinct groups: enriched high stereotypy, enriched low stereotypy, standard cage high stereotypy, and standard cage low stereotypy. This four-group, 2×2 design thus allowed us to avoid confounding behavioural outcome (high stereotypy/low stereotypy) with housing condition (standard/enriched). Apart from our studies, we are not aware of any published literature identifying neurobiological mechanisms that mediate the ameliorative or preventative effects of environmental complexity on stereotypies. We therefore review our studies in the three subsections below.

7.5.1. Neuronal metabolic activity

Our initial studies assessed whether environmental enrichment-related effects on the development of stereotyped behaviour in deer mice were associated with alterations in neuronal metabolic activity (Turner *et al.*, 2002). Neuronal activity was assessed using cytochrome oxidase (CO) histochemistry, an index of oxidative energy metabolism. CO indexes long-term changes in neuronal functional activity and has been shown to correlate with indices of activity-dependent plasticity (Poremba *et al.*, 1998; Wong-Riley *et al.*, 1998).

In terms of CO activity in the motor cortex, striatum, nucleus accumbens, thalamus and hippocampus, our findings revealed a clear and striking environment by behaviour interaction. All stereotyping animals had relatively low levels, regardless of their housing type. So too did the standard-housed mice which did not develop stereotypies; and these three groups did not significantly differ. However, the non-stereotyping, enriched-housed animals were distinct: they had relatively high CO activity in all of these brain regions. These results suggest that, in a model of spontaneous stereotypy, the enrichment-related prevention of stereotyped behaviour is associated with increased neuronal metabolic activity particularly in motor cortex and basal ganglia.

7.5.2. Dendritic morphology

Next, we evaluated whether the environmental enrichment-related effects on the development of stereotyped behaviour in deer mice were associated with alterations in dendritic morphology (Turner *et al.*, 2003). Dendritic alterations have been shown to be sensitive measures of experience-dependent changes in brain structure. Dendritic morphology was assessed in layer V pyramidal neurons of the motor cortex, medium spiny neurons of the dorsolateral striatum and granule cells of the dentate gyrus using Golgi-Cox histochemistry (Gibb and Kolb, 1998). These brain regions were selected based on our CO findings suggesting the importance of cortical-basal ganglia circuitry and relative lack of importance of limbic areas.

Once again, we found an environment by behaviour interaction, with the low stereotyping, enriched-housed animals differing from each of the three other groups, which in turn were statistically indistinguishable from each other. The enriched low-stereotypy mice exhibited significantly higher dendritic spine densities in the motor cortex (see Table 7.3), and also the striatum, but not hippocampus, compared with the other groups, suggesting that enrichment-related prevention of stereotyped behaviour is associated with increased dendritic spine density in neuroanatomical loci comprising cortical-basal ganglia circuitry.

7.5.3. Neurotrophic factors

As indicated previously, neurotrophins promote neuron survival and growth, and play an important role in use-dependent plasticity. Changes in neurotrophin expression would thus seem a likely candidate mechanism mediating the changes in neuronal metabolic activity and dendritic morphology observed in our enriched, low-stereotypy mice. Therefore, we evaluated whether environmental enrichment-related effects on the development of stereotyped behaviour in deer mice were associated with alterations in neurotrophin levels (Turner and Lewis, 2003). The motor cortex, striatum and hippocampus were dissected, and the levels of BDNF and NGF in each brain region were analysed using ELISA kits. There were no differences in either NGF or BDNF in either the motor cortex or the hippocampus. However, in the striatum, just as we predicted the enriched low-stereotypy mice exhibited significantly more BDNF (but not NGF) than the enriched high stereotypy and standard cage mice (see Fig. 7.1). These results provide evidence that the enrichment-related prevention of stereotyped behaviour in deer mice is associated with increased BDNF in the striatum.

Table 7.3. Number of dendritic spines per micron in motor cortex pyramidal neurons.

Rearing environment	Low stereotypy	High stereotypy
Environmental enrichment	1.86 ± 0.09^a	1.06 ± 0.06^b
Standard cage	1.00 ± 0.3^b	0.98 ± 0.04^b

Values with different letters (a and b) are significantly different.
NB. Similar results were found for the dorsolateral striatum (see Turner *et al.*, 2003).

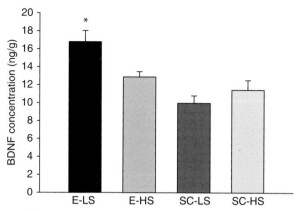

Fig. 7.1. Striatal BDNF protein concentrations (ng/g wet weight) from enriched low stereotypic (E-LS; $n = 8$), enriched high stereotypic (E-HS; $n = 7$), standard-caged low stereotypic (SC-LS; $n = 9$) and standard-caged high stereotypic (SC-HS; $n = 10$) deer mice (from Turner and Lewis, 2003). * = Signifies a significant difference from all other groups.

7.5.4. Summary

The studies in Section 7.5 thus represent our initial efforts to identify the brain changes mediating the effects of environmental complexity on stereotypies. As indicated in Section 7.4, environmental complexity induces a large number of alterations in brain function and structure. None the less, our findings do indicate particular structural and functional differences associated not just with enrichment, but specifically with the enrichment-related prevention of stereotypy. Moreover, these differences were found in brain areas that comprise part of the cortical-basal ganglia motor circuits. This is a particularly significant outcome since several lines of evidence support the importance of such circuitry in the expression of repetitive behaviour (as in Chapters 5 and 8, this volume). These lines of evidence are reviewed and pursued further in the following section.

7.6. Abnormal Repetitive Behaviour and Cortical-basal Ganglia Circuitry

Our environmental complexity findings, as surveyed above, suggest alterations in cortico-basal ganglia-cortical circuitry in the development of stereotypies. In this section, we therefore review this circuitry and show evidence from other work, especially clinical findings and drug studies, implicating cortical-basal ganglia circuitry in repetitive behaviour. Since our environmental complexity findings only indirectly indicate involvement of this circuitry, we will also provide here more direct evidence that cage stereotypies in rodents are associated with perturbations in the cortical-basal ganglia-cortical motor loop.

7.6.1. Cortical-basal ganglia circuitry

The motor circuit we hypothesize to mediate the expression of cage
stereotypy originates in the pre- and primary motor cortices, travels
through the direct (or striatonigral) pathway of the basal ganglia, and
ultimately terminates on thalamocortical relay neurons that stimulate
the supplementary motor cortex to provide positive feedback to selected
motor programmes in the primary motor cortex (see Olanow *et al.*, 2000
for review). This type of circuit is shown diagrammatically in Fig. 7.2, and

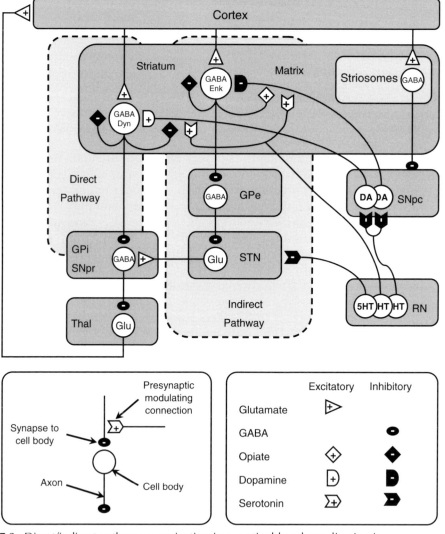

Fig. 7.2. Direct/indirect pathway organization in a cortical-basal ganglia circuit.

(*Legend continued on next page*)

discussed further by Garner in Box 7.2. It is modulated by midbrain input via the nigrostriatal pathway, as well as by projections from the frontal cortex (see Fig. 7.2).

We postulate that stereotypic behaviour is expressed as a consequence of abnormal facilitation of selected motor programmes due to imbalanced activity between the direct pathway, and the 'indirect' (or striatopallidal) basal ganglia pathway (see Chapter 5, this volume and Box 7.2). Simplistically, the direct pathway of the basal ganglia promotes movement, whereas the indirect pathway of the basal ganglia suppresses movement. So, for example, the impairment of voluntary movement associated with Parkinson's disease is thought to reflect relative overactivity of the indirect pathway, whereas the excess movement characteristic of Huntington's disease is thought to reflect relative overactivity of the direct pathway (Graybiel, 2000).

In the direct pathway, D1 dopamine receptors are colocalized with glutamate receptors on medium spiny neurons of the striatum that use gamma aminobutyric acid or GABA as a neurotransmitter (Steiner and Gerfen, 1998; Gerfen, 2000; Olanow *et al.*, 2000). These direct pathway neurones also express the neuropeptides dynorphin (see Fig. 7.2) and

---

Fig. 7.2. *Continued*
J.P. Garner

This figure shows the connections of a generic corticostriatal circuit loop, using the circuitry of the dorsal striatum as an example. Several corticostriatal loops – see Box 7.2 and Chapter 11, this volume – run between the basal ganglia areas of the frontal and motor cortex (including the orbitofrontal 'limbic' areas, prefrontal 'executive' areas, premotor 'motor planning' areas and primary motor areas). All are involved in controlling and sequencing behaviour. Although anatomically distinct, the general connections of each loop are the same. Distinct brain areas here are shown in dark grey shaded boxes. The cortical area in which a loop originates (e.g. premotor and primary motor cortex, for the 'motor' loop implicated in this chapter) is indicated in the figure. The striatum is the 'input' area of the basal ganglia, and is traditionally divided into the dorsal striatum (caudate and putamen) and ventral striatum (nucleus accumbens and part of the olfactory tubercle). Although this figure (and chapter) focuses on the dorsal striatum, note that a very similar circuit runs through the ventral striatum and ventral pallidum, and appears involved with sequencing behavioural aspects of emotion and reward (see Chapter 8, this volume). Note too that while the midbrain dopaminergic input to this illustrated circuit is via the nigrostriatal pathway from the substantia nigra pars compacta, SNpc, the midbrain dopaminergic input to the ventral striatal circuit is from the ventral tegmental area, via the mesoaccumbens pathway (see Chapter 8, this volume). Other areas are labelled with standard primate nomenclature: GPi, Globus pallidus internal segment; GPe, Globus pallidus external segment; SNpr, Substantia nigra pars reticulata; STN, Subthalamic nucleus; Thal, Thalamus; RN, Raphe nuclei. GPi and SNpr appear to be only arbitrarily divided by a white matter tract in primates, in the same way that the caudate and putamen are only arbitrarily divided by a white matter tract (Albin *et al.*, 1995). In rodents, these arbitrary divisions do not occur, and so the dorsal striatum is also called the caudatoputamen; and the equivalent of the GPi/SNpr is called the SNpr, and the equivalent of the GPe is called the Globus Pallidus. For clarity, modulatory circuitry involving acetylcholine, substance P and adenosine has been omitted, as have modulatory projections to the striosomes and frontal cortex. Projections whose innervation of particular cells is unknown or complex are depicted as projecting to the brain area rather than a particular cell (e.g. the inhibitory serotonin projection to STN).

substance P and A1 adenosine receptors (an issue we return to later). The D1 receptors are positively coupled to adenyl cyclase stimulation, and when activated, they increase the overall excitability of neurons (Aosaki *et al.*, 1998; Price *et al.*, 1999). Therefore, in medium spiny neurons of the direct pathway, dopamine acts to amplify the excitatory corticostriatal input, resulting in increased GABAergic inhibition of the major inhibitory output nuclei of the basal ganglia, the substantia nigra

---

Box 7.2. Direct–Indirect Pathway Organization, Modulation and Drug Effects

J.P. GARNER

Different corticostriatal loops – e.g. limbic, prefrontal and motor loops – sequence different components of behaviour (e.g. sequencing goals, behaviours, movements, etc.), depending on the cortical area involved (see Chapters 5 and 11, this volume; Alexander and Crutcher, 1990; Rolls, 1994; Albin *et al.*, 1995). The basic circuitry is mediated by glutamate (Glu) and GABA-releasing cells, which form two distinct branches within each loop: the indirect (or striatopallidal) pathway, and the direct (or striatonigral) pathway (see Fig. 7.2). The sum action of the direct pathway is to increase the activity of neurons in the thalamus, while the sum action of the indirect pathway is to inhibit neurons in the thalamus. Thus, under normal circumstances the direct pathway activates behaviour, and the indirect pathway inhibits behaviour, at the level of control that each loop is responsible for (Alexander and Crutcher, 1990; Albin *et al.*, 1995). Because inhibition by the indirect pathway prevents inappropriate behaviours and movements, down-regulation of the indirect pathway appears to be integral to unwanted movements (including stereotypy-like symptoms: Chapter 5, this volume) in human disorders (Albin *et al.*, 1995). Conversely, down-regulation of the direct pathway leads to a decrease in the activation of all behaviours (Albin *et al.*, 1995).

   These pathways are modulated by opiates (Dyn, dynorphin; Enk, enkephalin), dopamine (DA), serotonin (5HT) and several other neurotransmitters. The same general indirect/ direct pathway organization, and modulatory connections are present in birds, and probably all chordates (e.g. Reiner *et al.*, 1998). Dopaminergic projections from the SNpc to the dorsal striatum comprise the mesostriatal or nigrostriatal pathway (see Fig 7.2; also Box 7.3), and activate post-synaptic D1 receptors in the direct pathway, D2 receptors in the indirect pathway. In general terms, dopamine release thus increases the sensitivity of the entire system, increasing the likelihood of all behaviours being elicited (Robbins, 1997), through stimulating the direct pathway and inhibiting the indirect pathway. Opiate release mediates medium-term feedback, correcting overactivity in both the direct pathway (dynorphin) and indirect pathway (enkephalin) (Steiner and Gerfen, 1998; see also Fig. 7.2, and Lewis *et al.*, this chapter). The role of serotonin (5-HT) in the basal ganglia is particularly complex. One function appears to be promoting 'attentional' and context-dependent inhibitory processes (Robbins, 1997). In addition to dopaminergic drugs, opiate and serotonergic drugs can both induce and reduce stereotypies, and these effects often involve the modulation of dopamine transmission (Curzon, 1990; Waddington *et al.*, 1990; Cabib, 1993). Broadly speaking, stereotypy can thus be reduced by drugs that activate the indirect pathway or suppress the direct pathway; while stereotypy is selectively induced by drugs that suppress the indirect pathway. This relatively simplified picture is by no means the whole story, however – for instance the 'striosomes' (see Fig. 7.2) play an as yet poorly understood role in some drug-induced stereotypies (Canales and Graybiel, 2000; Graybiel *et al.*, 2000).

pars reticulata (SNpr) and the medial globus pallidus (GPi). This removal of inhibitory tone over the thalamocortical motor relay ultimately results in increased activity of the supplementary motor cortex neurons that provide positive feedback to active motor programmes.

In the indirect pathway, D2 receptors are colocalized with the glutamate receptors on striatal medium spiny neurons (Steiner and Gerfen, 1998; Gerfen, 2000; Olanow *et al.*, 2000). These neurons add-itionally express the neuropeptide enkephalin (see Fig. 7.2) and A2 adenosine receptors (again, an issue we return to later). These indirect pathway D2 receptors are negatively coupled to adenyl cyclase stimu-lation, and when activated they thus *reduce* the overall excitability of neurons (McPherson and Marshall, 2000; Nicola *et al.*, 2000; Pisani *et al.*, 2000). Therefore, in medium spiny neurons of the indirect pathway, dopamine acts to diminish the excitatory corticostriatal input, resulting in decreased GABAergic inhibition of the external globus pallidus (GPe), causing this nucleus to exert more inhibitory control over the subthalamic nucleus (STN). This increased inhibition of STN removes its excitatory influence on the inhibitory output nuclei of the basal ganglia, the GPi and SNpr, so that these become less activated, thus ultimately disinhibiting the thalamocortical relay neurons previously discussed. (NB. The importance of striatal D2 receptors in mouse strain differences and stress sensitization is discussed in Chapter 8).

7.6.2. Drug-induced repetitive behaviour and cortical–basal ganglia circuitry

Early experiments first established the importance of the basal ganglia in the mediation of repetitive behaviours by showing that dopamine or a dopamine agonist (e.g. apomorphine) injected into the corpus striatum induced stereotyped behaviour in rats (e.g. Ernst and Smelik, 1966; see also Boxes 7.2 and 7.3). Subsequent experiments stimulated other recep-tors in these pathways. For example, intrastriatal administration of a spe-cific type of glutamate receptor agonist, NMDA, also induces stereotypic behaviour that is often indistinguishable from dopamine agonist-induced stereotypy. Such stereotypy can be attenuated by intrastriatal administra-tion of the NMDA receptor antagonist CPP (Karler *et al.*, 1997). Notably, glutamatergic induction of stereotypic behaviour is not restricted to NMDA-sensitive glutamate receptors, but can also be influenced through modulation of other types of glutamate receptors in the dorsal striatum (Mao and Wang, 2000).

Given the capacity of intrastriatally administered glutamate agonists to induce stereotypy, it is not surprising that other intracortical manipula-tions enhancing the activity of excitatory corticostriatal projections also exacerbate the expression of stereotypy. For instance, administration of either the D2 antagonist sulpiride or the GABA antagonist bicuculline into the frontal cortex enhances the motor stimulatory effects of amphetamine

(Karler *et al.*, 1998; Kiyatkin and Rebec, 1999). Conversely, amphetamine-induced stereotypy can be attenuated via intracortical infusion of DA or GABAergic agonists (Karler *et al.*, 1998). It is thought that midbrain dopaminergic projection neurons regulate the excitability of these corticofugal efferents through activation of GABAergic cortical interneurons, as well as through direct interaction with cortical pyramidal neurons.

Experiments in which the expression of drug-induced stereotypy was shown to be sensitive to manipulations in the SNpr of the direct pathway and the STN of the indirect pathway also support the hypothesized role of these pathways in repetitive behaviours. Specifically, intranigral GABA agonist administration induces intense stereotypy in rats (Scheel-Kruger *et al.*, 1978), and administration of a serotonergic (5-HT$_2$) antagonist into the STN reduces stereotypy. These manipulations are expected to have altered either directly (intranigral GABA agonist administration) or indirectly (intra-STN 5HT$_2$ antagonist administration) inhibitory GABAergic tone over thalamocortical relay neurons (Brunken and Jin, 1993) such that manipulations disinhibiting thalamocortical projections induced stereotypy, whereas stereotypy was attenuated by manipulations increasing inhibitory tone in the thalamus (Barwick *et al.*, 2002). Similarly, direct injections of opiate agonists into the substantia nigra produce intense stereotypies in rats (Iwamoto and Way, 1977), presumably due to disinhibition of nigrostriatal dopaminergic projections, as this manipulation has been shown to elevate striatal dopamine release in mice (Wood and Richard, 1982).

Pharmacological studies have also consistently shown that manipulations which enhance activity of the direct (striatonigral) pathway or which inhibit activity of the indirect (striatopallidal) pathway induce motor stereotypy (Ernst and Smelik, 1966; Scheel-Kruger and Christensen, 1980; Chen *et al.*, 2000; Salmi *et al.*, 2000) (though see Box 7.2, and Chapter 5, this volume for a take on such data that emphasizes inhibition of the indirect pathway in stereotypy *per se*). Moreover, the intense stereotypy induced by psychostimulant drugs is associated with increased expression of IEGs in striatonigral neurons (Canales and Graybiel, 2000). Conversely, as reviewed in the next section, non-drug-induced stereotypies are attenuated via pharmacological manipulations that, in mice, inhibit activity of the direct pathway (Presti *et al.*, 2003) or, in humans, enhance activity of the indirect pathway (Remington *et al.*, 2001). Interestingly, transgenic animals that selectively overexpress ΔFosB (an IEG involved in the modulation of neuronal excitability) in striatonigral projection neurons exhibit excessive wheel-running, whereas wheel-running is significantly inhibited in animals that overexpress the gene in striatopallidal projection (indirect pathway) neurons (Werme *et al.*, 2002; see Box 4.2, Chapter 4 for discussion of wheel-running as a stereotypy). Patients with obsessive–compulsive disorder (OCD) have also been reported to have fewer striatal D2 dopamine receptors, suggesting a loss of activity of the indirect or striatopallidal pathway (Denys *et al.*, 2004). (The symptoms of OCD and their relationship with stereotypies are discussed further in Chapters 5 and 10, this volume.) These studies together support the hypothesis that repetitive behaviours are expressed as a result of imbalanced activity along the direct and indirect pathways of the basal ganglia and that this imbalance is characterized by a relative increase in striatonigral tone. These studies also support our preliminary conclusion that environmental complexity prevents stereotyped behaviour by normalizing cortical–basal ganglia activity, including potentially restoring the balance between the direct and indirect pathways.

It is important to note, however, that while rats are often used as the model in drug-induced stereotypy experiments, this is perhaps somewhat paradoxical given their negligible levels of spontaneous cage stereotypy (Chapter 4, this volume). Furthermore, drugs that induce stereotypy do not always enhance animals' spontaneous cage-induced stereotypies, and often elicit forms of stereotypy that are quite different in form (see e.g. Box 5.4, Chapter 5, this volume; see also Box 7.3, this chapter). For example, in our own studies, neither systemically nor intrastriatally administered apomorphine increased spontaneous cage stereotypies in deer mice, although other repetitive behaviours (e.g. stereotyped sniffing) were observed (Presti *et al.*, 2002, 2004). These results were also consistent with work done showing that apomorphine did not affect spontaneous stereotypies in bank voles, and nor did the NMDA antagonist MK-801 (Vandebroek and Odberg, 1997; Vandebroek *et al.*, 1998). Thus although cage- and drug-induced stereotypies share important similarities, they are not identical phenomena.

7.6.3. Non-drug-induced repetitive behaviour and cortical–basal ganglia circuitry: clinical findings and further animal studies

The view that spontaneous and persistent repetitive behaviour – including stereotypies but also other repetitive traits – is linked to dysfunction of the neural circuits that transmit information between the cortex and basal ganglia is supported by several lines of evidence. For example, functional magnetic resonance imaging (fMRI) data indicate that, in subjects with OCD, symptomatology was significantly correlated with alterations in activation of the orbitofrontal cortex (Adler *et al.*, 2000). Furthermore, anatomical MRI analysis has revealed alterations in putamen volume within the basal ganglia of individuals suffering from trichotillomania (repetitive hair-pulling) (O'Sullivan *et al.*, 1997). Similarly, magnetic stimulation and fMRI studies demonstrate increased cortical excitability and abnormal corticobasal ganglia activation, respectively, in individuals with Tourette's syndrome, a disorder characterized by repetitive motor tics (Berardelli *et al.*, 2003). A relationship between repetitive behaviours and the basal ganglia has also been found in autism and related disorders. Reiss *et al.* (1995) have shown that in Fragile X syndrome, dorsal striatal volume is significantly correlated with stereotypies; while in a recent MRI study of autism (a disorder that has abnormal repetitive behaviour as a diagnostic criterion), caudate volume was associated with compulsions, rituals,

---

Box 7.3. A Brief Further Note on Psychostimulant-induced Stereotypies

D. Mills and A. Luescher

It has long been known that large doses of dopaminergic drugs, such as amphetamine or apomorphine, induce stereotypic behaviour in rodents and primates (as reviewed in this chapter, and Chapter 8, this volume). Dopamine injected into the caudate nucleus of the basal ganglia also produces stereotypic behaviour in cats (Cools and van Rossum, 1970). The fact that such pharmacologically induced stereotypic behaviours are aggravated by stress, and some environmentally induced stereotypic behaviours are enhanced by amphetamine, seems to indicate some commonality (see e.g. Chapter 8). However, differences between the stereotypies produced by these very different processes have also been recognized, e.g. in pigs and rodents (Terlouw *et al.*, 1992a,b; also see Lewis *et al.*, this chapter, plus Box 5.4, Chapter 5), and care is thus needed when extrapolating the information from one to the other (Mills, 2003). For example, when inducing a stereotypic behaviour pharmacologically, the doses of drugs used mean that the consequent tissue levels of neurotransmitter might be many times higher than the natural endogenous level. Thus, the drugs may be operating by chemically flooding the system to produce an extreme response that only resembles the 'natural' condition at the phenomenological level (face validity) but not the mechanistic one (construct validity) (see also Chapter 10, this volume).

This problem aside, it is interesting to note that at least two different brain dopamine systems are implicated in drug-induced stereotypy. The nigrostriatal dopamine system containing the caudate nucleus (and focused on by this chapter) mediates mainly oral stereotypic behaviour, at least in rats. However, the mesolimbic dopamine system including the nucleus accumbens (as focused on in Chapter 8, this volume) is also affected, and this activates increased locomotion. In apomorphine-induced stereotypic behaviour in rats, such locomotion appears first (or at lower doses of apomorphine), while the oral stereotypic behaviours appear later (or at higher doses of apomorphine) (e.g. Teitelbaum *et al.*, 1990). At present, the relevance of the distinction between these two effects for captivity-induced stereotypies has not been fully established, although one hypothesis is presented in Chapter 8, this volume, and discussed further in Chapter 11.

---

difficulties with minor change and complex motor mannerisms (Sears *et al.*, 1999).

Clinical findings also implicate neurotransmitter systems associated with the anatomical loci discussed above in the mediation of stereotypic behaviour. Our studies of individuals with mental retardation have shown that stereotyped behaviour is associated with alterations in indirect estimates of central dopamine function. One such indirect estimate is spontaneous blink rate. For example, unmedicated Parkinson's disease patients have low spontaneous blink rates, which normalize with L-DOPA treatment (Karson, 1983). We demonstrated that individuals with mental retardation who exhibit stereotypies also show substantially lower spontaneous blink rates than age, sex and IQ-matched controls (Bodfish *et al.*, 1995). We also demonstrated lower plasma concentrations of the dopamine metabolite homovanillic acid (HVA) in the same population (Lewis *et al.*, 1996). The hypo-dopaminergic state that this suggests might at first sight seem paradoxical, but this result merely re-emphasizes how dysregulation of cortical–basal ganglia circuitry can result from any number of perturbations, and how the actions of dopamine itself vary in different parts of these systems. Finally, individuals with mental retardation and stereotypies were also found to have deficits in a postural control task, also indicative of basal ganglia dysfunction (Bodfish *et al.*, 2001). As we saw in the previous section, both dopaminergic and serotonergic systems are integral modulators of the cortical–basal ganglia circuit hypothesized to control abnormal repetitive behaviour. These neurochemical systems have been implicated in the aetiology of autism through several genetic analyses of candidate genes. Such analyses have revealed disruption of alleles associated with dopaminergic biosynthesis pathways (Smalley *et al.*, 2002) and receptor loci (Feng *et al.*, 1998), as well as polymorphisms of a locus regulating transcription of the serotonin transporter gene (Yirmiya *et al.*, 2001) in autistic individuals. Similarly, the concentrations of dopamine and serotonin metabolites in the cerebrospinal fluid (CSF) of individuals with Prader–Willi syndrome, a genetic disorder characterized by stereotypic skin-picking and hyperphagia, are significantly elevated relative to those in healthy controls (Akefeldt *et al.*, 1998). A central theme of these findings is that disruption of these monoaminergic neurotransmitter systems, so critical to the modulation of corticobasal ganglia–cortical circuitry, is associated with stereotypic behaviours in several distinct clinical populations.

Our studies of early socially deprived non-human primates further support the association of stereotypies and altered cortical–basal ganglia function. As we reviewed briefly in Section 7.2, stereotyped behaviour is a predictable consequence of early social deprivation in these species. We found that it was associated with dopamine receptor supersensitivity (Lewis *et al.*, 1990), and a loss of dopamine innervation in striatum and dopamine cells in the substantia nigra (Martin *et al.*, 1991) (see Chapter 6, this volume, for more details).

Returning to caged rodents, alpha-methyl-*para*-tyrosine (a catecholamine synthesis inhibitor that decreases dopamine levels) and L-DOPA (a catecholamine precursor that increases dopamine levels) have been shown to attenuate and increase, respectively, captivity-induced stereotypies in bank voles. Other activities, as measured by a general activity meter, were reported not to be altered. A dopamine-β-hydroxylase inhibitor, which decreases norepinephrine synthesis, had no significant effects on stereotyped behaviour in these animals, suggesting the specific importance of dopamine but not other catecholamines (Ödberg *et al.*, 1987).

In our own work, spontaneous jumping stereotypy in deer mice was attenuated selectively via intrastriatal administration of either the D1 dopamine receptor selective antagonist SCH23390 or the NMDA receptor-selective glutamate antagonist MK-801 (Presti *et al.*, 2003). Importantly, observational data indicated no significant drug-related changes in non-stereotypic motor behaviour. These results show that interruption of cortical projections to striatum by MK-801 or dopaminergic projections to striatum from substantia nigra can selectively reduce captivity-induced stereotypy via alterations in the direct pathway.

Furthermore, as described earlier in this section, expression of the striatal neuropeptides dynorphin and enkephalin index the activity of the direct and indirect pathways, respectively (see also Fig. 7.2 and Box 7.2). We therefore hypothesized that spontaneous stereotypy in deer mice would be associated with an imbalance in the activity of the pathways, favouring overactivity of the direct pathway. We measured the concentrations of these striatal neuropeptides in dorsolateral striatum in spontaneously high and low stereotypy deer mice from standard cages (Presti and Lewis, 2005). As predicted, results indicated significantly decreased leu-enkephalin content and significantly increased (dynorphin/enkephalin) content ratios in the high-stereotypy mice relative to low-stereotypy mice. Moreover, we saw a significant negative correlation between striatal enkephalin content (indexing indirect pathway activity) and frequency of stereotypy ($r = -0.40$), as well as a significant positive correlation between the (dynorphin/enkephalin) content ratio and frequency of stereotypy ($r = +0.42$) in these mice. These data are consistent with our hypothesis that spontaneous stereotypic behaviour is a consequence of relative hyperactivity along cortico-basal ganglia-cortical feedback circuits involving the direct (facilitative) pathway, but suggest that primary perturbations to the indirect (inhibitory) pathway give rise to such imbalanced activity. (See Chapter 5 for behavioural data from caged rodents further consistent with this corticostriatal hypothesis, and further discussion of the relative roles of these two pathways in stereotypy.)

7.7. Conclusions and Future Research Questions

The association between environmental complexity and stereotypy underscores the potency of experiential factors, particularly early in life, in shaping the development of brain and behaviour. The evidence for adverse outcomes following rearing in restricted or impoverished environments is overwhelming; it cuts across multiple domains of function and across species, including humans. Abnormal repetitive behaviours are an all too common outcome of such environments. The literature reviewed here indicates that increasing the complexity of the environment over standard captive housing conditions can markedly reduce the development of such behaviours – yet this literature is surprisingly sparse as to the mechanisms underlying these effects, especially given the voluminous work on environmental restriction and stereotypy. However, attempting to identify neural mediators of enrichment-related prevention of stereotypy has been the focus of several of our studies. Animals benefiting from an enriched environment (i.e. exhibiting little or no stereotypy) show increased neuronal metabolic activity, increased dendritic spine density and increased BDNF concentrations. Moreover, these effects are evident in the motor cortex and basal ganglia but not the hippocampus. This work, as well as our studies of intracerebral drug administration and striatal peptide concentrations, points to a pre-eminent role for cortical–basal ganglia circuitry in the development and maintenance of abnormal repetitive behaviour.

Much work remains to be done, however, to elaborate how environmental influences alter the brain to bring about a shift from behaviour characterized as periodic or regular, to behaviour that is complex, variable and thus adaptive for the organism and we end by giving just some examples of questions yet to be answered:

1. We opened this chapter by hypothesizing an intrinsic link between the complexity of an environment and the complexity of the behaviour shown by animals developing therein (Section 7.2). Yet an important line of research remains to still identify the key environmental factors that underlie the positive effects of enrichment on stereotypy: Complexity *per se*? The performance of species-typical behaviour patterns? Stress-reduction? It is likely not any single factor (e.g. increased space) but rather combinations of factors that permit expression of key species-typical behaviours. Moreover, the importance of these environmental factors is likely to vary across species and topographies (e.g. jumping in voles, cribbing in horses, etc.) of repetitive behaviour.

2. In Section 7.3, we showed that despite some intriguing evidence, there is still relatively little information as to whether true sensitive periods exist for either the induction of stereotypy by environmental restriction, or the prevention/reversal of stereotypy by increased environmental complexity. There is also as yet little information on how long is needed for such environments to have their effects; nor on what determines whether their effects will last if the environment is altered. For instance, little information is available regarding the persistence of enrichment effects on the brain after the organism is returned to a restricted environment, and while it might be argued that exposure to a more complex environment early in development should confer greater persistence of effects (or neuroprotection), this fascinating area still remains to be fully explored.

3. The impact of increased environmental complexity on CNS structure and function is well appreciated, and based on a long history of work. As we reviewed in Sections 7.4 and 7.5, environmental enrichment confers diverse advantages on intact animals (e.g. inhibition of apoptosis, increased hippocampal neurogenesis). It also ameliorates the effects of a variety of insults to the CNS (e.g. gene perturbations, traumatic brain injury, ischaemia). However, the possible relationship(s) between this long litany of effects on brain function, and the mediation of environmental effects on stereotypy, remains as yet little explored.

4. Likewise, in Section 7.4 we reviewed how enrichment-related alterations in CNS structure have diverse functional behavioural outcomes (e.g. improved spatial memory, decreased stress reactivity). Yet the possible relationship(s) between these specific outcomes and the prevention or attenuation of stereotypy remains uninvestigated.

5. Finally, no systematic attention has been directed toward the issue of individual differences in response to environmental complexity. As we have clearly shown, not all animals benefit equally from the larger, more complex environments, neither in terms of stereotypy (see Section 7.3) nor brain function (see Section 7.5). Conversely, while abnormal repetitive behaviours are an all too common outcome of barren environments, again not all animals are affected equally. Such outcomes thus likely depend on genotypic characteristics, but again little work has explored either the genetics of stereotypy or gene by environment interactions: further important lines of research to pursue in the future.

References

Adler, C.M., McDonough-Ryan, P., Sax, K.W., Holland, S.K., Arndt, S. and Strakowski, S.M. (2000) fMRI of neuronal activation with symptom provocation in unmedicated patients with obsessive compulsive disorder. *Journal of Psychiatric Research* 34, 317–324.

Akefeldt, A., Ekman, R., Gillberg, C. and Mansson, J.E. (1998) Cerebrospinal fluid monoamines in Prader–Willi syndrome. *Biological Psychiatry* 44, 1321–1328.

Albin, R.L., Young, A.B. and Penney, J.B. (1995) The functional anatomy of disorders of the basal ganglia. *Trends in Neurosciences* 18, 63–64.

Alexander, G.E. and Crutcher, M.D. (1990) Functional architecture of basal ganglia circuits – neural substrates of parallel processing. *Trends in Neurosciences* 13, 266–271.

Ames, A. (2001) The management of captive polar bears. PhD thesis, Open University.

Anderson, B.J., Gatley, S.J., Rapp, D.N., Coburn-Litvak, P.S. and Volkow, N.D. (2000) The ratio of striatal D1 to muscarinic receptors changes in aging rats housed in an enriched environment. *Brain Research* 872, 262–265.

Aosaki, T., Kiuchi, K. and Kawaguchi, Y. (1998) Dopamine D1-like receptor activation excites rat striatal large aspiny neurons *in vitro*. *The Journal of Neuroscience* 18, 5180–5190.

Barwick, V.S., Jones, D.H., Richter, J.T., Hicks, P.B. and Young, K.A. (2002) Subthalamic nucleus microinjections of 5-HT2 receptor antagonists suppress stereotypy in rats. *Neuroreport* 11, 267–270.

Berardelli, A., Curra, A., Fabbrini, G., Gilio, F. and Manfredi, M. (2003) Pathophysiology of tics and Tourette syndrome. *Journal of Neurology* 250, 781–787.

Berkson, G. (1968) Development of abnormal stereotyped behaviours. *Developmental Psychobiology* 1, 118–132.

Bodfish, J.W., Powell, S.B., Golden, R.N. and Lewis, M.H. (1995) Blink rate as an index of dopamine function in adults with mental retardation and repetitive behavior disorders. *American Journal of Mental Retardation* 99, 335–344.

Bodfish, J.W., Parker, D.E., Lewis, M.H., Sprague, R.L. and Newell, K.M. (2001) Stereotypy and motor control: differences in the postural stability dynamics of persons with stereotyped and dyskinetic movement disorders. *American Journal of Mental Retardation* 106, 123–134.

Bredy, T.W., Humpartzoomian, R.A., Cain, D.P. and Meaney, M.J. (2003) Partial reversal of the effect of maternal care on cognitive function through environmental enrichment. *Neuroscience* 118, 571–576.

Brunken, W.J. and Jin, X.T. (1993) A role for 5HT3 receptors in visual processing in the mammalian retina. *Visual Neuroscience* 10, 511–522.

Cabib, S. (1993) Neurobiological basis of stereotypies. In: Lawrence, A.B. and Rushen, J. (eds) *Stereotypic Animal Behaviour: Fundamentals and Applications to Welfare.* CAB International, Wallingford, UK, pp. 119–145.

Callard, M.D., Bursten, S.N. and Price, E.O. (2000) Repetitive behaviour in captive roof rats (*Rattus rattus*) and the effects of cage enrichment. *Animal Welfare* 9, 139–152.

Camel, J.E., Withers, G.S. and Greenough, W.T. (1986) Persistence of visual cortex dendritic alterations induced by postweaning exposure to a 'super-enriched' environment in rats. *Behavioral Neuroscience* 100, 810–813.

Canales, J.J. and Graybiel, A.M. (2000) A measure of striatal function predicts motor stereotypy. *Nature Neuroscience* 3, 377–383.

Carlson, M. and Earls, F. (1997) Psychological and neuroendocrinological sequelae of early social deprivation in institutionalized children in Romania. *Annals of the New York Academy of Sciences* 807, 419–428.

Caston, J., Devulder, B., Jouen, F. and Lalonde, R. (1999) Role of an enriched environment on the restoration of behavioral deficits in Lurcher mutant mice. *Developmental Psychobiology* 35, 291–303.

Cheal, M. (1987) Environmental enrichment facilitates foraging behavior. *Physiology and Behavior* 39, 281–283.

Chen, J.F., Beilstein, M., Xu, Y.H., Turner, T.J., Moratalla, R., Standaert, D.G., Aloyo, V.J., Fink, J.S. and Schwarzschild, M.A. (2000) Selective attenuation of psychostimulant-induced behavioral responses in mice lacking A(2A) adenosine receptors. *Neuroscience* 97, 195–204.

Comery, T.A., Shah, R. and Greenough, W.T. (1995) Differential rearing alters spine density on medium-sized spiny neurons in the rat corpus striatum: evidence for association of morphological

plasticity with early response gene expression. *Neurobiology of Learning and Memory* 63, 217–219.

Cools, A.R. and van Rossum, J.M. (1970) Caudal dopamine and stereotype behaviour of cats. *Archives Internationales de Pharmacodynamie et de Therapie* 187, 163–173.

Cooper, J.J., Ödberg, F. and Nicol, C.J. (1996) Limitations on the effectiveness of environmental improvement in reducing stereotypic behaviour in bank voles (*Clethrionomys glareolus*). *Applied Animal Behavior Science* 48, 237–248.

Cotman, C.W. and Berchtold, N.C. (2002) Exercise: a behavioral intervention to enhance brain health and plasticity. *Trends in Neuroscience* 25, 295–301.

Curzon, G. (1990) Stereotyped and other motor responses to 5-hydroxytryptamine receptor activation. In: Cooper, S.J. and Dourish, C.T. (eds) *Neurobiology of Stereotyped Behaviour.* Clarendon Press, Oxford, UK, pp. 142–168.

Dahlqvist, P., Zhao, L., Johansson, I.M., Mattsson, B., Johansson, B.B., Seckl, J.R. and Olsson, T. (1999) Environmental enrichment alters nerve growth factor-induced gene A and glucocorticoid receptor messenger RNA expression after middle cerebral artery occlusion in rats. *Neuroscience* 93, 527–535.

Dahlqvist, P., Ronnback, A., Risedal, A., Nergardh, R., Johansson, I.M., Seckl, J.R., Johansson, B.B. and Olsson, T. (2003) Effects of postischemic environment on transcription factor and serotonin receptor expression after permanent focal cortical ischemia in rats. *Neuroscience* 119, 643–652.

Davenport, R.K. and Menzel, E.W. (1963) Stereotyped behaviour of the infant chimpanzee. *Archives of General Psychiatry* 8, 99–104.

Davis, K. (1940) Extreme isolation of a child. *American Journal of Sociology* 45, 554–565.

Davis, K. (1946) Final note on a case of extreme isolation. *American Journal of Sociology* 52, 432–437.

Denys, D., van der Wee, N., Janssen, J., DeGeus, F. and Westenberg, H.G.M. (2004) Low level of dopaminergic D2 receptor binding in obsessive–compulsive disorder. *Biological Psychiatry* 55, 1041–1045.

Duffy, S.N., Craddock, K.J., Abel, T. and Nguyen, P.V. (2001) Environmental enrichment modifies the PKA-dependence of hippocampal LTP and improves hippocampus-dependent memory. *Learning and Memory* 8, 26–34.

Ehninger, D. and Kempermann, G. (2002) Regional effects of wheel running and environmental enrichment on cell genesis and microglia proliferation in the adult murine neocortex. *Cerebral Cortex* 13, 845–851.

Einon, D.F. and Sahakian, B.J. (1979) Environmentally induced differences in susceptibility of rats to CNS stimulants and CNS depressants: evidence against a unitary explanation. *Psychopharmacology* 61, 299–307.

Ernst, A.M. and Smelik, P.G. (1966) Site of action of dopamine and apomorphine on compulsive gnawing behaviour in rats. *Experientia* 22, 837–838.

Escorihuela, R.M., Tobena, A. and Fernandez-Teruel, A. (1995) Environmental enrichment and postnatal handling prevent spatial learning deficits in aged hypoemotional (Roman high-avoidance) and hyperemotional (Roman low-avoidance) rats. *Learning and Memory* 2, 40–48.

Feng, J., Sobell, J.L., Heston, L.L., Cook, E.H. Jr, Goldman, D. and Sommer, S.S. (1998) Scanning of the dopamine D1 and D5 receptor genes by REF in neuropsychiatric patients reveals a novel missense change at a highly conserved amino acid. *American Journal of Medical Genetics* 81, 172–178.

Fernandez-Teruel, A., Escorihuela, R.M., Driscoll, P., Tobena, A. and Battig, K. (1992) Differential effects of early stimulation and/or perinatal flumazenil treatment in young Roman low- and high-avoidance rats. *Psychopharmacology* 108, 170–176.

Fernandez-Teruel, A., Escorihuela, R.M., Castellano, B., Gonzalez, B. and Tobena, A. (1997) Neonatal handling and environmental enrichment effects on emotionality, novelty/reward seeking, and age-related cognitive and hippocampal impairments: focus on the Roman rat lines. *Behavioral Genetics* 27, 513–526.

Francis, D.D., Diorio, J., Plotsky, P.M. and Meaney, M.J. (2002) Environmental enrichment reverses the effects of maternal separation on stress reactivity. *The Journal of Neuroscience* 22, 7840–7843.

Fraser, D. (1975) The effect of straw on the behaviour of sows in tethered stalls. *Animal Production* 21, 59–68.

Freedman, D.A. (1968) On the role of coenesthetic stimulation in the development of psychic structure. *The Psychoanalytic Quarterly* 37, 418–438.

Frick, K.M. and Fernandez, S.M. (2003) Enrichment enhances spatial memory and increases synaptophysin levels in aged female mice. *Neurobiology of Aging* 24, 615–626.

Frick, K.M., Stearns, N.A., Pan, J.Y. and Berger-Sweeney, J. (2003) Effects of environmental enrichment on spatial memory and neurochemistry in middle-aged mice. *Learning and Memory* 10, 187–198.

Gerfen, C.R. (2000) Molecular effects of dopamine on striatal-projection pathways. *Trends in Neuroscience* 23, S64–S70.

Gibb, R. and Kolb, B. (1998) A method for vibratome sectioning of Golgi-Cox stained whole rat brain. *Journal of Neuroscience Methods* 79, 1–4.

Glass, L. and Mackey, M.C. (1988) *From Clocks to Chaos: The Rhythms of Life.* Princeton University Press, Princeton, New Jersey, 248 pp.

Goldberger, A. (1997) Fractal variability versus pathologic periodicity: complexity loss and stereotypy in disease. *Perspectives in Biology and Medicine* 40, 543–561.

Graybiel, A.M. (2000) The basal ganglia. *Current Biology* 10, R509–R511.

Graybiel, A.M., Canales, J.J. and Capper-Loup, C. (2000) Levodopa-induced dyskinesias and dopamine-dependent stereotypies: a new hypothesis. *Trends in Neurosciences* 23, S71–S77.

Greenough, W.T., Hwang, H.-M.F. and Gorman, C. (1985) Evidence for active synapse formation or altered postsynaptic metabolism in visual cortex of rats reared in complex environments. *Proceedings of the National Academy of Sciences USA* 82, 4549–4552.

Haemisch, A., Voss, T. and Gartner, K. (1994) Effects of environmental enrichment on aggressive behavior, dominance hierarchies, and endocrine states in male DBA/2J mice. *Physiology and Behavior* 56, 1041–1048.

Hamm, R.J., Temple, M.D., O'Dell, D.M., Pike, B.R. and Lyeth, B.G. (1996) Exposure to environmental complexity promotes recovery of cognitive function after traumatic brain injury. *Journal of Neurotrauma* 13, 41–47.

Hansen, S.W. (1992) Stress reactions in farm mink and beech marten in relation to housing and domestication. Ph.D. thesis, Cøpenhagen University, Cøpenhagen.

Hebb, D.O. (1949) *The Organization of Behavior: A Neuropsychological Theory.* John Wiley & Sons, New York.

Hockly, E., Cordery, P.M., Woodman, B., Mahal, A., van Dellen, A., Blakemore, C., Lewis, C.M., Hannan, A.J. and Bates, G.P. (2002) Environmental enrichment slows disease progression in R6/2 Huntington's disease mice. *Annals of Neurology* 51, 235–242.

Iwamoto, E.T. and Way, E.L. (1977) Circling behavior and stereotypy induced by intranigral opiate microinjections. *The Journal of Pharmacology and Experimental Therapeutics* 203, 347–359.

Johansson, B.B. (1996) Functional outcome in rats transferred to an enriched environment 15 days after focal brain ischemia. *Stroke* 27, 324–326.

Johansson, B.B. and Ohlsson, A.L. (1996) Environment, social interaction, and physical activity as determinants of functional outcome after cerebral infarction in the rat. *Experimental Neurology* 139, 322–327.

Kaplan, M.S. (2001) Environment complexity stimulates visual cortex neurogenesis:

death of a dogma and a research career. *Trends in Neuroscience* 24, 617–620.

Karler, R., Bedingfield, J.B., Thai, D.K. and Calder, L.D. (1997) The role of the frontal cortex in the mouse in behavioral sensitization to amphetamine. *Brain Research* 757, 228–235.

Karler, R., Calder, L.D., Thai, D.K. and Bedingfield, J.B. (1998) The role of dopamine in the mouse frontal cortex: a new hypothesis of behavioral sensitization to amphetamine and cocaine. *Pharmacology Biochemistry and Behavior* 61, 435–443.

Karson, C.N. (1983) Spontaneous eye-blink rates and dopaminergic systems. *Brain* 106, 643–653.

Keiper, R.R. (1969) Causal factors of stereotypies in caged birds. *Animal Behaviour* 17, 114–117.

Kempermann, G., Gast, D. and Gage, F.H. (2002) Neuroplasticity in old age: sustained fivefold induction of hippocampal neurogenesis by long-term environmental enrichment. *Annals of Neurology* 52, 135–143.

Kiyatkin, E.A. and Rebec, G.V. (1999) Modulation of striatal neuronal activity by glutamate and GABA: iontophoresis in awake, unrestrained rats. *Brain Research* 822, 88–106.

Klein, S.L., Lambert, K.G., Durr, D., Schaefer, T. and Waring, R.E. (1994) Influence of environmental enrichment and sex on predator stress response in rats. *Physiology and Behavior* 56, 291–297.

Kobayashi, S., Ohashi, Y. and Ando, S. (2002) Effects of enriched environments with different durations and starting times on learning capacity during aging in rats assessed by a refined procedure of the Hebb–Williams maze task. *Journal of Neuroscience Research* 70, 340–346.

Kolb, B. and Gibb, R. (1991) Environmental enrichment and cortical injury: behavioral and anatomical consequences of frontal cortex lesions. *Cerebral Cortex* 1, 189–198.

Latham, N. (2005) Refining the role of stereotypic behaviour in the assessment of welfare: stress, general motor persistence and early environment in the development of abnormal behaviour. Ph.D. thesis, Oxford University, Oxford, UK.

Lewis, M.H., Gluck, J.P., Beauchamp, A.J., Keresztury, M.F. and Mailman, R.B. (1990) Long-term effects of early social isolation in *Macaca mulatta*: changes in dopamine receptor function following apomorphine challenge. *Brain Research* 513, 67–73.

Lewis, M.H., Bodfish, J.W., Powell, S.B., Wiest, K., Darling, M. and Golden, R.N. (1996) Plasma HVA in adults with mental retardation and stereotyped behavior: biochemical evidence for a dopamine deficiency model. *American Journal of Mental Retardation* 100, 413–418.

Mallapur, A., Waran, N. and Sinha, A. (2005) Factors influencing the behaviour and welfare of captive lion-tailed macaques in Indian zoos. *Applied Animal Behaviour Science* 94, 341–352.

Mao, L. and Wang, J.Q. (2000) Distinct inhibition of acute cocaine-stimulated motor activity following microinjection of a group III metabotropic glutamate receptor agonist into the dorsal striatum of rats. *Pharmacology Biochemistry and Behavior* 67, 93–101.

Martin, L.J., Spicer, D.M., Lewis, M.H., Gluck, J.P. and Cork, L.C. (1991) Social deprivation of infant rhesus monkeys alters the chemoarchitecture of the brain. *Journal of Neuroscience* 11, 3344–3358.

McPherson, R.J. and Marshall, J.F. (2000) Substantia nigra glutamate antagonists produce contralateral turning and basal ganglia *Fos* expression: interactions with D1 and D2 dopamine receptor agonists. *Synapse* 36, 194–204.

Meehan, C.L., Garner, J.P. and Mench, J.A. (2004) Environmental enrichment and development of cage stereotypy in Orange-winged Amazon parrots (*Amazona amazonica*). *Developmental Psychobiology* 44, 209–218.

Mills, D.S. (2003) Medical paradigms for the study of problem behaviour: a critical review. *Applied Animal Behaviour Science* 81, 265–277.

Nicola, S.M., Surmeier, J. and Malenka, R.C. (2000) Dopaminergic modulation of neuronal excitability in the striatum and nucleus accumbens. *Annual Review of Neuroscience* 23, 185–215.

Ödberg, F.O. (1987) The influences of cage size and environmental enrichment on the development of stereotypies in bank voles (*Clethrionomys glareolus*). *Behavioral Processes* 14, 155–173.

Ödberg, F.O., Kennes, D., De Rycke, P.H. and Bouquet, Y. (1987) The effect of interference in catecholamine biosynthesis on captivity-induced jumping stereotypy in bank voles (*Clethrionomys glareolus*). *Archives of International Pharmacodynamics and Therapeutics* 285, 34–42.

Olanow, C.W., Obesa, J.A. and Nutt J.G. (2000) Basal ganglia, Parkinson's disease and levodopa therapy. *Supplement to Trends in Neurosciences* 23, S1–S126.

Olsson, I.A. and Dahlborn, K. (2002) Improving housing conditions for laboratory mice: a review of "environmental enrichment". *Laboratory Animals* 36, 243–270.

O'Sullivan, R.L., Rauch, S.L., Breiter, H.C., Grachev, I.D., Baer, L., Kennedy, D.N., Keuthen, N.J., Savage, C.R., Manzo, P.A., Caviness, V.S. and Jenike, M.A. (1997) Reduced basal ganglia volumes in trichotillomania measured via morphometric magnetic resonance imaging. *Biological Psychiatry* 42, 39–45.

Pham, T.M., Winblad, B., Granholm, A.C. and Mohammed, A.H. (2002) Environmental influences on brain neurotrophins in rats. *Pharmacology Biochemistry and Behavior* 73, 167–175.

Philbin, N. (1998) Towards an understanding of stereotypic behaviour of laboratory macaques. *Animal Technology* 49, 19–33.

Pinaud, R., Penner, M.R., Robertson, H.A. and Currie, R.W. (2001) Upregulation of the immediate early gene arc in the brains of rats exposed to environmental enrichment: implications for molecular plasticity. *Brain Research. Molecular Brain Research* 91, 50–56.

Pisani, A., Bonsi, P., Centonze, D., Calabresi, P. and Bernardi, G. (2000) Activation of D2-like dopamine receptors reduces synaptic inputs to striatal cholinergic interneurons. *Journal of Neuroscience* 20, 69.

Poremba, A., Jones, D. and Gonzalez-Lima, F. (1998) Classical conditioning modifies cytochrome oxidase activity in the auditory system. *European Journal of Neuroscience* 10, 3035–3043.

Powell, S.B., Newman, H.A., Pendergast, J. and Lewis, M.H. (1999) A rodent model of spontaneous stereotypy: initial characterization of developmental, environmental, and neurobiological factors. *Physiology and Behavior* 66, 355–363.

Powell, S.B., Newman, H.A., McDonald, T.A., Bugenhagen, P. and Lewis, M.H. (2000) Development of spontaneous stereotypy in deer mice: effects of early and late exposure to a more complex environment. *Developmental Psychobiology* 37, 100–108.

Presti, M.F. and Lewis, M.H. (2005) Striatal opioid peptide content in an animal model of spontaneous stereotypic behavior. *Behavioural Brain Research* 157, 363–368.

Presti, M.F., Powell, S.B. and Lewis, M.H. (2002) Dissociation between spontaneously-emitted and psychostimulant-induced stereotypy. *Physiology and Behavior* 75, 1–7.

Presti, M.F., Mikes, H.M. and Lewis, M.H. (2003) Selective blockade of spontaneous motor stereotypy via intrastriatal pharmacological manipulation. *Pharmacology, Biochemistry, and Behavior* 74, 833–839.

Presti, M.F., Gibney, B.C. and Lewis, M.H. (2004) Effects of intrastriatal administration of selective dopaminergic ligands on spontaneous stereotypy in mice. *Physiology and Behavior,* 80, 433–439.

Price, C.J., Kim, P. and Raymond, L.A. (1999) D1 dopamine receptor-induced cyclic AMP-dependent protein kinase phosphorylation and potentiation of striatal glutamate receptors. *Journal of Neurochemistry* 73, 2441–2446.

Raine, A., Mellingen, K., Liu, J., Venables, P. and Mednick, S.A. (2003) Effects of environmental enrichment at ages 3–5 years on schizotypal personality and antisocial behavior at ages 17 and 23 years. *American Journal of Psychiatry* 160, 1627–1635.

Rampon, C., Jiang, C.H., Dong, H., Tang, Y-H., Lockhart, D.J., Schultz, P.J., Tsien, J.Z. and Hu, Y. (2000a) Effects of environmental enrichment on gene expression in the brain. *Proceedings of the National Academy of Sciences USA* 97, 12880–12884.

Rampon, C., Tang, Y.P., Goodhouse, J., Shimizu, E., Kyin, M. and Tsien, J.Z. (2000b) Enrichment induces structural changes and recovery from nonspatial memory deficits in CA1 NMDAR1-knock-out mice. *Nature Neuroscience* 3, 238–244.

Rasmuson, S., Olsson, T., Henriksson, B.G., Kelly, P.A., Holmes, M.C., Seckl, J.R. and Mohammed, A.H. (1998) Environmental enrichment selectively increases 5-HT1A receptor mRNA expression and binding in the rat hippocampus. *Brain Research. Molecular Brain Research* 53, 285–290.

Reiner, A., Medina, L. and Veenman, C.L. (1998) Structural and functional evolution of the basal ganglia in vertebrates. *Brain Research Reviews* 28, 235–285.

Reis, G.F., Lee, M.B., Huang, A.S. and Parfitt, K.D. (2005) Adenylate cyclase-mediated forms of neuronal plasticity in hippocampal area CA1 are reduced with aging. *Journal of Neurophysiology* 93, 3381–3389.

Reiss, A.L., Abrams, M.T., Greenlaw, R., Freund, L. and Denckla, M.B. (1995) Neurodevelopmental effects of the FMR-1 full mutation in humans. *Nature Medicine* 1, 159–167.

Remington, G., Sloman, L., Konstantareas, M., Parker, K. and Gow, R. (2001) Clomipramine versus haloperidol in the treatment of autistic disorder: a double-blind, placebo-controlled, crossover study. *Journal of Clinical Psychopharmacology* 21, 440–444.

Robbins, T.W. (1997) Arousal systems and attentional processes. *Biological Psychology* 45, 57–71.

Rockman, G.E. and Borowski, T.B. (1986) The effects of environmental enrichment on voluntary ethanol consumption and stress ulcer formation in rats. *Alcohol* 3, 299–302.

Rolls, E.T. (1994) Neurophysiology and cognitive functions of the striatum. *Revue Neurologique* 150, 648–660.

Rosenzweig, M.R. and Bennett, E.L. (1996) Psychobiology of plasticity: effects of training and experience on brain and behavior. *Behavioral Brain Research* 78, 57–65.

Rosenzweig, M.R., Love, W. and Bennett, E.L. (1968) Effects of a few hours a day of enriched experience on brain chemistry and brain weights. *Physiology and Behavior* 3, 819–825.

Russell, R.L. and Pihl, R.O. (1978) The effect of dose, novelty, and exploration on amphetamine-produced stereotyped behavior. *Psychopharmacology (Berlin)* 60, 93–100.

Sahakian, B.J. and Robbins, T.W. (1975) Potentiation of locomotor activity and modification of stereotypy by starvation in adomorphine treated rats. *Neuropharmacology* 14, 251–257.

Saito, S., Kobayashi, S., Ohashi, Y., Igarashi, M., Komiya, Y. and Ando, S. (1994) Decreased synaptic density in aged brains and its prevention by rearing under enriched environment as revealed by synaptophysin contents. *Journal of Neuroscience Research* 39, 57–62.

Salmi, P., Sproat, B.S., Ludwig, J., Hale, R., Avery, N., Kela, J. and Wahlestedt, C. (2000) Dopamine D(2) receptor ribozyme inhibits quinpirole-induced stereotypy in rats. *European Journal of Pharmacology* 388, R1–R2.

Scheel-Kruger, J. and Christensen, A.V. (1980) The role of gamma-aminobutyric acid in acute and chronic neuroleptic action. *Advances in Biochemical Psychopharmacology* 24, 233–243.

Scheel-Kruger, J., Arnt, J., Braestrup, C., Christensen, A.V., Cools, A.R. and Magelund, G. (1978) GABA–dopamine interaction in substantia nigra and nucleus accumbens – relevance to behavioral stimulation and stereotyped behavior. *Advances in Biochemical Psychopharmacology* 19, 343–346.

Schoenecker, B., Heller, K.E. and Freimanis, T. (2000) Development of stereotypies and polydipsia in captive bank voles (*Clethrionomys glareolus*) and their laboratory-bred offspring. *Applied Animal Behaviour Science* 68, 349–357.

Schrijver, N.C., Bahr, N.I., Weiss, I.C. and Würbel, H. (2002) Dissociable effects of isolation rearing and environmental enrichment on exploration, spatial learning and HPA activity in adult rats. *Pharmacology, Biochemistry, and Behavior* 73, 209–224.

Sears, L.L., Vest, C., Mohamed, S., Bailey, J., Ranson, B.J. and Piven, J. (1999) An MRI study of the basal ganglia in autism. *Progress in Neuro-psychopharmacology and Biological Psychiatry* 23, 613–624.

Smalley, S.L., Bailey, J.N., Palmer, C.G., Cantwell, D.P., McGough, J.J., Del'Homme, M.A., Asarnow, J.R., Woodward, J.A., Ramsey, C. and Nelson, S.F. (2002) Evidence that the dopamine D4 receptor is a susceptibility gene in attention deficit hyperactivity disorder. *Molecular Psychiatry* 3, 427–430.

Smith, M.A., Bryant, P.A. and McClean, J.M. (2003) Social and environmental enrichment enhances sensitivity to the effects of kappa opioids: studies on antinociception, diuresis and conditioned place preference. *Pharmacology, Biochemistry, and Behavior* 76, 93–101.

Sørenson, G. and Randrup, A. (1986) Possible protective value of severe psychopathology versus lethal effects of an unfavourable milieu. *Stress Medicine* 2, 103–105.

Steiner, H. and Gerfen, C.R. (1998) Role of dynorphin and enkephalin in the regulation of striatal output pathways and behavior. *Experimental Brain Research* 123, 60–76.

Steward, O., Wallace, C.S., Lyford, G.L. and Worley, P.F. (1998) Synaptic activation causes the mRNA for the IEG Arc to localize selectively near activated postsynaptic sites on dendrites. *Neuron* 21, 741–751.

Teitelbaum, P., Pellis, S.M. and deVietti, T.L. (1990) Disintegration into stereotypy induced by drugs or brain damage: a microdescriptive behavioural analysis. In: Cooper, S.J. and Dourish, C.T. (eds) *Neurobiology of Stereotyped Behaviour.* Clarendon Press, Oxford, pp. 169–199.

Terlouw, E.M.C., DeRosa, G., Lawrence, A.B., Illius, A.W. and Ladewig, J. (1992a) Behavioural responses to amphetamine and apomorphine in pigs. *Pharmacology, Biochemistry and Behaviour* 43, 329–340.

Terlouw, E.M.C., Lawrence, A.B. and Illius, A.W. (1992b) Relationship between amphetamine- and environmentally-induced stereotypies in pigs. *Pharmacology, Biochemistry and Behaviour* 43, 347–355.

Thelen, E. (1980) Kicking, rocking, and waving: contextual analysis of rhythmical stereotypies in normal human infants. *Animal Behaviour* 29, 3–11.

Turner, C.A. and Lewis, M.H. (2003) Environmental enrichment: effects on stereotyped behavior and neurotrophin levels. *Physiology and Behavior* 80, 259–266.

Turner, C.A., Yang, M.C. and Lewis, M.H. (2002) Environmental enrichment: effects on stereotyped behavior and regional neuronal metabolic activity. *Brain Research* 938, 15–21.

Turner, C.A., King, M.A. and Lewis, M.H. (2003) Environmental enrichment: effects on stereotyped behavior and dendritic morphology. *Developmental Psychobiology* 43, 1–8.

van Keulen-Kromhout, G. (1978) Zoo enclosures for bears (Ursidae): their influence on captive behaviour and reproduction. *International Zoo Yearbook* 18, 177–186.

van Praag, H., Kempermann, G. and Gage, F.H. (1999) Running increases cell proliferation and neurogenesis in the adult mouse dentate gyrus. *Nature Neuroscience* 2, 266–270.

Vandebroek, I. and Ödberg, F.O. (1997) Effect of apomorphine on the conflict-induced jumping stereotypy in bank voles. *Pharmacology, Biochemistry, and Behavior* 4, 863–868.

Vandebroek, I., Berckmoes, V. and Ödberg, F.O. (1998) Dissociation between MK-801 and captivity-induced stereotypies in bank voles. *Psychopharmacology* 137, 205–214.

Vinke, C.M. (2004) Cage enrichments and the welfare of farmed mink. Ph.D. thesis, University of Utrecht, Utrecht, Netherlands.

Waddington, J.L., Molloy, A.G., O'Boyle, K.M. and Pugh, M.T. (1990) Aspects of stereotyped and non-stereotyped behaviour in relation to dopamine receptor subtypes. In: Cooper, S.J. and Dourish, C.T. (eds) *Neurobiology of Stereotyped Behaviour.* Clarendon Press, Oxford, pp. 64–90.

Wallace, C.S., Kilman, V.L., Withers, G.S. and Greenough, W.T. (1992) Increases in dendritic length in occipital cortex after 4 days of differential housing in weanling rats. *Behavioral and Neural Biology* 58, 64–68.

Werme, M., Messer, C., Olson, L., Gilden, L., Thoren, P., Nestler, E.J. and Brene, S. (2002) Delta FosB regulates wheel running. *Journal of Neuroscience* 22, 8133–8138.

Widman, D.R. and Rosellini, R.A. (1990) Restricted daily exposure to environmental enrichment increases the diversity of exploration. *Physiology and Behavior* 47, 57–62.

Wiedenmayer, C. (1997) Causation of the ontogenetic development of stereotypic digging in gerbils. *Animal Behaviour* 53, 461–470.

Wong, R. and Jamieson, J.L. (1968) Infantile handling and the facilitation of discrimination and reversal learning. *The Quarterly Journal of Experimental Psychology* 20, 197–199.

Wong-Riley, M.T., Huang, Z., Liebl, W., Nie, F., Xu, H. and Zhang, C. (1998) Neurochemical organization of the macaque retina: effect of TTX on levels and gene expression of cytochrome oxidase and nitric oxide synthase and on the immunoreactivity of Na+ K+ ATPase and NMDA receptor subunit I. *Vision Research* 38, 1455–1477.

Wood, P.L. and Richard, J.W. (1982) Morphine and nigrostriatal function in the rat and mouse: the role of nigral and striatal opiate receptors. *Neuropharmacology* 21, 1305–1310.

Würbel, H. (2001) Ideal homes? Housing effects on rodent brain and behaviour. *Trends in Neurosciences* 24, 207–211.

Yirmiya, N., Pilowsky, T., Nemanov, L., Arbelle, S., Feinsilver, T., Fried, I. and Ebstein, R.P. (2001) Evidence for an association with the serotonin transporter promoter region polymorphism and autism. *American Journal of Medical Genetics* 105, 381–386.

Young, D., Lawlor, P.A., Leone, P., Dragunow, M. and During, M.J. (1999) Environmental enrichment inhibits spontaneous apoptosis prevents seizures and is neuroprotective. *Nature Medicine* 5, 448–453.

8 The Neurobiology of Stereotypy II: the Role of Stress

S. Cabib

Universita 'La Sapienza', Dipartimento di Psicologia, via dei Marsi 78, Rome 00185, Italy

Editorial Introduction

In this chapter, Cabib contrasts the well-characterized responses of two mouse strains – DBA/2 and C57BL/6 – to sustained stress, and uses them to argue that some genotypes respond to certain types of stress with a suite of neurobiological changes ('stress sensitization') that have multiple behavioural effects, including stereotypy. Like Chapters 5 and 7, she portrays exaggerated, apparently functionless repeated behaviour as the products of enhanced behavioural activation by the brain's basal ganglia. However, unlike these chapters, her focus here is not on the dorsal striatum (DS) and its inputs and outputs, but instead on the ventral striatum (or nucleus accumbens (NAc)) and its inputs from the midbrain's ventral tegmental area (VTA). Furthermore, she emphasizes that the increased influence on behaviour of this subcortical part of the brain is not solely due to changes here, but also to a stress-induced decrease in the inhibitory influence usually exerted by the cortex.

Repeated exposure to uncontrollable aversive situations can cause profound, long-term changes in brain organization – effects that Cabib argues could underlie a new, more biologically meaningful definition of 'stress'. Such changes can occur even in adult animals whose brains are fully developed, and seem to fall into two classes. One type of change, readily seen in C57 mice for instance, is that stressors come to elicit increasingly minimal dopamine response from the NAc. This seems caused by enhanced dopamine release by the prefrontal cortex (pFC), which inhibits striatal activation. It is also correlated with a decreased density of post-synaptic striatal D2 receptors (receptors which, as we saw in the previous chapter, act to inhibit the indirect pathway and thence enhance direct pathway activity). Behaviourally, this acts to increase learned helplessness (e.g. causes less active struggling and more passive floating in 'forced swim' tests); to reduce responsiveness (including stereotypy) to dopamine agonists like amphetamine; and to reduce cage stereotypies. The second possible stress-induced change, in contrast, causes stress-induced NAc dopamine responses to become *increasingly* pronounced with repeated exposure, due to a wane in prefrontal inhibitory control and an increase in the density of post-synaptic striatal D2 receptors (thence leading to increased direct pathway output). A contributed box discusses how ventral tegmental area opioid receptors also play a role in this process. Isolated, food-deprived DBA mice illustrate this 'sensitization' well, plus the behavioural changes that follow it, namely reduced susceptibility to 'behavioural despair' in situations like the forced swim test; an enhanced susceptibility to the activating and stereotypy-inducing effects of amphetamine or similar compounds; and increased stereotypic climbing within the home-cage. This last emerges after 2 weeks of food restriction, and is especially seen before the expected arrival of food. A contributed box presents additional home-cage data backing this strain difference in cage stereotypy (but questioning the generality of this model across all mouse strains). From other literatures, Cabib then argues that further features are likely to characterize sensitization-induced cage stereotypy: a correlated general impulsivity; a tendency for the stereotypy to be elicited by a range of acute stressors (with a final contributed box nicely illustrating this with data from laboratory primates); and a developmental decrease in behavioural flexibility as the nigrostriatal systems more traditionally implicated in stereotypy (*cf.* Chapters 5 and 7, this volume) come to play a growing role in the behaviour's control.

GM

8.1. Introduction

In recent years, stress research has undergone a major evolution. Psychologists and psycho-neuroendocrinologists have increasingly applied cognitive concepts and theories to the interpretation of stress-related phenomena (Ursin and Eriksen, 2004). Moreover, technical developments in neuroscience have allowed collection of new data on the involvement of the central nervous system (CNS) in stress-related responses. Historically, the interpretation of stereotypy as a stress-related response had suffered from the lack of a clear, unitary definition of stress. Indeed, 'stress' had been used to indicate a general alarm reaction elicited by novel and unpredicted events; the physiological strain produced by prolonged exposure to extreme environmental conditions; and the defensive reaction of the whole organism against potentially dangerous stimuli. The concept of stress, therefore, seemed to offer little, if any, indication about the quality or quantity or temporal characteristics of stimuli or responses involved, making it of little help in understanding related phenomena, such as stereotypic behaviour. Now, however, thanks to recent advances in stress research, it is possible to re-evaluate the role of stress in the development and expression of stereotypies, using new concepts and data.

One simple way to circumvent the problem of defining stress might be to use the so-called hormonal stress response as the index of stressful situations. Indeed, activation of the 'HPA axis', the circuit connecting the brain areas of hypothalamus and pituitary with the adrenal glands, leading to the release of corticoid hormones, is considered 'the physiological stress response' by many researchers (e.g. Marinelli and Piazza, 2002). However, the HPA response is controlled 'upstream' by higher order brain structures such as the amygdala, hippocampus and frontal cortex in a complex and integrated way. Thus, the frontal cortex and hippocampus bear corticoid receptors whose activation inhibits HPA responses (Marinelli and Piazza, 2002), while the amygdala, more specifically its central nucleus, is necessary for the induction of adrenocortical activation (Goldstein *et al.*, 1996). Furthermore, the stress-related activation of such higher order brain areas is independent of HPA activation (Imperato *et al.*, 1991). This involvement of phylogenetically recent areas of the brain in the regulation of the hormonal stress response, and the independence of their response from HPA activation, supports the psychological view of a key role of cognitive evaluation in stress-related phenomena (see Ursin and Eriksen, 2004, for a recent review). Thus, brain stress responses depend on the organism's previous experience, the behavioural response allowed by the context, and the predictability of the stressful events (Cabib and Puglisi-Allegra, 1996a).

There is a general agreement that stress responses are essential physiological responses that do not represent health threats *per se*, but also, that if sustained they can lead to illness and disease (Ursin and Eriksen, 2004). Pathological alterations may thus derive from organisms' attempts to physiologically and neurologically adapt to maintain 'homeostasis' (Ursin and Eriksen, 2004). Brain systems have a number of different auto-regulatory mechanisms and retain a fairly large functional and even morphological neuroplasticity well into maturity (Perrotti *et al.*, 2004; Robinson and Kolb, 2004; see also Chapter 7, this volume). Thus, in conditions of prolonged, chronic pressure, brain functioning may change dramatically, in turn leading to profound behavioural alterations. Thus the term 'stress' might usefully be restricted to these neurobiological changes, as induced by aversive experiences that cannot be coped with behaviourally.

Here I therefore propose that some forms of stereotypy that develop in mature organisms derive from such brain 'adaptations' to stress-inducing circumstances. Note that I use the term 'adaptation' in the sense that neuroscientists use it, to mean changes in the brain that occur within the lifespan of individuals and are not directly related to learning. Such adaptation involves CNS re-organization, occurring via a series of complex reciprocal alterations; it is therefore relatively slow to develop, and can be long lasting (or even permanent). Also note that it can be either functional or pathological (see Box 1.4, Chapter 1, this volume). Adaptations to stress involve mainly the brain areas receiving dopamine projections from the mesencephalon (midbrain). In one form of adaptation, it leads to a progressive imbalance within the forebrain between cortical areas and subcortical areas such as the basal ganglia. Subsequently, within the basal ganglia, it also leads to imbalances between the dorsal striatum (DS) (Chapters 5 and 7, this volume) and the ventral striatum. The general dysfunctional outcome of these types of disturbance is a strong general tendency toward impulsivity, inflexibility and compulsivity that then renders animals susceptible to the stereotyped alteration of behavioural output.

In Section 8.2, I offer a brief introduction to the brain dopaminergic systems involved in stress-related responses, both short-term ones and long-term (i.e. adaptation). Section 8.3 reviews evidence on the relationship between brain dopaminergic dysfunctions and outcomes that are arguably pathological: impulsivity, inflexibility, compulsivity and stereotypy. Section 8.4 discusses the relationship between stress-induced altered dopaminergic functioning and the development and expression of stereotyped behaviour.

8.2. Brain Dopamine and Stress

Dopamine (DA) is a catecholamine neurotransmitter of the amine group. As we saw in the previous chapters (Chapters 5–7, this volume), it is one of the neurotransmitters most implicated in how animals respond to captivity. Within the brain, it has a rather selective distribution within two major systems: the 'mesocorticolimbic' and 'nigrostriatal' systems. These two systems have anatomical relationships and are activated by all so-called psychostimulant drugs (e.g. amphetamine; see e.g. Box 7.3, Chapter 7). However, they have been classically differentiated respectively as motivational (the mesocorticolimbic) and motor (the nigrostriatal) systems. Thus the mesocorticolimbic system is the key in turning 'liking' or 'disliking' into approach or avoidance (see Box 8.1), while the nigrostriatal pathway is the key in behavioural response selection and sequencing (see e.g. Chapter 7).

8.2.1. Mesocorticolimbic (motivational) dopamine and the responses to acute stress: coping versus failure

The mesocorticolimbic dopaminergic system is formed by projections of DA neurons that are located within the VTA of the mesencephalon (midbrain). VTA DA neurons send projections to three forebrain areas: (i) the cortical area known as the prefrontal cortex (pFC), giving rise to the 'mesocortical DA system'; (ii) the limbic system's amygdala and hippocampus (giving rise to the 'mesolimbic system'); and (iii) the ventral portion of the striatum, within the basal ganglia, known as the NAc, giving rise to the 'meso-accumbens' system. (See also Fig. 7.2 and Box 7.2, Chapter 7, this volume, for how this compares with the system involving the nigrostriatal pathway and DS.)

In rodents, sophisticated techniques have been used to quantify variations in neurotransmission in these specific brain areas in living, behaving animals. These reveal that very mild and short-lasting aversive stimuli, for example from stroking a rat's fur with a gloved hand, promote a selective increase of DA release within the amygdala (Inglis and Moghaddam, 1999); while stronger aversive stimuli enhance DA release in the pFC; and more prolonged aversive experiences (e.g. immobilization of the animal in a tightly fitting restraint apparatus) involve DA release within the NAc too (Puglisi-Allegra and Cabib, 1997). Thus as severity increases from mild to much more stressful, all three mesocorticolimbic

Box 8.1. Stress Sensitization and Exaggerated Reward-responses: the Role of VTA Opioids

B.M. Spruijt and R. van den Bos

The mesoaccumbens DA system seems to be involved in the cost–benefit analyses made of rewards and the effort animals will expend to obtain them (Salamone and Correa, 2002; see also Berridge and Robinson, 2003). Before obtaining rewards, such effort can be expressed as intense investigatory and locomotor anticipatory or 'appetitive' behaviour (e.g. Spruijt *et al.*, 2001). The magnitude of such behavioural responses is variable; for example, depending on species and test conditions, they may either decrease (cats: van den Bos *et al.*, 2003), or increase (e.g. rats: Von Frijtag *et al.*, 2000, 2002; van der Harst *et al.*, 2003a,b; van den Bos *et al.*, 2004; mink: Vinke *et al.*, 2004; cats: van den Bos *et al.*, 2003). Here, we argue that increased anticipatory responding is an opioid-mediated consequence of stress that underlies some stereotypies.

As Cabib reviews here, both stress and highly rewarding incentives can sensitize the mesoaccumbens DA system (also reviewed by Spruijt *et al.*, 2001). This could explain why some stressors, such as short-term social isolation (van den Berg *et al.*, 1999) or poor housing conditions (van der Harst *et al.*, 2003b), tend to increase rats' anticipatory activity before a standardized sucrose reward. This change in behaviour may reflect an increased tendency to seek rewards, to compensate for negative experiences (van der Harst *et al.*, 2003b), and opioids seem to play an important role. Thus stimulation of mu-opioid receptors in the ventral tegmental area (VTA: the origin of the mesoaccumbens DA system; see this chapter) enhances DA release in the nucleus accumbens (e.g. reviewed by Kas *et al.*, 2004). In contrast, expressions of reward-anticipation are blocked in rats by the mu-opioid antagonist naloxone (Spruijt *et al.*, 2001; Barbano and Cador, 2004), and in mice by genetic manipulations that cause a lack of mu-opioid receptors (Kas *et al.*, 2004). Thus the mesoaccumbens DA system is influenced by an opioid system that in turn seems important for this system's responsiveness to external stimuli. This opioid–dopamine interaction may be crucial for the development of stereotypies in stressful conditions, because when stress exaggerates behavioural responses to rewards, the repetition of such responses can potentially lead to stereotypy. For example, in mink and other carnivores, food-restriction increases the anticipatory behaviour to scheduled food, which often involves stereotypies (e.g. Vinke *et al.* 2002, 2004, and reviewed in Chapter 3, this volume). But note that food is not the only important reward. Some behaviour patterns are not controlled by homeostatic processes but are still vital for fitness, e.g. those relating to reproduction and self-maintenance (grooming etc.). It has been argued that such behaviours have inherently rewarding properties (e.g. Spruijt *et al.*, 1992, 2001). Following this rationale, these so-called 'ethological needs' are controlled by the mesolimbic opioid–dopamine system (Spruijt *et al.*, 1992, 2001). We therefore suggest that if some stressors exaggerate animals' 'wanting' of rewards, when no external rewards are immediately available, they may instead look for compensation by performing ethological needs. Thus depending on the species and circumstances, stressed individuals may display enhanced locomotion, sexual behaviour or self-maintenance such as grooming. Our hypothesis would help to explain why abnormal variants of such activities are so common in caged and stressed animals.

With repetition, such behaviours may then change in control: as this chapter discusses, sequences of behaviour can change from environmentally sensitive (i.e. flexible, reversible) to environmentally insensitive (i.e. irreversible, inflexible) due to events downstream of the mesoaccumbens system, e.g. in the nigrostriatal system (see also Toates, 2004). This change in control may explain why developing stereotypies are sensitive to both opioid and dopaminergic antagonists, while firmly established stereotypies are sensitive to the latter only (reviewed Mason, 1991; Willemse *et al.*, 1994). Overall, we thus highlight the central role of VTA opioids in Cabib's sensitization processes, a role we propose then increases – and ultimately renders stereotypic – anticipatory behaviour and/or the performance of 'ethological needs'.

pathways come to show enhanced DA release. However, experimental evidence indicates that DA release within the mesoaccumbens (NAcc) system, is actually under the inhibitory influence of DA release within the pFC (see McFarland and Kalivas, 2001 and Cabib et al., 2002 for reviews; also Box 5.4, Chapter 5, this volume). Thus neurotoxic lesions of the mesocortical system that reduce or eliminate stress-induced DA release here, facilitate mesoaccumbens DA responses (King et al., 1997). As we will see later, this potential inhibition from the pFC plays an important role in animals' behavioural responses to stress.

Experimental data support the view that mesocortical and meso-accumbens DA responses modulate behavioural responses, sometimes termed 'coping' responses, in stressful condition. The term 'coping' has several meanings; for example, in human research the term involves subjective feelings, and in ethological and stereotypy research it has typically meant having beneficial consequences (see Box 1.3, Chapter 1, this volume). Here, I intend the term coping to refer to behavioural efforts to master a situation. These may succeed (successful coping) or may not (failed coping). Escaping, hiding, freezing and fighting, as well as species-specific defensive displays during agonistic encounters, represent typical and simple coping responses. Pharmacological facilitation of DA trans-mission favours the expression of species-typical defensive reactions, and the expression of species-typical defensive responses in intact animals is accompanied by mesocortical and mesoaccumbens DA release. Thus, as an example, repeatedly defeated male mice as well as undefeated mice treated with a drug that enhances DA transmission display species-typ-ical defensive reactions in the presence of a non-aggressive conspecific (Puglisi-Allegra and Cabib, 1988; Belzung et al., 1991; Cabib et al., 2000b). Moreover, defeated mice exposed to non-aggressive conspecifics show enhanced mesoaccumbens and mesocortical DA release (Cabib et al., 2000b).

However, if animals cannot avoid, escape or control an aversive experience, i.e. they are unable to cope with the event successfully, the initial increase in DA release within the NAc changes into a decrease below resting levels (Puglisi-Allegra et al., 1991; Rossetti et al., 1993; Cabib and Puglisi-Allegra, 1994; Rada et al., 1998). To illustrate, in spec-ifically designed experiments, it was demonstrated that exposure to restraint produces a time-dependent biphasic alteration of DA release in the NAc: an initial increase of DA release is followed by a decrease below resting levels (Puglisi-Allegra et al., 1991; see Fig. 8.1). Furthermore, this effect was evident in both rats and mice, under different stressful condi-tions, and using different technical approaches (Puglisi-Allegra et al., 1991). It was proposed that the decrease of mesoaccumbens DA release was promoted by coping failure in unavoidable/uncontrollable situations where no behavioural coping was possible (Puglisi-Allegra et al., 1991; Imperato et al., 1993). Furthermore, as we will see below, it seems to be mediated by the inhibitory effect of pFC DA release on NAcc DA re-lease. Further support of this hypothesis comes from mice tested in a

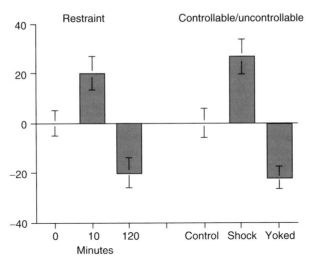

Fig. 8.1. Biphasic effects of different stressful experiences on dopamine release in the nucleus accumbens of mice of the DBA/2 strain. Time-dependent release during restraint is shown on the left-hand side; coping-dependent effects on the right. Controllable/uncontrollable indicates effects of 60 min of exposure to the shock-yoked condition (see text). Y-axis data are expressed as mean ($\pm$ SE) percent changes from basal (0 or control) tissue levels of 3-methoxytyramine (3-MT), a DA metabolite suitable for quantification of DA release *ex vivo*.

'shocked–yoked' situation. Here, pairs of animals are subjected to a series of electric shocks, with only one animal being able to interrupt shock delivery for both animals (Cabib and Puglisi-Allegra, 1994). Thus, in this experimental paradigm, both the shocked and yoked subjects receive exactly the same amount of shock, and for identical periods, but they experience it either with or without the possibility of being able to cope with the shock by turning it off. Mice exposed to this procedure show an increase of DA release in the NAc if they were allowed to control the shock experience (shocked condition), but a *decrease* of DA release if they were not allowed to exert any control (yoked condition; see Fig. 8.1).

So what mediates such effects? And what are their behavioural correlates? Recent results indicate that the inhibitory phase of mesoaccumbens DA response, seen in prolonged stressful conditions or where coping fails, is controlled by mesocortical DA release. The classic model used for these experiments is the so-called forced swimming or Porsolt's test (Porsolt *et al.*, 1977). Rodents, either mice or rats, are individually introduced into small tanks filled with warm water. The initial responses exhibited by individuals from both species are swimming and struggling to climb the walls, aimed at escaping. However, these attempts at active coping are soon abandoned when they fail, and the animals start floating in a rigid immobility that is called 'helplessness' or 'behavioural despair' (Porsolt *et al.*, 1977). Such helpless mice show a profound inhibition of DA release in the NAc. This reduction in mesoaccumbens activity is prevented by neurotoxic lesion of the mesocortical DA system (Ventura *et al.*, 2002), and moreover, mice or rats bearing such mesocortical DA lesions maintain active behavioural coping (Ventura *et al.*, 2002) instead of displaying helplessness. Furthermore, clinically effective anti-depressants reduce both behavioural helplessness, and mesocortical DA activation, and the inhibition of DA release in the NAc (Ventura *et al.*, 2002). Behavioural helplessness is not unique to the 'forced swim' test: it is also promoted in rodents by exposure to uncontrollable unavoidable shock, including the shocked–yoked paradigm previously described (Maier and Seligman, 1976; Overmier, 1988). This provides a further example of how a treatment that decreases mesoaccumbens DA release also causes a marked inhibition of behaviour.

Is such helplessness, functional or pathological? It has been hypoth-esized that this helplessness derives from the learned expectancy that there are no relationships between available responses and consequences and, for this reason, it is also called 'learned helplessness' (Maier and Seligman, 1976; Overmier, 1988; Levine and Ursin, 1991). It should be pointed out that although the helpless response is certainly functional in a situation that does not allow active coping, it is potentially extremely costly for the organism. For instance, it is easily generalized and may thus severely impair subsequent learning of escape/avoidance strategies in conditions that do allow this type of responses (Maier and Seligman, 1976; Overmier, 1988; Levine and Ursin, 1991). Moreover, helplessness has long-lasting negative effects on the expression of previously acquired appetitive responses, such as the consumption of palatable foods or intra-cranial auto-stimulation of selected brain areas (Cabib and Puglisi-Allegra, 1996a; for review), i.e. it is linked with anhedonia or the loss of pleasure. For this reason, learned helplessness is typically considered dysfunctional or pathological, and indeed these behavioural effects of unavoidable/inescapable experiences are among the most widely used models of human depressive syndromes (Maier and Seligman, 1976; Porsolt *et al.*, 1977; Overmier, 1988; Cabib and Puglisi-Allegra, 1996a).

8.2.2. Neurochemical adaptations during repeated stress

As we saw above, uncontrollable acute stress leads to an increase in mesoaccumbens DA transmission, followed by a decrease. However, the biphasic pattern of mesocorticolimbic DA response described above is elicited only by new experiences. When these experiences are then repeated, the pattern or response changes markedly. Indeed, it has been shown that repeated exposure to stressful restraint progressively elimin-ates the initial activation of mesoaccumbens DA release in both rats and mice, but without affecting the inhibitory phase of the stress response, i.e. the phase in which mesoaccumbens DA release becomes depressed (Imperato *et al.*, 1992, 1993; Cabib and Puglisi-Allegra, 1996b; see Fig. 8.2).

This change is an adaptation, not a mere artefact. For instance, it is not merely the result of an inability of the system's metabolism to sustain

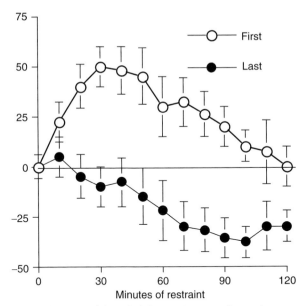

Fig. 8.2. Effects of repeated (ten daily) experiences of restraint stress on dopamine release in the rat nucleus accumbens. *Y*-axis data are expressed as mean percent changes ($\pm$ SE) from basal (0) DA outflow measured by intracerebral microdialysis on the first or last session of stress.

enhanced neurotransmitter synthesis, because the enhanced mesoaccumbens DA release that is elicited by termination of the restraint experience is always maintained (Imperato *et al.*, 1992). Thus it is suggested that the repeated exposure to the unavoidable experience specifically eliminates the initial, coping-related DA response (Imperato *et al.*, 1992). Interestingly – and perhaps unsurprisingly given the role of this system in behavioural helplessness – mesocortical DA activation is, in contrast, *not* reduced by repeated stress experiences (Cabib and Puglisi-Allegra, 1996b; see Fig. 8.3).

This adaptation of the mesocorticolimbic DA response as a result of repeated stress is affected strongly by gene–environment interactions. Thus, as can be seen in Fig. 8.3, mice of the inbred strain C57BL/6 (C57) are highly susceptible to inhibition of mesoaccumbens DA release in uncontrollable or unavoidable stressful conditions (Ventura *et al.*, 2001, 2002). In contrast, the pattern of response expressed by mice of the inbred strain DBA/2 (DBA) depends on their preceding experience. DBA mice living in standard conditions (group-housed and free-feeding, i.e. with *ad libitum* food), show a biphasic mesoaccumbens DA response to restraint (as was illustrated in e.g. Fig. 8.1), and then a progressive reduction of the initial, active phase and mesoaccumbens enhancement of DA release upon repeated restraint experience (Puglisi-Allegra *et al.*, 1991; Cabib and Puglisi-Allegra, 1996b). However, individually housed (though still free-feeding) DBA mice are characterized by a more rapid onset of the inhibitory phase in response to restraint; while individually housed and

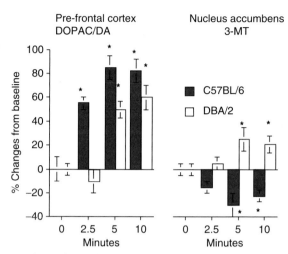

Fig. 8.3. Strain-dependent effects of restraint stress on mesocortical (left) and mesoaccumbens (right) dopamine release in the mouse. Data are expressed as mean (±SE) percent changes from basal (0) 3,4-dihydroxyphenylacetic acid (DOPAC)/dopamine (DA) ratios or tissue levels of 3-methoxytyramine (3-MT). DOPAC/DA is an index of released dopamine in the cortical areas for *ex vivo* analyses. *= statistical significance (P < 0.05) versus baseline (0).

food-restricted DBA mice show, in contrast, a prolonged activation of mesoaccumbens DA release (Cabib *et al.*, 2002). In parallel, mesocortical DA responses are enhanced in individually housed free-feeding DBA mice, but reduced in food-restricted DBA mice (Cabib *et al.*, 2002). Thus in DBAs, isolation on its own makes animals respond to restraint as though they repeatedly had been uncontrollably restrained previously, whereas the addition of food restriction has the opposite effect. This suggests that DBA and C57 mice differ in their general propensities to show CNS adaptation, as I consider below.

The plasticity within the mesocorticolimbic system, as evidenced by the adaptation of DA responses to repeated stress, may well involve DA receptors, especially DA receptors of the D2 type. Strain differences within mice would certainly be consistent with this. D2 receptors are expressed either by DA neurons or post-synaptic, non-DA neurons. The latter type of receptor mediates the effects of DA-typical transmission, while the former plays an auto-regulatory role by inhibiting DA synthesis and release, as well as inhibiting DA neuronal activity (Usiello *et al.*, 2000). Therefore, activation of the D2 post-synaptic receptors promotes the expression of DA-typical behaviours such as locomotion and stereotypies (see also Boxes 7.2 and 7.3, Chapter 7, this volume), while activation of the D2 auto-receptors inhibits these same behaviours. C57 and DBA strains differ in their behavioural responses to pharmacological stimulation of the D2 type of DA receptor (Puglisi-Allegra and Cabib, 1997). Thus, the behavioural profile promoted by D2 agonists (e.g. apomorphine) in group-housed free-feeding mice of the C57 strain reveals marked 'post-synaptic' effects, i.e. dose-dependent stimulation of locomotion and of stereotyped climbing. In contrast, the same agonists promote auto-receptor typical effects in mice of the DBA strain, i.e. an inhibition of spontaneous locomotion and climbing (Cabib *et al.*, 1997; Puglisi-Allegra and Cabib, 1997).

Evaluation of D2 densities within the basal ganglia of the two strains, using auto-radiographic techniques, further suggests a prevalence of the auto-receptor D2 in mice of the DBA strain, and in contrast the post-synaptic D2 receptors in C57 animals (Cabib *et al.*, 1998). Thus in standard-housed animals not exposed to experimental stressors, there are clear strain differences in the distribution of the D2 DA receptors. However, when animals were exposed to repeated daily restraint, this strain difference was abolished, due to opposite changes of D2 pre-synaptic and post-synaptic densities in the two strains (Cabib *et al.*, 1998). Thus with this type of repeated stress, DBA mice developed fewer of the auto-receptors involved in negative feedback, and more of the post-synaptic receptors involved in active DA responses. In contrast, with repeated restraint, C57s developed more auto-receptors, but reduced the post-synaptic receptors involved in active DA responses. Building on this, the behavioural responses to D2 agonists were then used for a genetic analysis involving recombinant inbred strains derived by C57 and DBA progenitors. The results confirmed the major regulatory role of gene–environment interactions on the D2-dependent phenotypes and identified a restricted number of loci involved in this regulation (Cabib *et al.*, 1997). These results indicate that two strains do not differ in their susceptibility to stress-induced neuroadaptation *per se*, but, rather in the type of neuroadaptive changes they undergo under environmental pressure.

8.2.3. Behavioural sensitization to psychostimulant drugs, and its inverse relationship with 'helplessness'-type DA adaptations

One long-known effect of exposure to repeated or chronic stressful experiences in laboratory rodents is 'behavioural sensitization'. This term refers to the enhancement of the behavioural effects of drugs of abuse, such that smaller doses are subsequently required to produce a given effect. Behavioural sensitization is also observed following repeated drug administration so that repeated psychostimulant dosing and stress share this enhancing effect (see Vanderschuren and Kalivas, 2000 for a review). The phenomenon is most studied in rodent models. It is supposed to be homologous to psychostimulant-induced psychoses in humans and to share neurobiological mechanisms with addiction and dependence (Laruelle, 2000; McFarland and Kalivas, 2001; Everitt and Wolf, 2002).

Interestingly, there are strain differences in this phenomenon that seem to parallel strain differences in responses of the DA system to repeated stress. Thus environmentally induced sensitization to the behavioural effects of the addictive psychostimulant amphetamine is observed in mice of the DBA strain, but not in C57 animals. This latter strain is also the most susceptible to the psychomotor and rewarding effects of amphetamine in standard living conditions (group-housed and free-feeding), but become less susceptible following food restriction or repeated restraint (Puglisi-Allegra and Cabib, 1997; Cabib *et al.*, 2000a; see Fig. 8.4). Thus, sensitivity to psychostimulants seems to be regulated by the same gene–environment interaction that controls the balance between D2 pre- and post-synaptic sensitivity (see above).

As we saw above, behavioural sensitization is not the only form of alteration induced by repeated or chronic stress: helplessness can also result. An enhanced predisposition to this type of helplessness, for instance in the forced swimming test, has been reported in food-restricted mice of the C57 strain (Alcaro *et al.*, 2002) and in individually housed DBA (Cabib *et al.*, 2002). In contrast, food-restricted mice of the DBA strain show a reduced propensity to helplessness (Cabib *et al.*, 1995; Alcaro *et al.*, 2002). The profile of behavioural adaptation thus appears to parallel the previously described neurochemical adaptation responses of these two strains. Thus, gene–environment interactions that facilitate mesoaccumbens DA transmission in stressful conditions (Cabib *et al.*, 2002) favour 'active coping' responses like escape, struggling and fighting (Alcaro *et al.*, 2002; Cabib *et al.*, 2002), and also promote behavioural sensitization to psychostimulant drugs (Cabib and Bonaventura, 1997; Cabib *et al.*, 2000a,b).

Interestingly, just as we might predict from the evidence above, the research on drug-induced behavioural sensitization points to an imbalance between mesocortical and mesoaccumbens DA transmission that

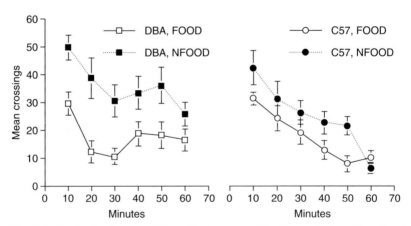

Fig. 8.4. Strain–feeding condition effects on sensitization as indexed by stimulant-induced activity. Effects of amphetamine (2 mg/kg) on locomotor activity (mean crossings ± SE) in individually-housed free-feeding (FOOD) and food-restricted (NFOOD) mice of the inbred strains DBA/2 (DBA) and C57BL/6 (C57).

favours the latter (Karler *et al.*, 1998; Prasad *et al.*, 1999; McFarland and Kalivas, 2001). In addition, repeated psychostimulant administration reduces immobility in the forced swimming test, and does so selectively in the DBA mouse strain susceptible to behavioural sensitization (Puglisi-Allegra and Cabib, 1997; Alcaro *et al.*, 2002). Box 8.1 further discusses the role of VTA opioids in helping to modulate this DA activity, and makes a suggestion as to why it is that particular behaviours are most prone to becoming intensely repetitive with chronic stress.

In conclusion, repeated stressful experiences and alteration of the living environment may promote opposite neurobehavioural adaptations, depending on complex gene–environment interactions. On one hand, there are CNS adaptations that favour mesoaccumbens DA transmission, active behavioural coping and the development of the phenomenon of behavioural sensitization. On the other, there are CNS adaptations that favour inhibition of DA transmission and helplessness.

8.2.4. Summary

The data reviewed above indicate that the mesocorticolimbic DA system is involved in both behavioural responses to acute stress as well as in CNS adaptations to prolonged stress. Within this system, the balance between mesocortical and mesoaccumbens DA transmission appears to play a major role in behavioural responses to long-term or uncontrollable stress. Thus, an imbalance that favours mesoaccumbens DA transmission is related to active coping responses (e.g. escape attempts), while an imbalance that favours mesocortical DA transmission is related to passive responses such as helplessness and also anhedonia. Finally, the balance between mesocortical and mesoaccumbens DA responses, and animals' behavioural responses to repeated stress are regulated by interactions between genotype and experience. Thus CNS adaptations promoted by particular environments may favour active coping or helplessness, depending on both genotype and the particular challenge in question.

Are such changes functional or dysfunctional? It is hard to say whether either meets the definition of pathology (see Box 1.4, Chapter 1, and Chapter 11). However, due to its proactive effects (i.e. its interference with subsequent expression of species-typical defensive responses as well as with learning of new defensive strategies), helplessness arguably represents a costly mechanism, and the underlying neurobiological adaptation may represent the substrate of a clinical depressive-like disturbance. None the less, the neurobiological adaptation underlying maintenance of active coping strategies may be pathological too, since it has been implicated in human drug addiction and schizophrenia. Building on this, in the section below, I discuss in more detail the pathological outcomes of the DA dysfunctions underlying behavioural sensitization; and also the role that these DA systems play in drug-induced stereotypies.

8.3. Pathological Outcomes of Dopamine Dysfunctions

In the earlier section, I pointed out that the dopaminergic disturbances involved in behavioural sensitization are used to model brain dysfunctions that underlie addiction and drug dependence. Addiction is a disorder involving compulsion and impulsivity; thus the neurobiological adaptation underlying sensitization may favour the development of compulsive–impulsive behavioural responses. This is one reason for wondering whether such changes might also be implicated in the spontaneous stereotypies of captive animals, which often seem to have these characteristics. Moreover, although classical behavioural sensitization is described as enhanced locomotor responses to psychostimulant drugs, sensitized rodents also show enhanced psychostimulant-induced stereotypies. That is, low doses of these drugs will be able to promote focused stereotypies normally observable only at very high doses (see previous paragraph on DA and stereotypies for details). In Section 8.3.1, I therefore review in more detail the brain circuits that might mediate sensitization-induced compulsion and impulsivity, plus those involved in drug-induced stereotypies and their interrelationships.

8.3.1. Sensitization and addiction to psychostimulant drugs

Different theories identify a range of different psychological and neurobiological substrates for addiction, and a full evaluation and comparison of such theories is beyond the scope of this chapter. This section will focus on evidence on the relationships between the brain neurochemical dysfunctions characteristic of sensitized animals and compulsive–impulsive alterations of behavioural output seen in studies using behavioural sensitization as a model of drug addiction.

Compulsive drug use is characterized by behaviour that may become dissociated from subjective measures of drug value (Robinson and Berridge, 1993), i.e. drugs are taken in a compulsive manner even when they yield little subjective reward. Such drug use is typically elicited by specific environmental cues (e.g. the contexts and drug 'paraphernalia' associated with experience of drug effects; e.g. Cardinal et al., 2002). It has therefore been hypothesized that exposure to addictive drugs promotes brain alterations that increase conditioned reward, that is, enhance the incentive value of stimuli associated with rewarding experience and reduce inhibitory control mechanisms over behaviour (Jentsch and Taylor, 1999; Robbins and Everitt, 1999). Indeed, Jentsch and Taylor (1999) have proposed that impulsivity resulting from frontostriatal dysfunction, plays an important role in addiction. Moreover, recent evidence further indicates that repeated drug exposure alters cortical cognitive function, and leads to the loss of inhibitory control mechanisms that normally help to regulate behavioural responses (Jentsch et al., 2002). This alteration of behaviour may well depend on the disruption of executive control provided by descending influences on striatal mechanisms from the pFC (Shallice and Burgess, 1996; Everitt *et al.*, 2001; Chapter 5, this volume).

As already discussed, the behavioural sensitization to psychostimulant drugs is associated with the loss of inhibitory DA tone in the pFC, leading to a loss of inhibitory control over NAc DA transmission (Karler *et al.*, 1998; Prasad *et al.*, 1999; McFarland and Kalivas, 2001). This condition may result from decreased DA release and/or decreased DA D2 receptor signalling in the pFC. Although it was originally imagined that such plasticity influenced mesoaccumbens DA release through excitatory inputs from pFC toward midbrain DA neurons, anatomical studies have shown that pFC afferents to the VTA do not synapse on mesoaccumbens DA neurons (Carr and Sesack, 2000). So, the route of communication between pFC and VTA DA neurons may be indirect (Everitt and Wolf, 2002). Whatever the exact mechanisms involved, it should be pointed out that the disruption of DA transmission within the pFC may also remove inhibitory control over other striatal regions too, notably the DS, the brain region particularly focused on by Garner (Chapter 5, this volume), and Lewis and colleagues (Chapter 7). Moreover, enhanced DA transmission within the NAc facilitates DA transmission in its adjacent (dorsal) domain, thus promoting a ventral-to-dorsal, NAc-to-DS progression (Haber *et al.*, 2000). The progressive involvement of DS DA transmission during such processes is considered to play a major role in the development of the 'inflexible' behavioural mode characteristic of addictive behaviour (Everitt and Wolf, 2002). Indeed, addictive behaviour is inflexible, because it persists despite considerable cost to the addict.

In conclusion, there is convincing evidence that corticostriatal dysfunctions related to behavioural sensitization to drugs might also be responsible of the compulsive, impulsive and inflexible characteristics of addictive behaviour. Furthermore, such changes may come to involve regions of the basal ganglia additional to the NAc, namely the DS focused in Chapters 5 and 7.

8.3.2. Brain dopamine and stereotyped behaviour: evidence from psychostimulants

The main evidence for an involvement of brain DA in stereotyped behavioural output comes from the observation that psychostimulant drugs – which enhance brain DA transmission – plus direct agonists of DA receptors, all promote stereotypies (Robbins *et al.*, 1990, for review; also see Chapter 7, this volume). Of course, DA agonists are not the only drugs that have this behavioural effect, since for example, serotonin (5HT) agonists are also known to induce stereotyped behaviour (see Curzon, 1990, for review; also Box 7.2, Chapter 7, this volume). However, the aim of this review is to identify the brain DA circuits that are specifically involved in the mediation of pharmacologically induced stereotypies and sensitization, and this is what I focus on below.

Addictive drugs such as psychostimulants stimulate motor activation in rodents. In rats, the initial response observable at low doses of the psychostimulant amphetamine is a general increase of exploration (psychomotor activation) (Robbins *et al.*, 1990). With increasing dosage, the pattern of effects undergoes major changes characterized by an increasing response rate within a progressively reduced number of response categories. The final output of this escalation is the expression of focused head and oral stereotypies at very high doses (Robbins *et al.*, 1990) (see also Boxes 7.2 and 7.3, Chapter 7).

Amphetamine increases DA release in the mesocortical and mesoaccumbens areas already discussed in this chapter. It also increases DA release in the nigrostriatal area (*cf.* Section 8.1 and Chapters 5 and 7). However, low to intermediate doses of the psychostimulant are more effective on mesoaccumbens than on nigrostriatal DA transmission, and result in locomotion. Higher doses progressively enhance nigrostriatal DA transmission, and it is this involvement that is paralleled by stereotypies. Thus, studies employing either local infusion of amphetamine, or neurotoxic lesions of the brain DA systems, have identified the mesoaccumbens DA system as responsible for stimulating the variable pattern of behavioural responses to low doses, and the nigrostriatal DA system as responsible for the focused stereotyped responses typical of the highest doses of psychostimulant (Robbins *et al.*, 1990). Moreover, direct stimulation of the projecting field of the nigrostriatal system appears to be a necessary but (importantly) not sufficient condition for the expression of DA-dependent stereotypies (Robbins *et al.*, 1990) (see also Boxes 7.2 and 7.3, Chapter 7, this volume). Based on these observations, it has been proposed that DA released by amphetamine within the NAc might be required to maintain the general stimulatory effect of these drugs and, at low doses, a parallel, moderate DA release in the DS might allow the independent, but coordinated, stimulation of somewhat distinct response elements permitting quite complex sequences of behaviour to be performed (*cf.* Box 7.2, Chapter 7). With increasing doses, the progressively greater activation of the DS may occlude the behavioural responses arising from the NAc. Therefore, amphetamine-induced stereotypies appear to depend on the competitive co-activation of mesoaccumbens and nigrostriatal DA systems.

As already discussed, the stereotyped responses to psychostimulants are susceptible to both drug- and environmentally induced sensitization (MacLennan and Maier, 1983; Reid *et al.*, 1998). Indeed, the expression of stereotyped responses at low doses of amphetamine in stress-sensitized rats is considered a problem related with the induction of sensitization in these animals, since it interferes with the right shift in the dose–response curve for the locomotor effects of psychostimulants. The latter observation is very relevant in light of the previously discussed hypothesis that one neurobiological mechanism of sensitization is a ventral-to-dorsal, NAc-to-DS progression of the DA response.

8.3.3. Summary

The neurobiological mechanisms underlying the phenomenon of behavioural sensitization involve a progressive imbalance between cortical and subcortical transmission in favour of the latter. This form of adaptation leads to a reduced inhibitory control over behavioural output that may be responsible for impulsive–compulsive responses. In addition, the progressive enhancement of nigrostriatal DA transmission may favour behavioural rigidity and inflexibility and is involved in facilitation of DA-dependent stereotypies (e.g. the characteristic head movements seen as a result of amphetamine). These observations support the hypothesis of a strong relationship between neurobiological adaptation underlying behavioural sensitization, the development of impulsive, compulsive and inflexible patterns and drug-induced stereotyped behavioural responses. Could similar processes occur during the development of cage stereotypies? This is what I discuss in Sections 8.4 and 8.5.

8.4. Stress and Dopamine in the Development and Expression of Stereotypy

In the previous sections of this chapter we have evaluated evidence pointing to:

1. The involvement of mesocorticolimbic DA transmission in behavioural responses to stressful experiences, with different patterns of adaptation corresponding to helplessness versus sustained, active coping attempts;
2. The involvement of dysfunctional mesocorticolimbic DA transmission in the phenomenon known as behavioural sensitization – the exaggerated response to psychostimulant drugs;
3. The profound effects of gene–environment interactions on mesocorticolimbic DA functioning, and on coping responses, seen even in mature organisms; and
4. The involvement of disturbed mesocorticolimbic DA transmission in the mediation of compulsive, impulsive and inflexible behavioural patterns (e.g. addiction), as well as in pharmacologically induced stereotypies.

In the following section, I use these points to create a connection between stress and the development and expression of stereotypies in caged animals.

8.4.1. Behavioural sensitization, stress and development of stereotyped behaviour

Environmentally induced enhanced behavioural responses to amphetamine, cocaine and morphine are observable following a range of different experimental procedures. However, all seem to have in common that they are stressful. Thus the effects of all these procedures are prevented by treatments that interfere with stress-induced enhancement of circulating glucocorticoids (Marinelli and Piazza, 2002). Glucocorticoids are the prototypical stress hormones (see Section 8.1); moreover, they modulate DA transmission within the NAc as well as the behavioural effects of different types of addictive drug (Marinelli and Piazza, 2002). Therefore, there is a wide agreement that environmental manipulations that promote the development of behavioural sensitization represent stressful situation for the animals.

None the less, as already discussed, the ability of such procedures to promote sensitization depends on the genetic background of the animals. Thus, individual housing promotes behavioural sensitization in rats (Marinelli and Piazza, 2002), but not in DBA mice (Cabib and Bonaventura, 1997); and repeated restraint and food restriction also promote sensitization in rats (Marinelli and Piazza, 2002) plus also mice of the DBA strain – but not C57 mice (Puglisi-Allegra and Cabib, 1997; Cabib *et al.*, 2000b). In the earlier sections I hypothesized that stereotypic behaviour may share the neurobiological substrates of behavioural sensitization. Since behavioural sensitization is dependent on specific gene–environment interactions, the development of stereotypies should depend on these interactions too.

Mice of the inbred strain C57 do not develop behavioural sensitization to the locomotor stimulant effects of amphetamine when individually-housed and food-restricted, while DBA mice do. However, mice from the latter strain do not show behavioural sensitization when individually-housed in a free-feeding condition (Cabib and Bonaventura, 1997; Cabib *et al.*, 2000a,b, 2002). These experiments were performed in mature male mice and required a relatively short-lasting period of differential housing (14 days). The in-cage behaviour exhibited by each strain during this period was also monitored daily, 30 min before food delivery. Mice of the C57 strain exposed to the food-restricted condition showed a progressive increase in non-stereotyped cage exploration. Such cage-cover climbing is part of cage exploration; thus free-feeding mice and food-restricted mice of the C57 strain will climb on the cage cover, move around the cage, dig into the ground, rear in the middle of the cage sniffing, and lean on the cage walls sniffing at specific spots, these behaviours all being expressed in a random sequence. Individually-housed, free-fed mice from both strains did not, in contrast, show any kind of behavioural change. Finally, DBA mice showed a progressive increase in *stereotyped* cage cover climbing (Cabib and Bonaventura, 1997). These stereotyping mice just climb up and down from the wire mesh of the cage-cover (Cabib and Bonaventura, 1997; see Fig. 8.5).

Interestingly, neither strain of mice showed stereotyped climbing when exposed to just 24 h of fasting (Cabib and Bonaventura, 1997): the response was only seen in DBA mice after 2 weeks of differential treatment. This suggests that this behavioural response was not solely elicited by the state of hunger *per se*, nor aimed at searching for food; but rather that the long-term nature of the experience was the key. Moreover, following 14 days of food restriction, mice of the DBA strain also showed an enhanced response (i.e. a sensitization) to the direct DA receptor agonist apomorphine, in a test for

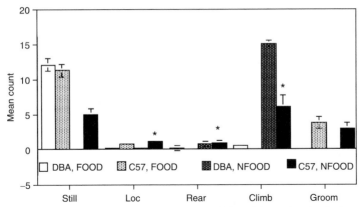

Fig. 8.5. Behavioural profile spontaneously exhibited by individually housed free-feeding (FOOD) and food-restricted (NFOOD) mice of the two inbred strains. Data refer to frequencies of different behavioural items registered in the 30 min preceding daily food delivery (2 min interval) (from Cabib and Bonaventura, 1997). *= Significant increase ($P < 0.01$) in comparison with free-feeding mice of the same strain.

pharmacologically induced climbing (Cabib and Bonaventura, 1997). This response is the prototypical DA-dependent behavioural stereotypy in mice (drug-induced stereotyped behaviour may be species-typical in rodents) and it is promoted by stimulation of the D2 post-synaptic receptors discussed earlier (Cabib *et al.*, 1997; Puglisi-Allegra and Cabib, 1997). In contrast, food-restricted mice of the C57 strain showed a marked reduction of apomorphine-induced climbing (Cabib and Bonaventura, 1997) (Fig. 8.6).

As already discussed, behavioural sensitization is the outcome of one of two possible adaptations promoted by the interaction between specific genotypes and specific (stressful) environmental conditions. Moreover, the neurobiological adaptation that underlies behavioural sensitization facilitates the maintenance of active behavioural coping in uncontrollable stressful conditions. In contrast, the other type of adaptation sensitizes the animals to behavioural helplessness. Interestingly, mice of the C57 strain appear more susceptible to the latter responses, and furthermore, this susceptibility is enhanced by food restriction (Alcaro *et al.*, 2002). In contrast, mice of the DBA strain are less susceptible to helplessness and its associated CNS changes, and this susceptibility is further reduced by food restriction (Alcaro *et al.*, 2002). These results support the view that genetic factors determine both coping styles in stressful situations and adaptation that favours maintenance of such coping styles in the face of repeated or chronic stressful conditions. Moreover, the neurobiological adaptation needed to maintain active coping renders the animals susceptible to impulsive, compulsive and inflexible behaviour, to enhanced sensitivity to the effects of addictive drugs, and to stereotypies within their home-cage. Note that this type of stereotypy does not necessarily represent a coping strategy *per se*, but is the 'side effect' of the genotype-dependent adaptation to particular types of stressful conditions. Box 8.2

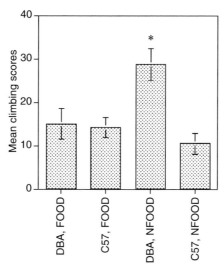

Fig. 8.6. Apomorphine (0.1 mg/kg)-induced climbing in individually housed free feeding (FOOD) and food restricted (NFOOD) mice of the inbred strains C57BL/6 (C57) and DBA/2 (DBA). * = Statistical significance ($P < 0.01$) versus all other groups.

further discusses strain-dependent differences in mouse stereotypy, to assess whether the differences reported here extend to other strains.

8.4.2. Stress, coping and the expression of environment-induced stereotypies

At this point, it should be clear that stressful environmental conditions can render some mature animals susceptible to developing stereotypic alterations of their behavioural outputs, depending on their genotype. This does not mean that these conditions immediately elicit stereotypic responses in the animals, but instead that they give rise to an underlying predisposition. Thus, just as behavioural sensitization is not expressed until drug challenge, even when promoted by prior environmental conditions, so too many prior conditions give rise to changes that are not manifest as stereotypy until the mice are exposed to particular eliciting factors.

The specific behavioural output expressed by a stress-sensitized organism thus depends on the stimuli that it encounters. Since the neuro-biological adaptation characteristic of sensitized subjects involves meso-corticolimbic DA transmission, all stimuli that activate this system are likely candidates as elicitors of stereotyped behaviour. As discussed, mesocortical DA transmission is activated by mildly aversive stimuli: in conditions of moderate aversive stimulation, pFC DA activation serves to control subcortical DA transmission. However, background sensitization promotes an imbalance that reduces cortical responses and facilitates subcortical activation. Thus sensitization may render the animal more susceptible to aversive stimuli, which will thus be more likely to elicit

Box 8.2. Strain Differences in the Cage Stereotypies of Laboratory Mice

G. Mason

Cabib convincingly proposes that stress-induced brain sensitization contributes to stereotypy. Her paradigm is a comparison of C57BL/6 ('C57') and DBA/2 ('DBA') mice in their reactions to long-term stress, sensitivity to drug sensitization and climbing responses to sustained food deprivation. However, she does not describe their behaviour in typical laboratory housing conditions: do the strain differences discussed here predict the degree of stereotypy shown night after night in their cages? The answer is yes, at least according to the one published study to systematically compare cage stereotypies across multiple strains. Nevison *et al.* (1999) kept *ad libitum*-fed male groups in two cage-types: standard barren cages merely lined with bedding; and semi-enriched cages additionally containing nesting and shelter. For 4 weeks, these were repeatedly videoed overnight. As predicted, DBAs showed more cage stereotypy than C57s. In barren cages, they spent a mean of 1.7% observations stereotyping, compared to the C57s' 0.5%; while in semi-enriched conditions these levels increased to 10.0% for DBAs and 1.3% for C57s. These data thus beautifully meet Cabib's prediction for her two focal strains (though the husbandry effect is rather unexpected; perhaps the barren conditions induced the 'behavioural despair' also discussed in this chapter?).

This study also collated data on four other strains, allowing further predictions to be tested – for if Cabib's framework has generality, the highest-stereotypy strains should also be least prone to 'behavioural despair' and most susceptible to stimulant sensitization. Nevison *et al.* (1999) found that CBA/Cas always showed the most cage stereotypies, and TOs and C57s, the least. The rank order of the intermediate stereotypers varied, however, being DBA > ICR(CD-1) > BALB/c in semi-enriched cages, but ICR(CD-1) = BALB/c > DBA in barren. Comparative data on drug sensitization have not been published for these strains. However, pre-pulse inhibition (PPI) data have been (Willott *et al.*, 2003). These are consistent with Cabib's prediction, although only if we assume that the subjects were housed in semi-enriched, not barren, cages (unspecified in the paper). Like schizophrenic humans and animals treated with DA agonists, DBAs showed low PPI: less than BALB/cs, which in turn showed less than C57s. Turning to models of depression, only two studies have investigated anhedonia/learned helplessness for three or more of these strains: Lucki *et al.* (2001) and Pothion *et al.* (2004). Both found the DBA versus C57 difference described by Cabib in this chapter; but other strain differences did *not* inversely correlate with cage stereotypy. Thus CBAs were most prone, *not* least, to anhedonia; while ICR(CD-1)s were more prone than BALBc/s to immobility in forced swim tests – *not* less prone or similar as their cage stereotypy would have predicted. Thus so far, this only partially supports Cabib's hypothesis beyond her two model strains – with three possible explanations. First, with data from so few strains, there may simply be insufficient replicates to test her idea reliably (Box 3.2, Chapter 3, illustrates how good comparative studies require a large N (here, of different strains) plus also need to control for phylogenetic relatedness). Second, the Nevison cage stereotypy data may have come from mice housed differently from the 'depression study' subjects. Housing-type clearly affects the magnitude – and even the rank order – of strain differences in cage stereotypy, and the extent to which similar variation occurs from one laboratory to the next is simply unknown. Last, it could be that, just as Cabib herself proposes here, stress-sensitization is only part of the full explanation: thus strain differences in mouse stereotypy are determined by multiple additional factors too (e.g. Chapter 4, this volume), which need factoring out before the hypothesis is tested further.

stereotypies. Box 8.3 gives one possible example of such a phenomenon in repeatedly stressed, individually-housed primates. On the other hand, the mesocorticolimbic systems are also activated by rewards (Kelly and Berridge, 2002), and by cues that predict rewards (Schultz, 2000). Thus the presence of food or environmental changes that predict food delivery may also activate the sensitized systems in an 'abnormal' way, eliciting stereotypies. Box 8.1 by Spruijt and van den Bos discusses this idea further, and Clubb and Vickery (Chapter 3, this volume) give some examples from carnivores where this may be occurring.

Box 8.3. Stress and the Performance of Primate Stereotypies

M.A. Novak, J.S. Meyer, C. Lutz and S. Tiefenbacher

Repeated or sustained uncontrollable stress is a risk factor for the development of some primate abnormal behaviour (*cf.* the argument developed in this chapter). For instance, the prevalence of self-injurious behaviour ('SIB', e.g. self-biting, self-slapping and head-banging) in rhesus monkeys is predicted by the age at which they are placed in those conditions, and also by the number of veterinary procedures experienced (e.g. venipunctures): for more details see Chapter 6. Furthermore, much converging evidence suggests that acute stress also triggers bouts of both stereotypies and SIB in laboratory primates. Indeed general stress or arousal more often seems the eliciting cue than the frustration of specific motivations (*cf.* Chapters 2 and 3, this volume). For instance, both stereotypies and SIB increase during increased sound levels (Berkson and Mason, 1964), and in the presence of a technician wearing black 'catch gloves' as if preparing for capture (Cross and Harlow, 1965). A tone previously paired with electric shock also increases SIB, compared to a similar but neutral tone (Gluck *et al.*, 1985). Monkeys with SIB often seem to self-bite in response to environmental change: for example, increased in SIB has been associated with temporary social separation or movement to a novel testing room (Suomi *et al.*, 1975; Lutz *et al.*, 2003), and even fairly routine husbandry events such as moving monkeys to a new cage or changing their location within their room (Pond and Rush, 1983). Self-biting also varied as a function of animal caretaker activity (Novak, 2003).

Physiological data further point to an effect of stress. When laboratory-housed monkeys were fitted with a heart-rate telemetry system, episodes of self-biting were preceded by a significant rise in heart rate (Novak, 2003). Additionally, individuals with high cerebrospinal fluid (CSF) levels of corticotrophin releasing factor (CRF), a key anxiety-related neuropeptide in the brain (Strome *et al.*, 2002), also had high rates of self-directed biting (Tiefenbacher *et al.*, 2002). Similarly, we found positive correlations between the amount of stereotypic behaviour in rhesus monkeys and both their morning plasma cortisol concentrations and levels of CRF in the CSF (unpublished results). Consistent with this, administering the anxiolytic drug diazepam also reduced SIB in a subset of monkeys showing this disorder (Tiefenbacher *et al.*, 2005), while fluoxetine reduces stereotypies in vervet monkeys (Hugo *et al.*, 2003) – an effect possibly mediated by its anxiolytic action (though see Chapter 10, this volume, for more on such pharmaceutical approaches). Primate stereotypies thus often seem responses to stress. One intriguing possible reason – in addition to the effects described by Cabib in this chapter – is that in some instances at least, performing these behaviours may help alleviate this stress: see our review (Chapter 6, this volume) for more details.

The mesocorticolimbic systems are not the only ones implicated. As already discussed, corticostriatal circuits can be sensitized, as characterized by a reduced frontocortical inhibitory control, enhanced mesoaccumbens DA responding and hyperactivity of the DS. These alterations are thought to further mediate the chain of impulsivity, compulsion and inflexibility. Thus, once activated by either aversive or rewarding stimuli, this chain may interfere with normal behavioural sequences. The specific morphology of the type of stereotypy thus elicited may then depend on species, current experience, past experience and current developmental stage.

As an example, stereotyped cage-cover climbing in food-restricted mice may possibly derive from food searching. Indeed, in standard conditions, food would be located in a hopper that forms part of the cage-cover. However, this does not mean that the stereotypy is an extremely motivated form of functional food searching. After all, as we saw above, animals fasted for 24 h do not show stereotyped climbing, and the stereotypy develops slowly over the first week of restriction. Furthermore, in the food-restriction paradigm, food is placed on the cage floor, not the cage-top food hopper (Cabib and Bonaventura, 1997). Instead, I suggest that the response represents the disinhibited expression of a previously functional response, even though this behaviour has no benefit in the fasting paradigm. As already discussed, the behaviour was quantified before daily food delivery and food was always delivered at the same time (Cabib and Bonaventura, 1997). So, this may be identified as a preprandial stereotypy activated by food expectancy. In non-sensitized animals this would lead to a mild activation of mesoaccumbens DA release controlled by the parallel activation of mesocortical DA response. This condition keeps behavioural output under control of current, as well as previously acquired, information (e.g. absence of food in the cage-cover hopper) allowing the expression of a complete pattern of cage exploration (as observed in C57 mice) – a response that may reflect attempts to escape the cage (Chapter 4, this volume). Instead, sensitized animals would respond with a large increase of mesoaccumbens DA release that is no longer controlled by the mesocortical response (Cabib *et al.*, 2002). This state would free the most long-established response (searching for food in the cage-lid food hopper) from inhibitory control by the ongoing experience (lack of reward) or the more recently acquired information (food is not there any more) (Ridley, 1994; Jentsh *et al.*, 2002). Finally, the activation of DA transmission in the DS, due to the enhanced mesoaccumbens activation, would focus the behavioural output on climbing by occluding all alternative responses (see previous section).

In conclusion, in a sensitized organism either mildly aversive or rewarding stimuli may activate mesoaccumbens DA transmission and then a chain of dysfunctional responses from the corticostriatal system. These responses alter ongoing, normal, behavioural output leading to the expression of stereotyped responses.

8.5. Discussion and Directions for Further Research

The hypothesis presented in this chapter is not meant to be exclusive: behavioural stereotypies might be best interpreted as multi-factorial phenomena (see e.g. Box 8.3), as are most behavioural phenotypes. Moreover, developmental processes (Chapters 6 and 7, this volume) may represent a major source of stereotypies in animals as well as in humans. None the less, neurobehavioural adaptations to stressful environmental changes in adulthood might help to explain the possibility of stereotypy development in mature individuals.

One important point in this hypothesis is that the neurobiological alterations we have discussed are not specific for stereotyped behaviour. In clinical terms they may determine a liability to disturbance in some specific genotype and in some specific conditions, while they are not responsible for the specific phenotype that is expressed. Thus, neurobiological mechanisms involved in the phenomenon of behavioural sensitization may be involved in the pathogenesis of schizophrenia, although they do not explain the syndrome. Indeed, sensitization of dopaminergic transmission within the striatum underlies the expression of 'positive symptoms' of schizophrenia (such as hallucinations) during the first psychotic episode and subsequent relapses (Laruelle, 2000). Interestingly, in the classic description of the pathogenesis of schizophrenia made by Bleuler (1950), stereotyped behaviour is said to appear before the clinical expression of the pathology. Moreover, disturbances in the relationship between cortical and subcortical functioning, the most relevant neurobiological alteration related to sensitization, have been implicated in different human disorders either involving or not involving stereotyped alteration of behavioural output (Ridley, 1994). If the neurobiological bases of sensitization represent a common substrate for different disturbances in very distant species (e.g. mouse cage stereotypies and human schizophrenia or addiction), they might well be involved in other stereotyped responses across other species and conditions. One possible test of this hypothesis would be behavioural sensitization, the enhanced species-typical behavioural response to psychostimulants, in stereotyping animals. Chapter 11 discusses such ideas further.

Overall, the neurobiological hypothesis presented here thus interprets animals' stereotypies as symptoms of a pathological adaptation of the CNS, and it implies the impact of pathogenically stressful experiences or environments on susceptible genotypes. However, the results obtained from comparing inbred strains of mice and different housing conditions also indicate that the absence of stereotyped responses does not mean that no pathological adaptations have occurred: a lack of stereotypy does not necessarily imply either normality or low stress. Therefore, overall, there is a need for more research aimed at investigating the outcomes of different types of environmentally induced neuroplasticity and change.

References

Alcaro, A., Cabib, S., Ventura, R. and Pug-lisi-Allegra, S. (2002) Genotype- and ex-perience-dependent susceptibility to depressive-like responses in the forced-swimming test. *Psychopharmacology (Berlin)* 164(2), 138–143.

Barbano, M.F. and Cador, M. (2004) Con-tribution of opioid and dopamine sys-tems to consummatory, motivational and anticipatory components of feeding behaviour. *FENS Meeting Lisbon, Portugal, 10–14 July* Abstract number A115.3.

Belzung, C., Cabib, S., Fabiani, L., Tolen-tino, P. and Puglisi-Allegra, S. (1991) LY 171555-induced hyperdefensiveness in the mouse does not implicate benzodi-azepine receptors. *Psychopharmacology (Berlin)* 103, 449–454.

Berkson, G. and Mason, W. (1964) Stereo-typed behaviors of chimpanzees: relation to general arousal and alternative activ-ities. *Perceptual and Motor Skills* 19, 635–652.

Berridge, K.C. and Robinson, T.E. (2003) Parsing reward. *Trends in Neuroscience* 26, 507–513.

Bleuler, E. (1950) *Dementia Precox.* Inter-national University Press, New York.

Cabib, S. and Bonaventura, N. (1997) Paral-lel strain-dependent susceptibility to en-vironmentally-induced stereotypies and stress-induced behavioral sensitization in mice. *Physiology and Behaviour* 61, 499–506.

Cabib, S. and Puglisi-Allegra, S. (1994) Op-posite responses of mesolimbic dopa-mine system to controllable and uncontrollable aversive experiences. *Journal of Neuroscience* 14, 3333–3340.

Cabib, S. and Puglisi-Allegra, S. (1996a) Stress, depression and the mesolimbic dopamine system. *Psychopharmacology* 128, 331–342.

Cabib, S. and Puglisi-Allegra, S. (1996b) Dif-ferent effects of repeated stressful experi-ences on mesocortical and mesolimbic dopamine metabolism. *Neuroscience* 73, 375–380.

Cabib, S., Zocchi, A. and Puglisi-Allegra S. (1995) A comparison of the behavioral effects of iminaprine, amphetamine and stress. *Psychopharmacology* 121, 73–80.

Cabib, S., Oliverio, A., Ventura, R., Lucch-ese, F. and Puglisi-Allegra, S. (1997) Brain dopamine receptor plasticity: testing a diathesis–stress hypothesis in an animal model. *Psychopharmacology (Berlin)* 132, 153–160.

Cabib, S., Giardino, L., Calza, L., Zanni, M., Mele, A. and Puglisi-Allegra, S. (1998) Stress promotes major changes in dopa-mine receptor densities within the mesoaccumbens and nigrostriatal sys-tems. *Neuroscience* 84, 193–200.

Cabib, S., D'Amato, F.R., Puglisi-Allegra, S. and Maestripieri, D. (2000a) Behavioral and mesocorticolimbic dopamine re-sponses to non-aggressive social inter-actions depend on previous social experiences and on the opponent's sex. *Behavioural Brain Research* 112, 13–22.

Cabib, S., Orsini C., Le Moal, M. and Piazza, P.V. (2000b) Abolition and reversal of strain differences in behavioral responses to drug of abuse after a brief experience. *Science* 289, 463–465.

Cabib, S., Ventura, R. and Puglisi-Allegra, S. (2002) Opposite imbalances between mesocortical and mesoaccumbens dopamine responses to stress by the same genotype depending on living condi-tions *Behavioural Brain Research* 129, 179–185.

Cardinal, R.N., Parkinson, J.A., Hall, J. and Everitt, B.J. (2002) Emotion and motiv-ation: the role of the amygdala, ventral striatum, and prefrontal cortex. *Neurosci-ence and Biobehavioural Review* 26, 321–352.

Carr, D.B. and Sesack, S.R. (2000) Projec-tions form the rat prefrontal cortex to the ventral tegmental area: target specifi-city in the synaptic associations with mesoaccumbens and mesocortical neurons. *Journal of Neuroscience* 20, 3864–3873.

Cross, H. and Harlow, H. (1965) Prolonged and progressive effects of partial isolation on the behavior of macaque monkeys. *Journal of Experimental Research in Personality* 1, 39–49.

Curzon, G. (1990) Stereotyped and other motor responses of 5-hydroxytryptamine receptor activation. In: Cooper, S.J. and Dourish, C.T. (eds) *Neurobiology of Stereotyped Behaviour.* Clarendon Press, Oxford, UK, pp. 142–168.

Everitt, B.J. and Wolf, M.E. (2002) Psychomotor stimulant addiction: a neural systems perspective. *Journal of Neuroscience* 22, 3312–3320.

Everitt, B.J., Dickinson, A. and Robbins, T.W. (2001) The neuropsychological basis of addictive behaviour. *Brain Research Reviews* 36, 129–138.

Gluck, J., Otto, M. and Beauchamp, A. (1985) Respondent conditioning of self-injurious behavior in early socially deprived rhesus monkeys (*Macaca mulatta*). *Journal of Abnormal Psychology* 94, 222–226.

Goldstein, L.E., Rasmusson, A.M., Bunney, B.S. and Roth, R.H. (1996) Role of the amygdala in the coordination of behavioral, neuroendocrine, and prefrontal cortical monoamine responses to psychological stress in the rat. *Journal of Neuroscience* 16, 4787–4798.

Haber, S.N., Fudge, J.L. and McFarland, N.R. (2000) Striatonigrostriatal pathways in primates form an ascending spiral from the shell to the dorsolateral striatum. *Journal of Neuroscience* 20, 2369–2382.

Hugo, C., Seier, J., Mdhluli, C., Daniels, W., Harvey, B.H., Du Toit, D., Wolfe-Coote, S., Nel, D. and Stein, D.J. (2003) Fluoxetine decreases stereotypic behavior in primates. *Progress in Neuropsychopharmacology and Biological Psychiatry* 27, 639–643.

Imperato, A., Puglisi-Allegra, S., Casolini, P. and Angelucci, L. (1991) Changes in brain dopamine and acetylcholine release during and following stress are independent of the pituitary-adrenocortical axis. *Brain Research* 538, 111–117.

Imperato, A., Angelucci, L., Casolini, P., Zocchi, A. and Puglisi-Allegra, S. (1992) Repeated stressful experiences differently affect limbic dopamine release during and following stress. *Brain Research* 577, 194–199.

Imperato, A., Cabib, S. and Puglisi-Allegra, S. (1993) Repeated stressful experiences differently affect the time-dependent responses of the mesolimbic dopamine system to stress. *Brain Research* 601, 333–336.

Inglis, F.M. and Moghaddam, B. (1999) Dopaminergic innervation of the amygdala is highly responsive to stress. *Journal of Neurochemistry* 72, 1088–1094.

Jentsch, J.D. and Taylor, J.R. (1999) Impulsivity resulting from frontostriatal dysfunction in drug abuse: implications for the control of behavior by reward-related stimuli. *Psychopharmacology (Berlin)* 146, 373–390.

Jentsch, J.D., Olausson, P., De La Garza, II, R. and Taylor, J.R. (2002) Impairments of reversal learning and response perseveration after repeated, intermittent cocaine administrations to monkeys. *Neuropsychopharmacology* 26, 183–190.

Karler, R., Calder, L.D., Thai, D.K. and Bedingfield, J.B. (1998) The role of dopamine in the mouse frontal cortex: a new hypothesis of behavioural sensitization to amphetamine and cocaine. *Pharmacology and Biochemistry of Behaviour* 61, 435–443.

Kas, M.J.H., van den Bos, R., Baars, A.M., Lubbers, M., Lesscher, H.M.B., Hillebrand, J.J.G., Schuller, A.G., Pintar, J.E. and Spruijt, B.M. (2004) Mu-opioid receptor knockout mice show diminished food-anticipatory activity. *European Journal of Neuroscience* 20, 1624–1632.

Kelly, A.E. and Berridge, K.C. (2002) The neuroscience of natural rewards: relevance to addictive drugs. *Journal of Neuroscience* 22, 3306–3311.

King, D., Zigmond, M.J. and Finlay, J.M. (1997) Effects of dopamine depletion in the medial prefrontal cortex on the stress-induced increase in extracellular dopamine in the nucleus accumbens core and shell. *Neuroscience* 77, 141–153.

Laruelle, M. (2000) The role of endogenous sensitization in the pathophysiology of schizophrenia: implications from recent brain imaging studies. *Brain Research Reviews* 31(2–3), 371–384.

Lucki, I., Dalvi, A. and Mayorga, A.J. (2001) Sensitivity to the effects of pharmacologically selective anti-depressants in different strains of mice. *Psychopharmacology* 155, 315–322.

Lutz, C., Marinus, L., Chase, W., Meyer, J. and Novak, M. (2003) Self-injurious behavior in male rhesus macaques does not reflect externally directed aggression. *Physiology and Behavior* 78, 33–39.

MacLellan, A.J. and Maier, S.F. (1983) Coping and stress-induced potentiation of stimulant stereotypy in the rat. *Science* 219, 1091–1093.

Maier, S.F. and Seligman, M.E.P. (1976) Learned helplessness: theory and evidence. *Journal of Experimental Psychology* 105, 3–46.

Marinelli, M. and Piazza, P.-V. (2002) Interaction between glucocorticoid hormones, stress and psychostimulant drugs. *European Journal of Neuroscience* 16, 387–394.

Mason, G.J. (1991) Stereotypies: a critical review. *Animal Behaviour* 41, 1015–1037.

McFarland, K. and Kalivas, P.W. (2001) The circuitry mediating cocaine-induced reinstatement of drug-seeking behavior. *Journal of Neuroscience* 21, 8655–8663.

Nevison, C.M., Hurst, J.L. and Barnard, C.J. (1999) Strain-specific effects of cage enrichment in male laboratory mice *(Mus musculus)*. *Animal Welfare* 8, 361–379.

Novak, M.A. (2003) Self-injurious behavior in rhesus monkeys: new insights into its etiology, physiology, and treatment. *American Journal of Primatology* 59, 3–19.

Overmier, J.B. (1988) Psychological determinants of when stressors stress. In: Hellhammer, D.H., Florin, I. and Weiner, H. (eds) *Neurobiological Approaches to Human Disease. Neuronal Control of Bodily Function: Basic and Clinical Aspects*, Vol. 2. Hans Huber, Toronto, pp. 236–259.

Perrotti, L.I., Hadeishi, Y., Ulery, P.G., Barrot, M., Monteggia, L., Duman, R.S. and Nestler, E.J. (2004) Induction of DeltaFosB in reward-related brain structures after chronic stress. *Journal of Neuroscience* 24, 10594–10602.

Pond, C. and Rush, H. (1983) Self-aggression in macaques: five case studies. *Primates* 24, 127–134.

Porsolt, R.D., Le Pichon M. and Jalfre, M. (1977) Depression: a new animal model sensitive to antidepressant treatments. *Nature* 266, 730–732.

Pothion, S., Bizot, J.C., Trovero, F. and Belzung, C. (2004) Strain differences in sucrose preference and in the consequences of unpredictable chronic mild stress. *Behavioural Brain Research* 155, 135–146.

Prasad, B.M., Hochstatter, T. and Sorg, B.A. (1999) Expression of cocaine sensitization: regulation by the medial prefrontal cortex. *Neuroscience* 88, 765–774.

Puglisi-Allegra, S. and Cabib, S. (1988) Pharmacological evidence for a role of D2 dopamine receptors in the defensive behavior of the mouse. *Behavioural Neural Biology* 150, 98–111.

Puglisi-Allegra, S. and Cabib, S. (1997) Psychopharmacology of dopamine: the contribution of comparative studies in inbred strains of mice. *Progress in Neurobiology* 51, 637–661.

Puglisi-Allegra, S., Imperato, A., Angelucci, L. and Cabib, S. (1991) Acute stress induces time-dependent responses in dopamine mesolimbic system. *Brain Research* 554, 217–222.

Rada, P.V., Mark, G.P. and Hoebel, B.G. (1998) Dopamine release in the nucleus accumbens by hypothalamic stimulation-escape behavior. *Brain Research* 782, 228–234.

Reid, M.S., Ho, L.B., Tolliver, B.K., Wolkowitz, O.M. and Berger, S.P. (1998) Partial reversal of stress-induced behavioral sensitization to amphetamine following metyrapone treatment. *Brain Research* 783, 133–142.

Ridley, R.M. (1994) The psychology of per-
serverative and stereotyped behaviour.
Progress in Neurobiology 44, 221–231.

Robbins, T.W. and Everitt, B.J. (1999) Drug
addiction: bad habits add up. *Nature* 398,
567–570.

Robbins, T.W., Mittelman, G., O'Brien, J. and
Winn, P. (1990) The neurophysiological
significance of stereotypy induced by
stimulant drugs. In: Cooper, S.J. and
Dourish, C.T. (eds) *Neurobiology of
Stereotyped Behaviour.* Clarendon Press,
Oxford, UK, pp. 25–63.

Robinson, T.E. and Berridge, K.C. (1993) The
neural basis of drug craving: an incentive-
sensitization theory of addiction. *Brain
Research Reviews* 18, 247–291.

Robinson, T.E. and Kolb, B. (2004) Structural plasticity associated with exposure
to drugs of abuse. *Neuropharmacology* 47
(Suppl. 1), 33–46.

Rossetti, Z.L., Lai, M., Hmadan, Y. and
Gessa, G.L. (1993) Depletion of mesolimbic dopamine during behavioral despair:
partial reversal by chronic imipramine.
European Journal of Pharmacology 242,
313–315.

Salamone, J.D. and Correa, M. (2002) Motivational views of reinforcement: implications for understanding the behavioral
functions of nucleus accumbens dopamine. *Behavioural Brain Research* 137,
3–25.

Schultz, W. (2000) Multiple reward signals
in the brain. *Nature Reviews Neuroscience* 1, 199–207.

Shallice, T. and Burgess, P. (1996) The domain of supervisory processes and temporal organization of behaviour.
*Philosophical Transactions of the Royal
Society, London B Biological Sciences*
351, 1405–1411.

Spruijt, B.M., van Hooff, J.A.R.A.M. and Gis-
pen, W.H. (1992) Ethology and neurobiology of grooming behavior. *Physiology
Reviews* 72, 825–852.

Spruijt, B.M., van den Bos, R. and Pijlman
F. (2001) A concept of welfare based
on reward evaluating mechanisms in
the brain: anticipatory behaviour as an indicator for the state of reward systems.

Applied Animal Behaviour Science 72,
145–171.

Strome, E.M., Wheler, G.H.T., Higley, J.D.,
Loriaux, D.L., Suomi, S.J. and Doudet,
D.J. (2002) Intracerebroventricular corti-
cotropin-releasing factor increases limbic
glucose metabolism and has social context-dependent behavioral effects in non-
human primates. *Proceedings of the
National Academy of Sciences* 99,
15749–15754.

Suomi, S.J., Eisele, C.D., Grady, S.A. and
Harlow, H.F. (1975) Depressive behavior
in adult monkeys following separation
from family environment. *Journal of Abnormal Psychology* 84, 576–578.

Tiefenbacher, S., Gabry, K.E., Novak, M.A.,
Pouliot, A.L., Gold, P.W. and Meyer, J.S.
(2002) Central levels of CRF and NPY in
male rhesus monkeys with self-injurious
behavior. *Society for Neuroscience*, Or-
lando, Florida.

Tiefenbacher, S.T., Fahey, M.A., Rowlett,
J.K., Meyer, J.S., Pouliot, A.L., Jones,
B.M. and Novak, M.A. (2005) The efficacy
of diazepam treatment on the management of acute wounding episodes in captive rhesus macaques. *Comparative
Medicine* 55, 387–392.

Toates, F. (2004) Cognition, motivation,
emotion and action: a dynamic and vulnerable interdependence. *Applied Animal Behaviour Science* 86, 173–204.

Ursin, H. and Eriksen, H.R. (2004) The cognitive activation theory of stress. *Psychoneuroendocrinology* 29, 567–592.

Usiello, A., Baik, J.H., Rouge-Pont, F.,
Picetti, R., Dierich, A., LeMeur M.,
Piazza, P.V. and Borrelli, E. (2000) Distinct functions of the two isoforms of
dopamine D2 receptors. *Nature* 408,
199–203.

van den Berg, C.L., Pijlman, F.T.A., Koning,
H.A.M., Diergaarde, L., van Ree, J.M. and
Spruijt, B.M. (1999) Isolation changes the
incentive value of sucrose and social behaviour in juvenile and adult rats. *Behavioural Brain Research* 106, 133–142.

van den Bos, R., Meijer, M.K., Renselaar, J.P.,
van der Harst, J.E. and Spruijt, B.M. (2003)
Anticipation is differently expressed in

rats (*Rattus norvegicus*) and domestic cats (*Felis silvestris catus*) in the same Pavlovian conditioning paradigm. *Behavioural Brain Research* 141, 83–89.

van den Bos, R., van der Harst, J., Vijftigschild, N., Spruijt, B., van Luijtelaar, G. and Maes, R. (2004) On the relationship between anticipatory behaviour in a Pavlovian paradigm and Pavlovian-to-instrumental transfer in rats (*Rattus norvegicus*). *Behavioural Brain Research* 153, 397–408.

van der Harst, J.E., Fermont, P.C.J., Bilstra, A.E. and Spruijt, B.M. (2003a) Access to enriched housing is rewarding to rats as reflected by their anticipatory behaviour. *Animal Behaviour* 66, 493–504.

van der Harst, J.E., Baars, A.M. and Spruijt, B.M. (2003b) Standard housed rats are more sensitive to rewards than enriched housed rats as reflected by their anticipatory behaviour. *Behavioural Brain Research* 142, 151–156.

Vanderschuren, L.J. and Kalivas, P.W. (2000) Alterations in dopaminergic and glutamatergic transmission in the induction and expression of behavioral sensitization: a critical review of preclinical studies. *Psychopharmacology (Berlin)* 151, 99–120.

Ventura, R., Cabib, S. and Puglisi-Allegra, S. (2001) Opposite genotype-dependent mesocorticolimbic dopamine response to stress. *Neuroscience* 104, 627–631.

Ventura, R., Cabib, S. and Puglisi-Allegra, S. (2002) Genetic susceptibility of mesocortical dopamine to stress determines liability to inhibition of mesoaccumbens dopamine and to behavioral 'despair' in a mouse model of depression. *Neuroscience* 115, 999–1007.

Vinke, C.M., Eenkhoorn, C.N., Netto, W.J., Fermont, P.C.J. and Spruijt, B.M. (2002) Stereotypical behaviour and tail biting in farmed mink (*Mustela vison*) in a new housing system. *Animal Welfare* 11, 231–245.

Vinke, C.M., van den Bos, R. and Spruijt, B.M. (2004) Anticipatory activity and stereotypical behaviour in American mink (*Mustela vison*) in three housing systems differing in the amount of enrichments. *Applied Animal Behaviour Science* 89, 145–161.

Von Frijtag, J.C., Reijmers, L.G.J.E., van der Harst, J.E., Leus, I.E., van den Bos, R. and Spruijt, B.M. (2000) Defeat followed by individual housing results in long-term impaired reward- and cognition-related behaviours in the rat. *Behavioural Brain Research* 117, 137–146.

Von Frijtag, J.C., van den Bos, R. and Spruijt, B.M. (2002) Imipramine restores the long-term impairment of appetitive behavior in socially stressed rats. *Psychopharmacology* 162, 232–238.

Willemse, T., Mudde, M., Josephy, M. and Spruijt, B.M. (1994) The effect of haloperidol and naloxone on excessive grooming behaviour of cats. *European Neuropsychopharmacology* 4, 39–45.

Willott, J.F., Tanner, L., O'Steen, J., Johnson, K.R., Bogue, M.A. and Gagnon, L. (2003) Acoustic startle and pre-pulse inhibition in 40 inbred strains of mice. *Behavioural Neuroscience* 117, 716–727.

9

Environmental Enrichment as a Strategy for Mitigating Stereotypies in Zoo Animals: a Literature Review and Meta-analysis

R. SWAISGOOD[1] AND D. SHEPHERDSON[2]

[1]Center for Conservation and Research for Endangered Species, Zoological Society of San Diego, PO Box 120551, San Diego, CA 92112, USA; [2]Oregon Zoo, 4001 SW Canyon Rd, Portland, OR, 97221, USA

Editorial Introduction

Zoo animals provided some of the earliest and best described cases of stereotypic behaviour, examples including the repetitive pacing of large carnivores (Chapter 2, this volume) or the monotonous swaying of elephants. Because stereotypic behaviour often dismays the public, as well as causing concern about welfare and stress, zoos have taken the lead in finding practical ways to reduce it. Usually these attempts are through what is known as 'environmental enrichment' – typically alterations to the enclosure, or additions of particular objects or stimuli, made with the broad aim of increasing welfare. In this chapter, Swaisgood and Shepherdson present an overview of the likely causes of stereotypy in zoo animals, bringing together many of the ideas of the previous chapters: the frustration of specific motivated behaviours; a general lack of physical complexity and/or sensory stimulation; and stress (though, rightly or wrongly, they downplay the role of major CNS changes in zoo animals). They show how in practice this has correspondingly led to enrichments designed to offer specific behavioural opportunities (often foraging-related), or to make enclosures more naturalistic, complex or stimulating (although relatively few to date have aimed to reduce stress *per se*). But how well does such enrichment work? After all, truly recreating natural environments, or offering genuine opportunities to, say, hunt, will often be hard if not impossible, and furthermore, the real psychological needs of most exotic species are as yet unknown.

Happily, studies of enrichment and its effects on stereotyping zoo animals have accumulated to the point where they are ripe for meta-analysis, which the authors of this chapter have done to great effect. They pool and analyse around 20 papers, dealing with about 100 individuals across multiple taxa. Their findings include that those animals with the most time-consuming stereotypies attract the most complex, multi-component 'everything-but-the-kitchen-sink' forms of enrichment. While this makes it hard to assess exactly what aspects of it 'work', this is an appropriate practical response and also seems effective, with the behaviour of these highly stereotypic animals being just as successfully reduced as is that of much less stereotypic individuals. Indeed overall, Swaisgood and Shepherdson find an impressive degree of success, enrichments typically reducing stereotypy-performance by half. It is notable, however, that not a single case eliminated the behaviour altogether, making it even more pressing to ascertain what 'works' and what does not. Are some types of environmental enrichment better than others, for example? Surprisingly, not obviously so, though the authors highlight multiple possible reasons for this. Enrichments that are implemented for a long time, however, do seem more effective than more short-lived forms. Boxes in this chapter further discuss the role of enrichment in the captive breeding of pandas; the practical realities of implementing enrichment programmes in zoos; the effects of enrichment on laboratory primates; and alarmingly, how the relative effectiveness of different enrichments can depend on precisely how their influence on stereotypy is measured.

The authors end with a very thoughtful guide to future work. Research in zoos is typically hampered by the small number of animals available, and also by the difficulty of providing sufficient scientific control in an environment where the primary goal, rightly, is to eliminate rather than merely understand stereotypies. The chapter therefore ends with suggestions as to how research on stereotypies in zoo animals could be improved – a topic potentially of great fundamental as well as practical value.

JR and GM

9.1. Introduction

The focus of this chapter is on lessons learned from studies of enrichment and stereotypy in the zoo community. Zoos have a rich history of pioneering work in this area, yet modern zoo research has not kept pace – and perhaps cannot keep pace – with research on lab and farm species. We in the zoo community have access to a diversity of species, making for interesting comparative studies (see Chapters 1 and 3, this volume), but our studies are often compromised by issues of limited sample sizes and little experimental control. Zoo enrichment practitioners also face great challenges in understanding the welfare of such a diverse array of wild animals. However, a renewed effort in this arena will pay off considerably, for no other community has so much unrealized potential.

Most modern zoos aim to present animals to their visitors in a way that conveys information about natural history and fosters positive attitudes to animals and their natural environments. Today the emphasis is on using this connection to encourage visitors to make 'environmentally friendly' choices that will benefit wildlife in the future. Zoos can also take part directly in the conservation of wildlife by augmenting and re-establishing wild populations with individuals reared or bred in zoos (e.g. Hutchins and Conway, 1995). Thus, zoos need to maintain a wide

diversity of species with an emphasis on genetically sustainable, pheno-
typically normal populations of rare and endangered species (e.g. see Box
9.1). Given this background, many zoos are highly motivated to tackle
stereotypies when they occur in their animal collections for several key
reasons.

First, concern for animal well-being is a fundamental ethic of modern
zoos, and stereotypic behaviour is an indicator of potential welfare problems. Second, stereotypies may indicate stress, and the deleterious effects
of stress on reproduction and health have been widely documented (e.g.
Moberg, 2000; Shepherdson *et al.*, 2004). Successful reproduction is central to the ability of zoos to display and reintroduce endangered species.

Box 9.1. Enrichment and Captive Breeding Programmes for Endangered Species:
the Case of the Giant Panda

R. SWAISGOOD

In today's zoo, and the conservation community at large, a prominent goal of maintaining wild
animals in captivity is to support efforts to conserve endangered species. The purpose of these
captive breeding programmes is to create a genetic reservoir as an insurance policy should the
species become extinct or genetically compromised in the wild. Because poor animal well-
being can be a significant obstacle to reproduction, much effort is made to monitor signs of
poor adjustment to captive conditions, including assessment of stress and abnormal behaviour,
and to determine what environmental provisions are necessary to sustain the species physically and psychologically (Shepherdson, 1994; Swaisgood, 2004b). The highly endangered
giant panda makes a good case study, illustrating just how integral such efforts are for
successful captive breeding. Notorious for their reluctance to breed in captivity, earlier efforts
focused on the role of olfactory communication in promoting libido (Swaisgood *et al.*, 2000).
However, it soon became apparent that communication alone would not solve all breeding
problems. Many pandas suffered from relatively severe stereotypies, which appeared linked to
reproductive problems. Assuming that behavioural needs were therefore not being met,
researchers set out to devise an enrichment programme with the short-term goal of reducing
stereotypies and the long-term goal of increasing reproduction (Swaisgood *et al.*, 2005b).
Initial tests showed that simple novel objects commanded the attention of pandas living in
impoverished conditions, and performed equally well compared with enrichments devised to
create opportunities to work for food. These enrichments led to a significant reduction in
stereotypies and signs of feeding anticipation, and also promoted behavioural diversity (Swaisgood *et al.*, 2001). These tests also shed light on the motivation underlying stereotypy performance. The novel objects served no biological goal other than the opportunity to *perform*
behaviours and/or gather information, yet the effects of this enrichment continued into the
aftermath of the enrichment interaction, suggesting that motivation to perform stereotypies was
influenced, i.e. effects were not just the result of taking up time to perform stereotypies. From
these initial insights researchers continued to expand the enrichment programme, often
adapting it to the needs of individual animals that still failed to reproduce (Zhang *et al.*,
2004). Eventually, the programme included enclosure redesign, dietary change (particularly
increased bamboo feeding) and other changes in basic husbandry practices: all this to much
effect, for this breeding centre in China grew from about 25 to nearly 80 animals in the space of
a few years post-implementation of this holistic programme (Swaisgood *et al.*, 2005a,b).

Third, the educational role of zoos can be compromised by animals displaying stereotypies: stereotypic behaviour is not a good representation of 'natural' behaviour, and is frequently perceived negatively by zoo visitors. Finally, abnormal behaviours may reduce the chances of an animal surviving in the wild, or indicate other problems that may negatively impact an animal's chance of surviving in the wild after release (Shepherdson, 1994; Vickery and Mason, 2003). Since stereotypies are rarely, if ever, seen in the wild, the key to resolving them clearly must reside in making appropriate changes to the zoo environment. These kinds of changes are usually encompassed by a general philosophy of animal husbandry termed 'environmental enrichment', a term now almost synonymous with efforts to improve animal well-being in the zoo community. The purpose of this chapter is to review the effectiveness of environmental enrichment in reducing stereotypic behaviour in zoo animals. In Sections 9.2 and 9.3, we discuss enrichments and how they might tackle stereotypies, before analysing their success in practice in Section 9.4.

9.2. What is Environmental Enrichment?

Environmental enrichment is a loosely defined term that describes actions taken to enhance the well-being of captive animals by identifying and providing key environmental stimuli (Shepherdson, 1998). Its conceptual roots can be traced back at least to the beginning of the last century, when primatologist Yerkes (1925) emphasized that captive animals should be given opportunities for both play and 'work' activities comparable to those performed by wild animals. The concepts subsequently grew in sophistication with the work of zoo biologists like Heini Hediger, Desmond Morris, Hal Markowitz and many others working in related fields. Their important contributions have been described elsewhere in greater detail (see Shepherdson, 1998).

Environmental enrichment aims to both pre-empt and cure stereotypies and other welfare problems. In practice, however, it is often reactive, and targeted at individuals or groups with overt behavioural problems. Examples range from naturalistic foraging tasks to objects introduced for manipulation, play and exploration, novelty and sensory stimulation. Social stimulation, and even training by humans (e.g. Kastelein and Wiepkema, 1988), are often described as enrichment. Renovating old and 'sterile' exhibits and the construction of new exhibits, with the design goal of providing enhanced opportunities for the expression of natural behaviour patterns, are also forms of enrichment (e.g. Little and Sommer, 2002). The activities performed in zoos in the name of environmental enrichment thus cover a multitude of innovative, imaginative and ingenious techniques, devices and practices, and the processes of developing and monitoring these enrichments are ever being improved (see Box 9.2).

Box 9.2. Enriching with SPIDER

J. Barber **and** D. Shepherdson

Environmental enrichment in zoos and aquaria has historically been a fairly unstructured, grassroots movement, with news of apparent successes spread by word-of-mouth, presentations at zoo-oriented conferences, newsletters like *Shape of Enrichment* (www.enrichment.org), and other web-based means (e.g. www.enrichmentonline.org). Enrichment goals were often rather unclear; 'success' poorly defined; enrichment choices somewhat arbitrary (often based on precedent rather than specific end results) and evaluation very subjective, with little done to report successes or failures consistently. Recent years, however, have seen a growing call for a more formal, goal-oriented approach (e.g. Mellen *et al.*, 1998; Mellen and MacPhee, 2001; Shepherdson, 2002). Indeed, in North America, legislation from the U.S. Department of Agriculture in 1991 required facilities with non-human primates to develop formal enrichment plans, and this prompted the American Zoo and Aquarium Association (AZA) to extend this requirement to all animals in AZA-accredited collections. Using guidelines like the 'SPIDER' model, the AZA proposes that this process involve: Setting goals; Planning; Implementing; Documenting; Evaluating and Readjusting (e.g. www.animalenrichment.org).

(S) *Setting goals* ideally involves knowing which behaviours are important for an animal's health and psychological well-being. Without data on what captive wild animals really need, this typically comes from understanding the animals' natural history, 'gut feelings' and general observations. Goals might be to provide opportunities to perform natural behaviour patterns (e.g. providing polar bears with hay to make day-nests, analogous to the tundra beds used by some bears in the wild); and/or to increase cognitive stimulation (perhaps via 'non-natural approaches' like computer terminals, or the human–animal communication involved in training) and/or to reduce stereotypy.

(P) Enrichment suitability needs careful consideration by keepers, curators, veterinarians and nutritionists during *planning*. The impact that enrichment may have on animal welfare is considered, but also safety, cost, space availability and visitor reactions – since these influence what is practical. Many exhibits were not initially designed to maximize animal welfare by current standards, and so during planning/approval there can be conflict between what is in the best interest of the animals, and the constraints of what is possible (e.g. not having space to house a whole pride of lions) or suitable (e.g. not wanting to feed carnivores whole carcasses in front of visitors).

(I) *Implementation* is the next step. Many factors impact the effectiveness of enrichment, including where and when the enrichment is provided, how long it is given, and to which animals (factors which remain relatively understudied). To be effective, enrichment needs to be varied, and scheduled in advance.

(D) *Documenting* is important to record the animal's response to enrichment initiatives. Simple records of whether an animal uses the enrichment are currently the commonest form of documentation. More scientific information on changes to the animals' time-budgets, and long-term physiological and behavioural data, are recorded less often, given the time investment and resources needed. The development of multi-institutional databases is an essential next step for the future.

(E) *Evaluating* effectiveness uses such records and the clearly defined goals (see 'S') to determine the extent to which an enrichment is successful. Formal, objective assessments remain, however, an underutilized aspect of all enrichment programmes.

Continued

Box 9.2. *Continued*

(R) The final step requires *readjustments* so that enrichment initiatives remain (or become) effective. For example, enrichment can succeed or fail for many reasons – depending on which animal it is given to, in the presence of which conspecific, where, or at what time. Some may simply fail to evoke the desired behavioural response. Creativity, knowledge of the individual animals, and a clear understanding of behavioural goals are the keys to success here.

As a process, SPIDER ideally guides the implementation of knowledge on a day-to-day basis, to realize the true value of enrichment. It is a long way from the scientific experiments reported in Chapters 2–8 of this book, but still represents a major advance, and perhaps could even yield future data for further meta-analyses like those presented in this chapter.

Positive effects of appropriate enrichment are frequently found. In research laboratories, animals reared in more enriched environments show a variety of brain changes, demonstrate improved learning ability, and are less emotionally reactive and more exploratory with novel objects and places (e.g. Renner and Rosenzweig, 1987; and Chapter 7, this volume). Animals living in enriched environments may also exhibit lower levels of pituitary-adrenal activation, and other indices of stress (Dantzer and Mormede, 1981; Carlstead *et al.*, 1993). Finally, enriched environments can promote a more diverse species-typical behavioural repertoire and a concomitant reduction in stereotypies (see previous chapters, and reviews in Shepherdson *et al.*, 1998). This suggests a clear role for enrichment in maintaining wild species in captivity, promising to increase successful mating and rearing of offspring, and to promote the development of more behaviourally competent candidates for reintroduction to the wild.

9.3. Reducing Stereotypic Behaviour with Environmental Enrichment: Principles Underlying the Practice

As discussed in this book, there are a number of potential causal factors in the development of stereotypic behaviour. Although all could be invoked to explain cases of stereotypy in zoos, some are probably more relevant than others. Some stereotypies in zoo animals doubtless stem from CNS pathology due to factors such as abnormal development (*cf.* Chapters 6 and 7, this volume); while others may occur as a direct consequence of veterinary health problems (e.g. dermatitis-induced stereotypic grooming, Virga, 2003; see also Chapter 10). However, zoo stereotypic behaviours are typically considered to be induced by an animal's current environment, and via the following potential causal factors:

1. Frustrated motivations to perform specific behaviours
Hughes and Duncan (1998) proposed that animals may suffer, and develop stereotypies, in situations where they are motivated to perform behaviours but are frustrated from performing them (see also Box 1.1 in Chapter 1; and Chapters 2–4, this volume). They focused on appetitive behaviour, i.e. behaviour patterns that precede and enable an animal to acquire a

certain resource or attain a particular state (such as access to a conspecific, a resting place, a cooler microclimate or food). However, frustrated consummatory behaviour, e.g. ingestion, can also be important. For instance, high levels of feeding motivation associated with insufficient nutrient intake can, quite independently of opportunities to perform foraging behaviours, lead to stereotypy performance (e.g. Rushen, 2003; Chapter 2, this volume). In this case, both the stereotypy and the behaviours directed to effective enrichments often seem to be analogues of the specific frustrated natural behaviour (e.g. Shepherdson *et al.*, 1993; Swaisgood *et al.*, 2001; see also Box 1.1, Chapter 1).

2. *Paucity of behavioural opportunities*
Zoo environments have traditionally provided few challenges, leaving animals with large amounts of free time and no appropriate behaviour with which to fill that time (e.g. Carlstead, 1996). This may cause or enhance stereotypies either through reduced behavioural competition (due to few competing activities; see e.g. Chapters 2 and 4 on 'channelling'), or by encouraging animals to perform stereotypies to self-stimulate, e.g. to increase arousal to some optimal level (see Berkson and Mason, 1964; Mason, 1991).

3. *Lack of sensory stimulation*
Zoo environments may offer relatively little to *perceive* as well as to *do*; furthermore, the stimulation they provide may be very unvarying and predictable (e.g. Chapter 7, this volume). Animals in such stimulus-poor environments may either reduce activity and stimulus-seeking behaviour, or seek out stimulation, again perhaps through stereotypy (e.g. Carlstead, 1996). Furthermore, repeated behaviours may become more stereotyped in form if the environment does not cause them to be varied (again see Chapters 2 and 4 on 'channelling').

4. *Stress*
Stress is defined as: (i) the animal's perception of a threat that challenges internal homeostasis; and (ii) the behavioural and physiological adjustments that the organism undergoes to avoid or adapt to the stressor and return to homeostasis (Moberg and Mench, 2000; and Chapter 8, this volume). Several aspects of the zoo environment may be a source of stress to zoo animals. Humans, nearby predator species (e.g. Carlstead *et al.*, 1993), confinement with potentially aggressive conspecifics (Wielebnowski *et al.*, 2002a), and ambient noise levels (Owen *et al.*, 2004) all cause signs of stress. This may be exacerbated if the animal has no control over its exposure to stress, nor any other means of coping. Stress may result when animals do not have control over salient environmental factors, either reinforcing or aversive, i.e. where access to resources and even stimulus feedback is no longer *contingent* upon behaviour (e.g. Sambrook and Buchanan-Smith, 1997; Markowitz and Aday, 1998). Increasing the degree of control animals have over their environments may thus be one

mechanism by which more complex environments tend to reduce stress and result in psychologically healthier animals (Wemelsfelder, 1993). Cabib (Chapter 8, this volume) discusses one way in which stress may cause stereotypies. A further, functional hypothesis for a link between stress and stereotypies is that paradoxically, animals under stress may stereotype to reduce arousal to an optimal level (e.g. Mason, 1991; Carlstead, 1996; Mason and Latham, 2004; also Chapter 6, this volume).

Based on these putative causal factors, a number of enrichment 'strategies' have evolved to guide efforts to reduce stereotypic behaviour. These strategies can be categorized as follows:

1. *Mimicking nature*
Almost a philosophical stance, this principle has long played a major role in zoo enrichment (e.g. Hutchins *et al.*, 1984). The aim is to stimulate natural behaviours or try to mimic specific environmental factors important in the wild (dens, food items, social groupings, increased space, etc.). The essence of this approach is that since species have evolved over many generations to survive and thrive in their wild habitat, mimicking nature should satisfy their motivational needs (see Section 9.3(1), earlier). Mimicking nature could also affect stereotypic behaviour through the other mechanisms listed above. However, this is a 'scatter shot' approach: it does not really seek to address a specific frustrated motivation, but rather hopes that providing a more natural environment will satisfy psychological needs whatever they are. Veasy *et al.* (1996) outline the drawbacks of an overly simplistic application of this concept, especially if the implicit assumption is that everything in the wild is good for well-being. Since this is obviously not the case, we are generally left with little guidance as to which characteristics or behaviours of the wild environment should be mimicked and which should not, and therefore enrichment practitioners typically resort to informed intuition.

2. *Increasing the physical complexity of the environment*
Specifically increasing physical, and temporal, complexity – without necessarily mimicking nature – may add biologically relevant information to an animal's enclosure, resulting in increased opportunities for exploration and sensory stimulation, and perhaps alleviating sub-optimally low arousal (Renner and Rosenzweig, 1987; Carlstead, 1996). It can thus provide hitherto absent behavioural opportunities through increased diversity of substrates and physical structures. By providing a context within which an animal can learn to increase its chance of achieving a desired goal through the performance of appropriate behaviour, it can potentially provide greater contingency or control (thence reducing stress).

3. *Increasing sensory stimulation*
If sensory stimulus deprivation is suspected, then modifications to the environment can be used to increase sensory stimulation *per se*

(Carlstead, 1996). Changes to the environment may be similar to those for increasing complexity, but greater emphasis is placed on stimulating the five senses. For example, scents can be introduced, noises made, visual complexity added (even video images have been tried), food varied in texture and/or taste, and substrates added to provide varied tactile feedback. Social interactions may also function to increase sensory stimulation (e.g. via allogrooming, olfactory stimuli, etc.).

4. *Meeting specific frustrated motivations*
This principle is essentially a refinement of mimicking nature, but with the primary objective being to elicit the performance of specific behaviours. In practice, foraging behaviours are the most frequently targeted in this category. These are appetitive behaviours judged particularly likely to be important because they are essential for daily survival in the wild (thence likely to be highly motivated), and most species spend most of their waking hours foraging in nature. However, zoo animals often are fed highly prepared diets that do not allow them to employ natural search, acquisition and processing behaviours; and many observations do link feeding and foraging with stereotypies (e.g. Falk, 1977; Carlstead, 1996; Rushen, 2003; Young, 2003; and Chapters 2 and 3, this volume).

5. *Removing sources of stress or providing coping options*
In many cases the form and timing of the behaviour itself may indicate the source of stress or frustration, and in these cases enrichment may consist of removing the stressor or providing the animal with a means of coping with the stressor. For example, leopard cats stressed by the close proximity of predators had lower corticoid and pacing levels when provided with appropriate hiding places (Carlstead *et al.*, 1993; see also Chapter 3, this volume).

6. *Providing enrichments that give the animal control*
Markowitz (e.g. Markowitz and Aday, 1998) is best known for pioneering this approach to enrichment. Over the years he and his co-workers have engineered numerous devices that re-establish contingency between behaviour and various consequences. As an example, food delivery can be made contingent upon the animal chasing artificial prey. This approach has been extended to include numerous other control options in the captive environment (also reviewed in Sambrook and Buchanan-Smith, 1997).

These categories of enrichment strategy are neither exhaustive nor mutually exclusive. A complex environment is likely, for example, to contain more options for control and coping with stress, to increase sensory stimulation, and to promote opportunities to perform behaviours that meet specific motivational needs. To the extent that the increased complexity is 'natural', complex environments may also mimic nature. Furthermore, multiple strategies often are implemented together (in an 'everything-but-the-kitchen-sink' approach). Thus isolating and understanding the effects of these different strategies is in practice a daunting task. Yet if we are really to understand how enrichment works and explain the differential efficacy between enrichment strategies, this must be done. In the literature analysis that follows, we have made a first attempt to do just this.

If implementing enrichments involves some challenges (see Box 9.1), then research on enrichments involves even more. Zoos are not primarily research institutions, and controlled experiments are challenging both practically and ethically (Swaisgood *et al.*, 2003b). Animals have to be available to display to the public, and to undergo veterinary procedures and many other events that may jeopardize a closely controlled experimental design. Sample sizes are often small, and animals within an enclosure are not statistically independent. Also, it may not be ethical to continue with conditions that are clearly not helping or, conversely, to return to baseline conditions when the experimental condition has proved effective (see Chapter 10 for similar issues facing vets dealing with companion animals). However, in spite of these problems, many attempts have been made and published to evaluate the effectiveness of enrichment in zoo animals. It is these we utilize below.

9.4. What can we Learn about Enrichment and Stereotypy from an Analysis of Published Zoo-based Research?

9.4.1. Our literature review and analysis: methods

A review of the literature can provide insights into how zoo practitioners tackle stereotypy with enrichment, and move us one step closer to understanding what works in practice. Here we share results from a quantitative analysis of zoo-based published literature, and draw additional insights from the specifics of some case studies.

For this analysis, we reviewed all papers published between 1990 and 2003 in three peer-reviewed journals: *Animal Welfare, Applied Animal Behaviour Science* and *Zoo Biology*, where most peer-reviewed zoo enrichment studies are published. Where appropriate, we refer to studies published elsewhere, but the statistical analysis stems from only these papers, to avoid possible biases associated with the 'file drawer' approach (Lipsey and Wilson, 2001). We included only publications meeting the following criteria: (i) the animals were studied in two different situations that varied in terms of enrichment quality (i.e. control versus enriched); (ii) the performance of stereotypies was quantitatively measured and mean values were reported; and (iii) the study was conducted at a zoological park, aquarium or conservation breeding centre. Studies at biomedical research facilities were not included in this analysis (though see Box 9.3 for the types of insight they can generate). We found 18 publications meeting these criteria, but our sample size was increased to 23 because some papers included more than one study or reported results from two or more species. We refer to these 23 samples as 'studies'. This

Box 9.3. The Effects of Enrichment in Biomedical Facilities: Some Insights into their Effects on Laboratory Primates' Stereotypies

M.A. Novak, J.S. Meyer, C. Lutz, S. Tiefenbacher, J. Gimpel and G. Mason

In biomedical research laboratories, non-human primates are often kept in groups or pairs in indoor facilities, and more rarely, in individual cages, a housing treatment particularly linked with stereotypy (Chapter 6, this volume). In these various conditions, which environmental enrichments reduce abnormal behaviour, and how might they act? Non-human primate data yield two take-home messages relevant to this chapter. First, very different forms of enrichments can be similarly effective at reducing abnormal behaviour. Second, sometimes some abnormal behaviours but not others are affected by a given enrichment.

To illustrate the first issue, in individually housed rhesus monkeys, self-directed abnormal behaviour can be reduced by exposure to a 'musical feeder box' (Line *et al.*, 1990), placement in an enriched playpen (Bryant *et al.*, 1988) and adding a conspecific companion (Eaton *et al.*, 1994); while stereotypic pacing and other whole-body stereotypies can be decreased by providing a foraging/grooming board (Bayne *et al.*, 1991), a food puzzle feeder (Novak *et al.*, 1998), or substantially larger novel cages (Draper and Bernstein, 1963; Paulk *et al.*, 1977). Similarly in group-housed rhesus monkeys, adding foraging opportunities (by scattering food in litter) to a standard cage, or increasing space allowance and climbing opportunities, both effectively reduce stereotypic pacing, and to equal degrees (Gimpel, 2005). Together, such findings suggest either that very diverse enrichments share a common important feature, such as stress-reduction (Chapter 8, this volume), or instead that different enrichments act to tackle stereotypies in varying, quite different ways (see this chapter for a list of likely means).

Turning to the issue of differential effectiveness on different forms of abnormal behaviour, in three of the above studies of single-housed primates, enrichments successfully reduced whole-body stereotypies, but failed to affect the self-directed abnormalities such as self-injurious behaviour (Paulk *et al.*, 1977; Bayne *et al.*, 1991; Novak *et al.*, 1998). Similarly in group-housed rhesus monkeys, while extra space or foraging substrates reduced pacing, they failed to reduce body-rocking or self-directed behaviours (Gimpel, 2005). This finding could indicate that different types of environmental deficit are responsible for different types of abnormal behaviour, in which case trying yet further diverse types of enrichment should prove effective (see above for successful attempts to reduce self-directed stereotypies). Alternatively, it could indicate that some abnormal behaviours are inherently harder than others to tackle, perhaps because they are pathologies induced by early experience rather than direct products of the current environment. Note, however, that even such pathologies may eventually slowly respond to changes in housing, when these changes are sustained and appropriate (Chapter 6, this volume).

small number of peer-reviewed studies may seem disappointing, but this again speaks of the difficulty of carrying out first-rate research in a zoo environment.

Before proceeding, some caveats on the analysis are necessary. First, some of our data are not fully independent, for example, when two or more species are given the same enrichment test at the same facility. Second, our sample does not come from a random sample of zoo animals, but instead emphasizes carnivores (6 ursids, 5 felids) and primates (5), with 6 others coming from disparate taxa (3 seals, 1 elephant, 1 giraffe and 1 conure). It seems likely that these species are either more prone to stereotypy or that zoo biologists are more concerned about their welfare – a form of 'targeting' that we return to later in this chapter. Third, the sample sizes for these studies are small (median = 4, range = 1–11). Fourth, a further shortcoming is that many studies used several different forms of enrichment simultaneously: often a veritable laundry list of enrichments is used. Such approaches can offer little insight into the underlying basis for stereotypy performance and its alleviation – they often address all four putative causal factors underlying stereotypy, and indeed this is probably precisely why such tactics are chosen. And herein lies the conundrum for zoo-based research on enrichment: zoo practitioners, generally speaking, are looking out for the welfare of their own animals first and foremost, and often the solution to 'problem behaviours' must be found quickly. These priorities do not lend themselves to the kind of careful experimental designs that tease out the various hypotheses to explain stereotypy and the effects of enrichment. Thus in our sample, 13 of 23 studies suffered from the use of many diverse enrichments, making our task of identifying 'what works and what does not' exceedingly difficult.

To obtain a comparable dependent variable, we took the following approach. Data reported in the original articles were in the form of simple percent time stereotyping before and after enrichment, which we used to calculate the percent change in stereotypy to standardize the measure (see Box 9.4 for some relevant measurement issues here). Thus, a reduction in stereotypy from 4% time to 2% time would be equivalent to a reduction from 10% to 5%. This method is analogous in form to calculation of 'effect size' using standard literature meta-analysis techniques (Lipsey and Wilson, 2001), with the exception that we did not correct for bias arising from varying sample sizes in different studies. The formula for this correction requires knowledge of the standard error, which was reported in only a subset of the papers in our review. However, because paper sample sizes in our analysis were small and varied little, giving each study equal weight in the analysis probably introduces little bias. Although some authors reported different forms of stereotypy (e.g. locomotor, oral, repetitive movements), many reported results on composite stereotypies – an amalgamation of several stereotypic forms. While clearly much is to be learned by an analysis of stereotypic forms that may vary in underlying motivation (e.g. Mason, 1993; see also Chapters 2–4, this volume), our data therefore did not permit this. Thus, all forms of stereotypy were lumped for analysis.

Our independent variable of interest – enrichment type – categorized enrichment according to the following dimensions. First, to evaluate whether enrichment targeted at feeding motivation was more effective, we categorized enrichment as feeding only, non-feeding only or a mix of both. Examples of non-feeding enrichment include enclosure modifications and provision of manipulable objects. Within the category of feeding enrichment, we classified the enrichment further by what kind of foraging behaviour was promoted – searching, extracting or processing. Second, we tried to determine which enrichment strategies worked better than others. We originally discussed six enrichment strategies or principles used in

Box 9.4. Evaluating Stereotypy Frequency in Enrichment Studies: Different Methods Lead to Different Conclusions

S. Vickery

Studies to determine the effect of environmental enrichment on stereotypies can differ in how they evaluate stereotypy frequency. For example, stereotypy might be measured as a proportion of all observations made across all hours ('Method 1') (e.g. Carlstead *et al.*, 1991; Shepherdson *et al.*, 1993; Grindrod and Cleaver, 2001); as a proportion of observations made during the period immediately after providing the enrichment ('Method 2') (e.g. Bloomstrand *et al.*, 1986; Powell, 1995; Tepper *et al.*, 1999; Swaisgood *et al.*, 2001: in these studies, observation periods ranged between 30 min and 2 h after providing enrichments); or as a proportion of all observations, controlling for the time an animal spends interacting with the enrichment ('Method 3') (e.g. Swaisgood *et al.*, 2001; Vickery, 2003). But do these three methods lead to the same conclusions?

Data were collected during an enrichment experiment involving 14 individually caged bears (8 Asiatic black bears (*Ursus thibetanus*: 4♂, 4♀), and 6 Malayan sun bears (*Helarctos malayanus*: 4♂, 2♀)), and analysed using these three different methods (Vickery, 2003). The enrichment experiment comprised five stages: 'Pre' – a 10-day pre-enrichment baseline; 'Ob1' – a 7-day object-only enrichment stage (e.g. cage furniture, heavy-duty plastic containers, straw); 'Ob+fd' – a 7-day stage of object (as 'Ob1') plus food enrichment (e.g. whole coconuts, hidden food); 'Ob2' – an exact replicate of 'Ob1' and 'Post' – a 10-day post-enrichment baseline. During each stage, the bears' behaviour was observed between 0700 and 1800 h by scan-sampling from observation hides.

The three different methods of evaluating changes in stereotypy led to quite different conclusions as to the enrichments' effectiveness. When the measure used was total daily stereotypy (Method 1), only object enrichments had a near significant effect, and only during their first week of provision (GLM for 'Ob1' versus baseline: $F_{1,11} = 4.51$; $P = 0.057$). When only stereotypy in the first hour after enrichment presentation was considered (Method 2), only foraging enrichments were effective (GLM for 'Ob+fd' versus baseline: $F_{1,11} = 9.09$; $P = 0.012$). When stereotypy was assessed controlling for time spent using the enrichments (Method 3), no effects at all were found. Therefore, these three different methods of evaluating stereotypy clearly measure quite different things, and rank the enrichments differently. Differences between Methods 1 and 2, and Method 3 probably arose because only the latter controls for effects due to time occupation/behavioural substitution, while differences between Methods 1 and 2 are probably due to Method 2 overemphasizing the efficacy of enrichments that are used most when first introduced. However, it is not clear which of these methods reveals the most about an animal's welfare; and these results also suggest that it will often not be valid to compare enrichment studies or attempt to draw general conclusions from them if their methods of quantifying stereotypy frequency differ.

practice: mimicking nature; increasing environmental complexity or sensory stimulation; enabling specific motivated behaviours; reducing stress and giving animals control or contingency. These were based on the motivational principles discussed in Section 9.3. Of these categories, only four were amenable to analysis; however, two – (5) removing the source of stress and (6) providing enrichments that give animals control – were excluded because they occurred too infrequently and/or it was difficult to determine the degree to which they applied to a given study.

We evaluated each of the 23 studies, ranking them on a four-point scale for the four remaining categories:

1. *Mimicking nature*
By far the most frequent principle alluded to in the zoo enrichment literature is the need to model captive environments after the stimuli and behavioural opportunities present in nature. We tried to estimate the extent to which the captive environment mimicked what would be found in the wild for the species.

2. *Increasing environmental complexity*
Often enrichment practitioners attempt to make the environment more complex, varying the number and types of objects, stimuli and behavioural opportunities, without paying much attention to what is 'natural'. Indeed, practitioners may hope that artificial items serve as a functional analogues for more natural ones (Forthman-Quick, 1984; Swaisgood *et al.*, 2003a). Complexity is a difficult concept (Sambrook and Buchanan-Smith, 1997). One method is to sum the number of 'features' it takes to describe the enclosure, but this only yields a facade of quantification. We took a more subjective approach, summing the enrichment changes that appeared to increase biologically relevant aspects of the environment. On a practical level these aspects were the only ones regularly described by authors. Indeed given the problem of quantifying complexity, Sambrook and Buchanan-Smith (1997) conclude that 'a subjective estimate is probably sufficient' (p. 208).

3. *Increasing sensory stimulation*
This category overlaps somewhat with (2), but the emphasis is on providing multiple stimuli for different sensory modalities. Several studies scored high in (2) but low in (3), suggesting redundancy was not too problematic in practice.

4. *Enabling highly motivated natural behaviours*
Most examples here dealt with attempts to increase foraging behaviour. For analysis, we evaluated studies based on how much animals were challenged to work for food, in terms of effort expended by the animals (see also our previous paragraph on feeding and non-feeding enrichments).

We analysed the effects of enrichment type with a two-way ANOVA, with enrichment category as one factor and taxonomic group (carnivore, primate, other) included as a blocking variable to reduce any statistical noise or confounding effects of phylogeny (see e.g. Box 3.2, Chapter 3, this volume). Our results are given below. Note that because of the limitations to our data-set, these are presented in the spirit of exploration, hopefully driving future hypothesis-generation and testing.

9.4.2. Results I: what kinds of enrichments were used?

As we noted in our methods, zoo enrichment practitioners often used a diverse array of enrichments. Below we describe briefly some of these enrichments before proceeding to the analysis.

In one type of enrichment commonly used, major permanent changes to exhibits were made or altogether new exhibits were built. Often live vegetation such as grass, bushes and trees were planted extensively or other naturalistic items such as logs, stumps, branches, stones, rock ledges, artificial vines, wading pools, substrate (e.g. leaves, dirt) and so forth were added. These sorts of enhancements allow animals to climb, explore, dig, dirt-bathe and generally stimulate sensory input and provide opportunities for physical exercise and locomotor play. They also give animals some basic choices, for example, to seek sun or shade for thermoregulation, to take cover from visitors or other nearby animals or to seek a spot where they can view potential threats from a safe position above ground (see Forthman et al., 1995; Poole, 1998).

The second common type of enrichment was the provisioning of manipulable objects, usually on a temporary basis. Different objects, such as plastic balls, fresh branches, egg cartons, cardboard boxes and so forth are rotated through the enclosures on an irregular basis. Feeding enrichments, the last main type used in zoos, were similarly diverse. Generally speaking, the design attempted to encourage one or more aspects of natural foraging behaviour. Some were designed to increase the amount of time the animal spent searching for the food, for example, by scattering the food around or hiding it under or on top of objects in the enclosure. For predators, live or simulated prey were used to stimulate capture behaviour. A number of clever methods were used to increase opportunities for animals to extract their food. Small food items were placed in puzzle feeders – hollow wooden or plastic objects with holes – forcing the animal to manipulate the feeder to get the food to fall through the holes. Food may also be frozen in ice blocks, requiring the animal to bite away the ice to get to the food. In one case, chimpanzees were given honey pots with 20 different tools such as string and wire brushes, used to extract the honey in a task designed to mimic the 'ant-fishing' behaviour seen in nature (Celli et al., 2003). Finally, some enrichments were designed to enhance the handling and processing behaviours occurring just before consumption, for example, by adding browse or whole carcasses to the diet.

As mentioned briefly in the methods, many zoo enrichment practitioners used a wide variety of these enrichments simultaneously. This 'everything-but-the-kitchen-sink' approach can produce dramatic results. In one example (Grindrod and Cleaver, 2001), enrichments for common seals included bottles, boxes, balls, trays, buoys, wood blocks, ice blocks, mirrors, water spray, fountains, music, floating mats, islands, fish pulled across the water and self-propelled bottles containing fish. Most of these were given with and without fish hidden in or around the enrichment. It is not surprising that these enrichments effected a reduction in stereotypy from about 65% to 23% of the animals' time. Clearly this strategy worked well in this case. It is also an example of the creativity and energy used to recreate environments for zoo animals. But what aspects of it specifically worked? It is difficult to determine. Was it mostly due to the wood blocks? Were hidden fish the key? This is a good example of excellent enrichment, but with a relatively poor transfer of information to other facilities hoping to improve the welfare of their animals. Short of reproducing the entire programme, one does not learn *how* to use enrichments to achieve the desired effect.

Our hunch had been that such pragmatic 'everything-but-the-kitchen-sink' approaches were most often adopted when there was a large problem to be solved. This was confirmed in our data-set when we determined whether multiple and diverse environmental changes were made simultaneously (e.g. was the whole enclosure renovated and altered in many ways?) or whether just a single strategy was taken (e.g. was a single or a set of similar feeder devices or manipulable objects added?). In the studies where many and diverse changes were made, animals performed stereotypies for significantly more time pre-enrichment ($N = 13$, mean $= 26.4\%$) than in studies where only few changes were made ($N = 10$, mean $= 7.9\%$, $F_{19} = 4.8$, $P = 0.04$). This suggests that enrichment practitioners throw more enrichment at animals displaying greater problems.

9.4.3. Results II: how well did enrichments work?

The most important question to begin with is, simply, does zoo enrichment work? Our cross-study analysis of effect size shows that – at least in these published studies – zoo enrichment works and it works quite well (Fig. 9.1; repeated measures ANOVA: $F_{1,20} = 20.6$, $P = 0.0002$ with Greenhouse–Geisser adjustment to correct for correlation of repeated measures). Among carnivores, primates and other species, a reduction in stereotypy performance of between 50% and 60% was observed following the onset of the enrichment programme. However, it is also essential to point out that in no case was stereotypy completely abolished.

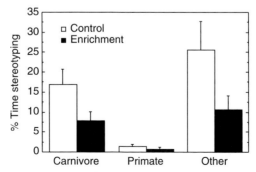

Fig. 9.1. Zoo enrichments significantly reduce stereotypy performance in various species. *Y*-axis values represent the mean of study means (for all individuals included in the study) for percent time spent performing stereotypies.

One possible concern is that these observed effects may be merely short term, and that once the novelty has worn off, the stereotypy levels will return to pre-enrichment levels. If this is the case, our results will overstate the real effectiveness of enrichment. To test this notion, we reasoned that studies of longer duration should give subjects more time for habituation. However, we found no such negative relationship between study duration and percent change in stereotypy following enrichment. In fact, we found the opposite i.e. that longer studies were associated with greater enrichment efficacy, although this effect was marginally non-significant ($r^2 = 0.17$, $P = 0.06$; median $= 4.5$ months; range $= 0.5$–16 months). These results indicate that the enrichment effects were fairly robust to habituation, perhaps because so many enrichments were comprehensive in scope (see earlier).

9.4.4. Results III: did some kinds of enrichment work better than others?

9.4.4.1. Feeding versus non-feeding enrichments

Of equal importance is to determine what *kind* of enrichments work. Although this may vary with taxon (see below), it is instructive to take a first look across all the species in our sample to see if any certain class of enrichment stands out. This is justifiable because most of the theories for why enrichment reduces stereotypy apply to most species. This may provide insight into generalized underlying factors associated with stereotypy performance in zoo animals.

As discussed above, this analysis was compromised by the fact that in many studies, several different forms of enrichment were given. One comparison we can make without such obfuscation is between enrichments that were based on feeding, non-feeding or a combination of both. Each study can be clearly classified into one of these three categories. Given the predominance of the concept of feeding motivation in stereotypy performance and enrichment, one might predict that enrichments with food are more effective than those without.

In our sample, however, this was not the case (Fig. 9.2). Feeding, non-feeding or a combination of the two were all equally effective at reducing stereotypies, in each case by 56–58% ($F_{3,15} = 0.48$, $P = 0.63$). Although taxonomic group was included in the model as a blocking variable to reduce taxon effects on the results, it is possible that some taxa were more likely to receive feeding enrichment than others. However, Table 9.1 suggests that this is not the case, with carnivores, primates and other orders being equally likely to 'attract' feeding enrichment from their caretakers. It may still be the case, however, that caretakers are tailoring their enrichments to suit the problem they are trying to solve – an issue we return to below.

9.4.4.2. Feeding enrichment: a closer look

Analysis of more specific details of enrichment strategies may uncover some patterns. To examine this possibility, we further categorized each

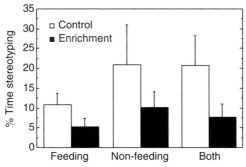

Fig. 9.2. Enrichments with and without food did not differ in terms of their effectiveness in reducing stereotypy performance.

study of feeding enrichment according to the type of behaviour promoted: search, extract, process or mixed. It is also possible – in most cases – to determine whether feeding enrichment involved a change in the actual diet or just a change in the presentation of food.

Examination of Fig. 9.3 suggests that the type of feeding task may affect its success at reducing stereotypy levels, with extraction behaviours being less successful at reducing stereotypies. However, all extraction tasks were given to primates (Table 9.1), so feeding task is confounded with taxonomic order. From this we can see that in the three cases in which feeding enrichments were given to primates – all extraction tasks – they performed poorly. By contrast, in the remaining studies where primates were given non-feeding enrichments (major exhibit changes), there was almost 90% reduction in stereotypies. However, these studies with primates are particularly difficult to interpret because pre-enrichment levels of stereotypy fell in the range of 1–3% of the total activity budget, far less than for other groups (see Figs 9.1 and 9.4): thus taxon and prior stereotypy level were confounded. Similarly, all search tasks were given to carnivores, thus again confounding feeding task with taxonomic group.

Table 9.1. The frequency with which different taxonomic groups received feeding and non-feeding enrichment or a combination of both.

	Feeding	Non-feeding	Both	Total	
Carnivore	8	3	0	11	
Primate	3	1	2	6	
Other	2	0	4	6	
Total	13	4	6	23	
	Search	Extract	Process	Mixed	Total
Carnivore	4	0	1	2	7
Primate	0	3	0	0	3
Other	0	0	2	0	2
Total	4	3	3	2	12

Thus overall, it is unsurprising that the statistical model which controls for such taxonomic effects, does not even approach significance for the effects of feeding task on enrichment efficacy ($F_{2,7} = 0.53$, $P = 0.61$).

To understand fully how feeding enrichments influenced stereotypy performance, it is necessary to know whether these enrichments involved a dietary change or whether diet remained unchanged while opportunities for feeding behaviours changed. The observed effects of any feeding enrichment may result from improved nutrition, meeting ethological needs to forage or both. Unfortunately, the authors did not always state whether food used in enrichment involved a change in diet, so it was impossible to test this. To distinguish between nutritional and ethological needs hypotheses, it will be necessary for authors to report changes in diet. Even so, there will be many instances where nutritional and behavioural changes will occur together for many feeding enrichments. For example, Stoinski *et al.* (2000) replaced the normal feed for elephants, Bermuda hay, with an equal dry weight of natural browse, which dramatically changes both diet and foraging behaviours (*cf.* Chapter 2, this volume). This experimental constraint will be common in most studies designed to increase the processing component (see below) of feeding behaviour, whereas increasing search and extraction time can be easily manipulated without changing diet. Thus, the role of processing behaviour, such as handling and mastication, will be more difficult to distinguish from nutritional needs. Future studies should strive to change the amount of processing work without changing nutritional content.

9.4.4.3. The role of enrichment strategies

As outlined in Section 9.3, enrichment could address four potential causes of stereotypy – sustained motivations to perform specific behaviours; the paucity of behavioural opportunities; a lack of sensory stimulation; and stress. As a result of this, several enrichment strategies or principles

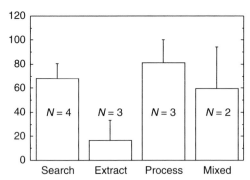

Fig. 9.3. Enrichments promoting different forms of foraging behaviour did not differ in terms of their effectiveness in reducing stereotypy performance. Although extraction of food appears least effective here, this effect can be explained more parsimoniously by taxonomic effects (see text). (The *y*-axis = % change in stereotypy.)

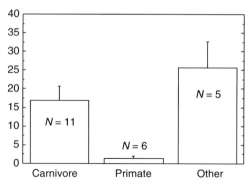

Fig. 9.4. Baseline proportion of time spent performing stereotypies before enrichment manipulations for different taxonomic groups.

are used in practice: mimicking nature; increasing environmental complexity or sensory stimulation; enabling specific motivated behaviours; reducing stress and giving animals control or contingency. Of these six categories, we could assess the first four.

There are several caveats. Although we tried to be as objective and consistent as possible when ranking studies according to these principles, there is inevitably a subjective component to how studies were scored. Since the studies were not designed to manipulate just one of these principles/strategies, we cannot compare a relatively pure manipulation of ethological needs with a relatively pure manipulation of physical complexity. Therefore, the tests that follow compare one type of enrichment to the combined effects of all other forms of enrichment in the sample, not to control or to baseline conditions. Surprisingly, our analysis found no association between the evaluated scale of any of the four enrichment principles and the percent change in stereotypy (ANCOVA with taxon as blocking variable: $F_{1,18} < 0.74$, $P > 0.41$). Even studies rated highly across several principles did not fare better than those with low ratings. The fact that none of the tests even approached significance suggests that these measures of enrichment properties have little bearing on how effective enrichment is at reducing stereotypy. Alternatively, researchers could have tailored the enrichments to the specific problems seen in their study subjects, an issue we return to later.

9.4.5. Results IV: taxonomic effects on enrichment and stereotypy revisited

Do some taxonomic groups spend more time performing stereotypies than others? Clearly, different species have markedly different evolutionary histories, biological make-up, life history characteristics and behavioural temperaments. To the extent that these characteristics are shared by related species, we might find that some groups are prone to perform stereotypy at higher levels (see e.g. Chapter 3, this volume, on carnivores).

It certainly appears that taxa differ in their typical forms of stereotypy, as we saw in Chapter 1, Fig. 1.2. For this analysis, we examined only the data for pre-enrichment periods, and found that taxon did significantly affect stereotypy performance ($F_{3,20} = 6.0$, $P = 0.009$; Fig. 9.4), with primates being the least stereotypy-prone in our sample. Among the carnivores, cats spent more time stereotyping than bears (22.6% versus 12.4%) (see Chapter 3 for a more detailed analysis of species-differences within the Carnivora).

One might speculate that the social housing prevalent for primates makes them less vulnerable to stereotypy (Chapter 6, this volume), but the 'other' group also contained mostly socially housed animals that also displayed high stereotypy levels. Further speculation on potential taxon-specific causes of stereotypy seems imprudent with this limited data-set, though.

By contrast, these taxonomic groups did not differ in their response to enrichment. We found no differences in the percent change in stereotypy levels from pre-enrichment control periods to enrichment ($F_{3,19} = 0.29$, $P = 0.75$; Fig. 9.5). Enrichment effected a 50–60% reduction in stereotypy performance for each of the three taxonomic groups. Thus, we conclude that enrichment as currently used is effective regardless of taxonomic affiliation, and also that pre-enrichment stereotypy levels do not determine how effective enrichment will be.

9.5. Conclusions and Directions for Future Research

What lessons can we learn from this first, and somewhat preliminary, systematic analysis of the zoo literature on stereotypy and enrichment?

First, our results suggest that environmental enrichment does effectively reduce stereotypy in zoo animals, at least in those studies that are published in refereed journals. The good news was that in our sample, enrichment on average halved the time spent performing stereotypies.

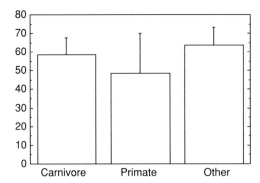

Fig. 9.5. The percent reduction in stereotypy performance following enrichment in different taxonomic groups.

Clearly zoo enrichment practitioners are doing something right, and one cannot help but be impressed by the creativity and diversity of effort found in these enrichment programmes.

With such dramatic reductions, this result also offers great promise, by suggesting that most stereotypies in zoos are not yet neurally 'hard-wired' and that they are still responsive to positive environmental change – even though subjects were typically adult animals with pre-existing stereotypies (*cf.* e.g. Chapter 7; see Mason and Latham, 2004; see also Chapter 10, this volume).

Furthermore, as far as we could tell, such effects were not just transient and just due to novelty alone. Indeed longer-lasting enrichments tended to have greater, not lesser, effects. This could be because longer-lasting enrichment studies may have been better planned and executed, given the long-term investment, or perhaps because some stereotypies are resilient to enrichment and the cumulative effects of long-term enrichment produce stronger effects.

However, a third finding, lest we be too joyful in our successes, was that in no case was stereotypy completely abolished. This may indicate that the enrichments used were always insufficient to tackle all underlying motivations to perform stereotypies. Perhaps we still do not understand fully the underlying motivations for stereotypy in zoo animals. Or alternatively, could stereotypies be both easily influenced by environmental context yet *not* be fully extinguishable because of some neural habit or a form of CNS dysfunction (*cf.* Chapters 5–8, this volume)? The answer to this question seems unknown. Stereotypy abolishment, though rare, has been reported (Mason, 1991; see also Chapter 10, this volume), but there has been no systematic effort to evaluate the context in which abolishment occurs. Large-scale change in basic husbandry and environment is the most likely candidate to reduce stereotypies to nil in zoo environments. In one instructive example, a male giant panda was moved from a small, relatively barren pen to a larger, more naturalistic pen under completely different feeding and animal care management, including a supplemental enrichment programme. Before this environmental transformation, this male spent more than one-third of his time pacing, but ceased pacing permanently (for at least 7 years) after the change (R. Swaisgood, unpublished data). However, another male underwent a similar transition to a new and improved environment, yet persisted in his tongue-flicking stereotypy. Are some forms of stereotypy more easily abolished than others? (See Box 9.3 for similar examples from laboratory primates.) Or do individual temperament, developmental history, etc. explain these divergent anecdotes? Again, these zoo data highlight some important unanswered research questions.

Our fourth main finding was that no type of enrichment seemed to be more successful than any other. For example, we could find no evidence that feeding enrichment was any more effective at reducing stereotypies than non-feeding enrichment; and there are within-study examples, too, where the two approaches were equally effective at reducing stereotypy (e.g. Swaisgood *et al.*, 2001). We also hypothesized that the type of feeding behaviour (e.g. searching, extracting, processing) promoted by feeding enrichment might determine its efficacy, and again our analysis failed to discover a relationship (although species differences in feeding behaviour may have influenced this result). Given the prominence of the surmised role that feeding motivation and frustrated foraging play in stereotypy performance, this finding is rather surprising. We made one last attempt to see if any universal factors were responsible for the efficacy of enrichment, by examining the importance of four prominent enrichment principles (mimicking nature, environmental complexity, sensory stimulation, foraging challenge). However, again we came up empty-handed, with enrichment strategies ranking high in one principle proving no better than strategies ranking high in other principles. It is rather disappointing that from this review, we have learned so little about 'which enrichments work and which don't'. So, what reasons could there be for this?

First, we should acknowledge that confounding variables and limited sample sizes greatly compromised our analyses. Thus, to give just one example, we could conclude little about the taxon-specific effects of different types of foraging enrichment. Or perhaps methods of scoring stereotypy affected the results, and thence added noise, as discussed in Box 9.4. Future surveys with larger sample sizes, and more equitable representation across taxonomic groups, would be required to investigate these issues adequately. A second possible reason is that none of the variables examined here are truly important in determining successful enrichment. We are reluctant to jump to this conclusion, but an alternative possibility could be merely that our schemes and scoring systems – sometimes based on rather limited data – failed to encapsulate or rank the properties of enrichment that really are differentially important to stereo-typing animals. A third explanation for the apparently similar success of diverse enrichments is that all the variables examined here were equally valuable. Perhaps the most parsimonious conclusion is that more than one factor is causal in stereotypy performance. Thus animals may well be kept in conditions in which multiple factors give rise to stereotypies; for instance, conditions could both thwart opportunities to perform foraging behaviours *and* exploratory, stimulus-seeking behaviours. The box in this chapter (Box 9.3) on laboratory primates, also illustrates how diverse enrichments may be similarly effective at reducing stereotypies, presumably working via different processes. Indeed, in a further study, not reviewed by them, rhesus monkeys housed individually in small cages showed reduced corticoid levels and abnormal behaviours even when just given simple manipulable devices (Line *et al.*, 1991). Thus, in the most extreme situations, one might argue that 'something, anything' added to a stimulus-poor environment may have equally meaningful effects on stereotypy (and perhaps other indices of welfare).

Finally, it is plausible that enrichment practitioners selected enrichment intelligently based on the factors apparently operating in their own study animals. Thus enrichment practitioners may have tailored enrichment to individuals and existing conditions (see Mellen and MacPhee, 2001), as well as to taxonomic group. Perhaps they successfully identified which animals needed feeding enrichment (e.g. those pacing at the door waiting to be fed), distinguishing them from those that needed more general enhancements in their environment (e.g. those kept in small, barren enclosures observed pacing at the cage boundary or displaying escape-related stereotypies). Such creative and appropriate tailoring of enrichment may obscure any universal patterns. If the enrichments chosen are tailored for maximum appropriateness, this could well explain why they all seemed to be similarly effective, and would clearly obscure any differential enrichment efficacy that might be seen if their use was more random. Indeed in our literature sample, we occasionally came across authors who explicitly acknowledged such tailoring of enrichment to the observed problem. For example, Baxter and Plowman (2001) provided giraffes with high-fibre meadow hay *because* the animals performed oral stereotypies, believed to be caused by frustrated feeding motivation (*cf.* Chapter 2, this volume). Furthermore, in our analyses we showed, for example, that 'everything-but-the-kitchen-sink' approaches were most likely to be adopted when stereotypy levels were high; and that food extraction enrichments were given to primates more often than to other taxa – again suggesting that the enrichments used in each study were selected on a tailored, case-by-case basis.

Finally, our fifth and last main finding from the survey was that different taxa varied in the amount of pre-enrichment stereotypy. Zoo primates showed low levels, a result nicely complementing the zoo and laboratory comparisons given in Chapter 6, this volume. Carnivores, especially felids, showed high levels. As it stands we cannot say why – this could reflect biological differences between mammalian taxa, differences in typical housing or both – but future zoo research is ideally placed to investigate further.

So what can we conclude about the value of zoo work to the scientific understanding of stereotypies? It is clear that there are many obstacles. For example, practices like enrichment 'tailoring', and the use of simultaneously presented, multiple diverse enrichments, probably mean that while zoo researchers will continue effectively helping their animals, they will often contribute relatively little to understanding these behavioural phenomena. However, zoo data can still have real value. For one, the sheer diversity of species available for study at zoos could add tremendously to our understanding of phylogenetic effects on stereotypy, and allow the novel test of hypotheses (*cf.* Clubb and Mason, 2003; Chapter 3, this volume). Thus zoo researchers have access to individuals with varying evolutionary histories, life history strategies, foraging strategies, ranging patterns, social systems and vastly different temperaments and perhaps behavioural needs. Furthermore, zoo animals often differ too in aspects of their early experience (see Box 7.1, Chapter 7, this volume).

So, which of these characteristics are associated with different forms of stereotypy and how do they respond to different housing arrangements, husbandry practices and enrichments? With an improved empirical arsenal, zoo enrichment practitioners stand ready to make major inroads in this arena. Zoo studies are also valuable because they highlight the research questions that impede the successful elimination of stereotypy. For example, how does one prevent the development of stereotypies in captive environments in the first place? Is prevention truly better than cure? Although zoos contain plenty of animals that have never performed stereotypies, no one to date has systematically analysed the factors responsible for stereotypy *development*. Future emphasis should be placed on how to abolish stereotypies altogether, something that did not occur in our sample, and which – as we saw earlier in this discussion – raised a number of questions about the causation of stereotypy.

This literature review and meta-analysis has thus yielded some new insights into stereotypies and enrichment, but also has been instructive in highlighting some of the shortcomings of zoo enrichment research and pointing to new directions that merit attention in the future. We would like to see a zoo literature that enables a better understanding of the *causation* and *motivations* underlying stereotypies, eventually allowing enrichment strategies to be shaped out of a *predictive* theoretical framework. If this can be achieved, the zoo community can greatly enhance its contributions to the larger fields of animal welfare and applied ethology. Below we enumerate several suggestions to facilitate this:

1. Study one or a few enrichments at a time, or otherwise allow the separate effects of each type of enrichment to be measured.
2. Test specific predictions from motivational models attempting to explain stereotypy – for example by collecting additional data on stress levels (e.g. Swaisgood *et al.*, 2001; Wielebnowski *et al.*, 2002a; Mason *et al.*, in press).
3. Analyse the effects of enrichment when the subjects are not directly interacting with the enrichment, to determine whether it affects the motivation to perform stereotypies or just occupies animals' time (*cf.* Box 9.4).
4. Conduct studies aimed at understanding the environmental and biological factors underlying the development of stereotypies, for example by also looking at non- or low-stereotyping animals.
5. Conduct multiple-institution studies that increase sample size and expand the generalizability of findings to various captive environments (*cf.* some equine studies; Chapter 2, this volume). For experimental examples in zoos see e.g. Mellen *et al.* (1998); Wielebnowski *et al.* (2002b) and Shepherdson *et al.* (2004), and for multi-institutional questionnaire surveys see Bashaw *et al.* (2001).
6. Describe the form of stereotypies in detail, and analyse the effects of enrichments on different stereotypy forms separately. If individual stereotypy forms occur too infrequently to merit separate analysis, at least provide descriptive statistics (e.g. in table form) that may be used later in meta-analyses.

7. Examine the long-term effects of enrichments to rule out possible novelty effects, understand how to avoid habituation, and reveal why long-term enrichments may be more effective.

8. Design studies to determine the circumstances in which stereotypies are abolished, not just diminished (even if this means using/reporting anecdotes).

9. Report negative and non-significant findings on the effects of enrichment on stereotypy, even if this means 'hitching' them on to more publishable positive results. Currently, only successful enrichment studies are likely to be published.

10. Include low-stereotypers too. In evaluating our results, our sample probably represents some of the worst cases of stereotypy in the zoo community. It could be instructive (and potentially also valuable for welfare – see Chapter 11, this volume), if less 'alarming' subjects also merited large enrichment research projects.

11. Conduct further meta-analyses, incorporating non-journal data. To illustrate, a better idea of the full amount of effort going into enrichment is found in the numerous publications in unrefereed journals and conference proceedings (e.g. *Shape of Enrichment*, *AZA Proceedings*; see also Box 9.2).

Overall, we recognize that, at times, the goals of research and promoting well-being may run counter to one another in zoos. While on the one hand we advocate controlled studies to understand the individual effects of particular enrichments, we also acknowledge that the goal of optimal well-being and reproduction will only be achieved through more holistic approaches, incorporating many varieties of enrichment into basic husbandry practices (Mellen and MacPhee, 2001; Swaisgood, 2004a). On the other hand, holistic approaches will continue to involve educated guesswork and 'everything-but-the-kitchen-sink' tactics, unless we fully understand the causes of zoo animals' stereotypies.

Acknowledgements

We thank Lauren Brudney, Megan Owen and Staci Wong for their assistance with the literature review.

References

Bashaw, M.J., Tarou, L.R., Maki, T.S. and Maple, T.L. (2001) A survey assessment of variables related to stereotypy in captive giraffe and okapi. *Applied Animal Behavior Science* 73, 235–247.

Baxter, E. and Plowman, A.B. (2001) The effect of increasing dietary fibre on feeding, rumination and oral stereotypies in captive giraffes (*Giraffa camelopardalis*). *Animal Welfare* 10, 281–290.

Bayne, K., Mainzer, H., Dexter, S., Campbell, G., Yamada, F. and Suomi, S. (1991) The reduction of abnormal behaviors in individually housed rhesus monkeys (*Macaca mulatta*) with a foraging/grooming board. *American Journal of Primatology* 23, 23–35.

Berkson, G. and Mason, W.A. (1964) Stereotyped behaviors of chimpanzees: relation to general arousal and alternative activities. *Perceptual and Motor Skills* 19, 635–652.

Bloomstrand, M., Riddle, K., Alford, P. and Maple, T.L. (1986) Objective evaluation of a behavioral enrichment device for captive chimpanzees (*Pan troglodytes*). *Zoo Biology* 5, 293–300.

Bryant, C., Rupniak, N. and Iversen, S. (1988) Effects of different environmental enrichment devices on cage stereotypies and autoaggression in captive cynomolgus monkeys. *Journal of Medical Primatology* 17, 257–269.

Carlstead, K. (1996) Effects of captivity on the behavior of wild mammals. In: Kleiman, D.G., Allen, M.E., Thompson, K.V. and Lumpkin, S. (eds) *Wild Mammals in Captivity*. University of Chicago Press, Chicago, Illinois, pp. 317–333.

Carlstead, K., Seidensticker, J. and Baldwin, R. (1991) Environmental enrichment for zoo bears. *Zoo Biology* 10, 3–16.

Carlstead, K., Brown, J.L. and Seidensticker, J. (1993) Behavioral and adrenocortical responses to environmental change in leopard cats (*Felis bengalensis*). *Zoo Biology* 12, 321–331.

Celli, M.L., Tomonaga, M., Udono, T., Teramoto, M. and Nagano, K. (2003) Tool use tasks as environmental enrichment for captive chimpanzees. *Applied Animal Behavior Science* 81, 171–182.

Clubb, R. and Mason, G. (2003) Captivity effects on wide-ranging carnivores. *Nature* 425, 473–474.

Dantzer, R. and Mormede, P. (1981) Pituitary-adrenal consequences of adjunctive activities in pigs. *Hormones and Behavior* 15, 386–394.

Draper, W. and Bernstein, I. (1963) Stereotyped behavior and cage size. *Perceptual and Motor Skills* 16, 231–234.

Eaton, G., Kolley, S., Axthelm, M., Iliff-Sizemore, S. and Shigi, S. (1994) Psychological well-being in paired adult female rhesus (*Macaca mulatta*). *American Journal of Primatology* 33, 89–99.

Falk, J.L. (1977) The origin and functions of adjunctive behavior. *Animal Learning and Behavior* 4, 325–335.

Forthman, D.L., McManamon, R., Levi, U.A. and Bruner, G.Y. (1995) Interdisciplinary issues in the design of mammal exhibits (excluding marine mammals and primates). In: Gibbons, E.F., Durrant, B.S. and Demarest, J. (eds) *Conservation of Endangered Species in Captivity: an Interdisciplinary Approach*. State University of New York Press, Albany, New York, pp. 377–399.

Forthman-Quick, D.L. (1984) An integrative approach to environmental engineering in zoos. *Zoo Biology* 3, 65–77.

Gimpel, J. (2005) The effects of housing and husbandry on the welfare of laboratory primates. Ph.D. thesis, University of Oxford, Oxford, UK.

Grindrod, J.A.E. and Cleaver, J.A. (2001) Environmental enrichment reduces the performance of stereotypic circling behaviour in captive common seals (*Phoca vitulina*). *Animal Welfare* 10, 53–63.

Hughes, B.O. and Duncan I.J.H. (1988) The notion of ethological 'need', models of motivation and animal welfare. *Animal Behaviour* 36, 1696–1707.

Hutchins, M. and Conway, W.G. (1995) Beyond Noah's ark: the evolving role of modern zoos and aquariums in field conservation. *International Zoo Yearbook* 34, 117–130.

Hutchins, M., Hancocks, D. and Crockett, C. (1984) Naturalistic solutions to behavioral problems of captive animals. *Zoologische Garten* 54, 28–42.

Kastelein, R.A. and Wiepkema, P.R. (1988) The significance of training for the behaviour of Stellar sea lions (*Eumetopias jubata*) in human care. *Aquatic Animals* 14, 39–41.

Line, S.W., Clarke, A.S., Markowitz, H. and Ellman, G. (1990) Responses of female rhesus macaques to an environmental enrichment apparatus. *Laboratory Animals* 24, 3–220.

Line, S.W., Markowitz, H., Morgan, K.N. and Strong, S. (1991) Effects of cage size and environmental enrichment on behavioral and physiological responses of rhesus macaques to the stress of daily events. In: Novak, M. and Petto, A. (eds) *Through the Looking Glass: Issues of Psychological Well-being in Captive Non-human Primates*. American Psychological Association, Washington, DC, pp. 160–179.

Lipsey, M.W. and Wilson, D.B. (2001) *Practical Meta-analysis*. Sage Publications, Thousand Oaks, California.

Little, K.A. and Sommer, V. (2002) Change of enclosure in langur monkeys: implications for the evaluation of environmental enrichment. *Zoo Biology* 21, 549–559.

Markowitz, H. and Aday, C. (1998) Power for captive animals: contingencies and nature. In: Shepherdson, D.J., Mellen, J.D. and Hutchins, M. (eds) *Second Nature: Environmental Enrichment for Captive Animals*. Smithsonian Institution Press, Washington, DC, pp. 47–58.

Mason, G.J. (1991) Stereotypies: a critical review. *Animal Behaviour* 41, 1015–1037.

Mason, G.J. (1993) Forms of stereotypic behaviour. In: Lawrence, A. and Rushen, J. (eds) *Stereotypic Animal Behaviour*. CAB International, Wallingford, UK, pp. 7–40.

Mason, G.J. and Latham, N. (2004) Can't stop, won't stop: is stereotypy a reliable animal welfare indicator. *Animal Welfare* 13, S57–S69.

Mellen, J. and MacPhee, M.S. (2001) Philosophy of environmental enrichment: past, present, and future. *Zoo Biology* 20, 211–226.

Mellen, J.D., Hayes, M.P. and Shepherdson, D.J. (1998) Captive environments for small felids. In: Shepherdson, D.J., Mellen, J.D. and Hutchins, M. (eds) *Second Nature: Environmental Enrichment for Captive Animals*. Smithsonian Institution Press, Washington, DC, pp. 184–201.

Moberg, G.P. (2000) Biological response to stress: implications for animal welfare. In: Moberg, G.P. and Mench, J.A. (eds) *The Biology of Animal Stress: Basic Principles and Implications for Animal Welfare*. CAB International, Wallingford, UK, pp. 1–22.

Moberg, G.P. and Mench, J.A. (2000) *The Biology of Animal Stress: Basic Principles and Implications for Animal Welfare*. CAB International, Wallingford, UK.

Novak, M.A., Kinsey, J.H., Jorgensen, M.J. and Hazen, T.J. (1998) Effects of puzzle feeders on pathological behavior in individually housed rhesus monkeys. *American Journal of Primatology* 46, 213–227.

Owen, M., Swaisgood, R.R., Czekala, N.M., Steinman, K. and Lindburg, D.G. (2004) Monitoring stress in captive giant pandas (*Ailuropoda melanoleuca*): behavioral and hormonal responses to ambient noise. *Zoo Biology* 23, 147–164.

Paulk, H., Dienske, H. and Ribbens, L. (1977) Abnormal behavior in relation to cage size in rhesus monkeys. *Journal of Abnormal Psychology* 86, 87–92.

Poole, T.B. (1998) Meeting a mammal's psychological needs: basic principles. In: Shepherdson, D.J., Mellen, J.D. and Hutchins, M. (eds) *Second Nature: Environmental Enrichment for Captive Animals*. Smithsonian Institution Press, Washington, DC, pp. 83–94.

Powell, D.M. (1995) Preliminary evaluation of environmental enrichment techniques for African lions. *Animal Welfare* 4, 361–370.

Renner, M.J. and Rosenzweig, M.R. (1987) *Enriched and Impoverished Environments: Effects on Brain and Behavior*. Springer-Verlag, New York.

Rushen, J. (2003) Changing concepts of farm animal welfare: bridging the gap between applied and basic research. *Applied Animal Behaviour Science* 81, 199–214.

Sambrook, T.D. and Buchanan-Smith, H.M. (1997) Control and complexity in novel object enrichment. *Animal Welfare* 6, 207–216.

Shepherdson, D.J. (1994) The role of environmental enrichment in the captive breeding and reintroduction of endangered species. In: Mace, G., Olney, P.J.S. and Feistner, A. (eds) *Creative Conservation: Interactive Management of Wild and Captive Animals*. Chapman & Hall, London, pp. 167–177.

Shepherdson, D.J., (1998) Tracing the path of environmental enrichment in zoos. In: Shepherdson, D.J., Mellen, J. and Hutchins, M. (eds) *Second Nature: Environmental Enrichment for Captive Animals.* Smithsonian Institution Press, Washington, DC, pp. 1–12.

Shepherdson, D. (2002) Realizing the vision: improving zoo animal environments through enrichment. *Communique* June, 5–6.

Shepherdson, D.J., Carlstead, K., Mellen, J.D. and Seidensticker, J. (1993) The influences of food presentation on the behavior of small cats in confined environments. *Zoo Biology* 12, 203–216.

Shepherdson, D.J., Mellen, J.D., Hutchins, M. (1998) Second Nature: Environmental Enrichment for Captive Animals. Smithsonian Institution Press, Washington, DC.

Shepherdson, D.J., Carlstead, K.C. and Wielebnowski, N. (2004) Cross-institutional assessment of stress responses in zoo animals using longitudinal monitoring of faecal corticoids and behaviour. *Animal Welfare*, 13, S105–S113.

Stoinski, T.S., Daniel, E. and Maple, T.L. (2000) A preliminary study of the behavioral effects of feeding enrichment on African elephants. *Zoo Biology* 19, 485–493.

Swaisgood, R.R. (2004a) Captive breeding. In: Bekoff, M. (ed.) *Encyclopedia of Animal Behavior.* Greenwood Publishing Group, Westport, Connecticut, pp. 883–888.

Swaisgood, R.R. (2004b) What can captive breeding do for conservation and what can behavior research do for captive breeding? *The Conservation Behaviorist*, 2, 3–5.

Swaisgood, R.R., Lindburg, D.G., Zhou, X. and Owen, M.A. (2000) The effects of sex, reproductive condition and context on discrimination of conspecific odours by giant pandas. *Animal Behaviour* 60, 227–237.

Swaisgood, R.R., White, A.M., Zhou, X., Zhang, H., Zhang, G., Wei, R., Hare, V.J., Tepper, E.M. and Lindburg, D.G. (2001) A quantitative assessment of the efficacy of an environmental enrichment programme for giant pandas. *Animal Behaviour* 61, 447–457.

Swaisgood, R.R., Ellis, S., Forthman, D.L. and Shepherdson, D.J. (2003a) Improving well-being for captive giant pandas: theoretical and practical issues. *Zoo Biology* 22, 347–354.

Swaisgood, R.R., Zhou, X., Zhang, G., Lindburg, D.G. and Zhang, H. (2003b) Application of behavioral knowledge to giant panda conservation. *International Journal of Comparative Psychology* 16, 65–84.

Tepper, E.M., Hare, V.J., Swaisgood, R.R., Lindburg, D.G., Quan, Z.G., Ripsky, D.M., Hawk, K. and Hilpman, G. (1999) Evaluating enrichment strategies with giant pandas at the San Diego zoo. In: Hare, V.J., Worley, K.E. and Myers, K. (eds) *Proceedings of the Fourth International Conference on Environmental Enrichment.* Shape of Enrichment, San Diego, pp. 226–239.

Veasy, J.S., Waran, N.K. and Young, R.J. (1996) On comparing the behaviour of zoo housed animals with wild conspecifics as a welfare indicator. *Animal Welfare* 5, 13–24.

Vickery, S.S. (2003) Stereotypy in caged bears: individual and husbandry factors. PhD thesis, University of Oxford, Oxford, UK.

Vickery, S.S. and Mason, G.J. (2003) Behavioral persistence in captive bears: implications for reintroduction. *Ursus* 14, 35–43.

Virga, M. (2003) Behavioral dermatology. *Veterinary Clinics in North American Small Animal Practices* 33, 231–251.

Wemelsfelder, F. (1993) The concept of animal boredom and its relationship to stereotyped behaviour. In: Lawrence, A. and Rushen, J. (eds) *Stereotypic Animal Behaviour: Fundamentals and Applications to Welfare.* CAB International, Wallingford, UK, pp. 65–95.

Wielebnowski, N., Ziegler, K., Wildt, D.E., Lukas, J. and Brown, J.L. (2002a) Impact of social management on reproductive, adrenal and behavioural activity in the cheetah (*Acinonyx jubatus*). *Animal Conservation* 5, 291–301.

Wielebnowski, N.C., Fletchall, N., Carlstead, K., Busso, J.M. and Brown, J.L. (2002b) Noninvasive assessment of adrenal activity associated with husbandry and behavioral factors in the North American clouded leopard population. *Zoo Biology* 21, 77–98.

Yerkes, R.M. (1925) *Almost Human*. Century, New York.

Young, R.J. (2003) *Environmental Enrichment for Captive Animals*. Blackwell Science, Oxford, UK.

Zhang, G., Swaisgood, R.R. and Zhang, H. (2004) Evaluation of behavioral factors influencing reproductive success and failure in captive giant pandas. *Zoo Biology* 23, 15–31.

10 Veterinary and Pharmacological Approaches to Abnormal Repetitive Behaviour

D. Mills[1] and A. Luescher[2]

[1]Department of Biological Sciences, University of Lincoln, Riseholme Park, Lincoln, LN2 2LG, UK; [2]Animal Behavior Clinic, Purdue University, VCS, LYNN, 625 Harrison Street, West Lafayette, IN 47907-2026, USA

Editorial Introduction

Like the zoo personnel of the previous chapter, practising veterinarians are often charged with the challenging job of 'curing' stereotypic behaviour, and of doing so despite negligible fundamental information on the motivational or neurophysiological bases of the behaviours in question, and little good evidence as to the best treatments. So, faced with a tail-chasing dog, a pet cat that has licked its belly bald, or a much-loved horse that stubbornly crib-bites, what do vets do? What is fascinating about the holistic approach described by Mills and Luescher is the way that it picks up and mirrors many of the themes reviewed in this volume. First, is the animal showing perfectly normal responses to an abnormal environment? For example, is it being exposed to frightening, agonistic or motivationally conflicting situations that elicit perfectly relevant (if exaggerated, inappropriate and undesirable) species-typical responses? Or has it learned associations between particular activities and the arrival of a reward (e.g. tail-chasing prompts the owner to give it a toy)? In these instances, educating the owner into changing their expectancies, their own behaviour, and/or the circumstances to which the animal is exposed, may be the only treatments needed. Indeed multiple, diverse changes in husbandry may well be appropriate, from changing the ways the animal is fed to giving it more exercise, especially if the problem is severe or its specific motivational basis cannot be ascertained. Furthermore, because the animals involved are typically more tractable than those of Chapter 9, these approaches can be supplemented with other behavioural techniques such as habituation and conditioning through graded exposure to stimuli or the reinforcement of alternative, substitute responses.

But what if the animal itself is psychologically abnormal in some way? This can be a challenging judgement to make, but Mills and Luescher argue we can reach given sufficient evidence on a subject's previous experience and history, its genetic (breed) predispositions, the extent to which multiple problematic aspects of behaviour co-occur and the degree to which a problematic behaviour impairs an animal's physical health or its relationships with other individuals. In such

© CABI 2006. *Stereotypic Animal Behaviour:*
Fundamentals and Applications to Welfare, 2nd edn (eds G. Mason and J. Rushen)

instances, any environmental change that reduces general stress might be called for, but veterinarians also have another potential means of effecting treatment: the use of pharmacological compounds such as serotonin reuptake inhibitors (SRIs). After all, these approaches are a well-used tool of doctors faced with human patients with psychological or psychiatric problems (albeit that unlike doctors, vets cannot use reported feelings and cognitions to help refine their diagnosis).

Mills and Luescher do an excellent job in describing the pros and cons of the various treatment approaches that vets can use, including a detailed overview of the use of many pharmacological compounds. It is also striking that, unlike Chapter 9, they provide some examples of complete success. However, the authors also give important warnings and caveats, especially about pharmacological approaches; and they make strong recommendations for some surprisingly basic, much-needed research into the effectiveness and the potential for side effects of such compounds. They also return us to another theme that has recurred throughout the book: how heterogeneous is stereotypic behaviour, and are there fundamental sub-types that should be distinguished? As is argued powerfully in a contributed box, if we treat one specific type of pathology as though it is another, not only may our treatments fail, but we may compromise animal welfare further in the process. Furthermore, as this chapter argues, terms like 'stereotypy', 'obsessive–compulsive disorder (OCD)' and 'compulsive disorder (CD)' either do or should have distinct meanings, and this should be borne in mind despite the challenge of distinguishing and diagnosing them in animals. Mills and Luescher also recommend that we use the broad catch-all term 'stereotypic behaviour' for all apparently abnormal, apparently functionless, repetitive behaviours – from the most clockwork-like forms of pacing to the flexible wood-chewing of Chapter 2 or hair-plucking of Chapter 5. In this, they do a useful service since this gives us an item of vocabulary that we perhaps need, given our current state of knowledge, which does not imply a known or unitary cause, an issue of definition we return to in Chapter 11.

GM

10.1. Introduction

10.1.1. Chapter overview

Abnormal repetitive behaviour is not unique to farms, laboratories and zoos – it also occurs in the animals that humans keep for companionship. These behaviours (Table 10.1 gives examples), and the strategies of the veterinarians charged with alleviating them, are our subject here. This chapter thus describes how 'stereotypies' and other abnormal repetitive behaviours are approached in veterinary clinical practice, our aims being to describe the challenges faced when tackling these behaviours; to refine how key terms are used within the field; and to identify the issues around which empirical data are still lacking.

We begin with a historical perspective of veterinary behavioural medicine, and an introduction to the 'systems view' of problem behaviours that shapes our own work (this section). We then propose a framework

Table 10.1. Abnormal Repetitive Behaviours typically seen in certain breeds of dog and cat (for images, see our section of the book's website).

Breed/type	Common Abnormal Repetitive Behaviour
Doberman Pinscher	Flank-sucking
English Bull Terrier	Spinning in tight circles/sticking head under or between objects and freezing
Staffordshire Bull Terrier	Spinning in tight circles
German Shepherd	Tail-chasing (Overall and Dunham, 2002)
Australian Cattle Dog	Tail-chasing (Blackshaw *et al.*, 1994; Hartigan, 2000)
Miniature Schnauzer	Checking hind-end
Border Collie	Visual fixations, e.g. shadow staring
Large breed dogs	Persistent licking, causing granulomas
Siamese/Burmese cat	Wool sucking (Luescher *et al.*, 1991)

to differentiate between forms of abnormal repetitive behaviour. Repetitive behaviour problems are often labelled 'compulsive' in the veterinary literature, but historically the difference between compulsive disorders (CDs) and stereotypies has been unclear, and so we propose a distinction (Section 10.2). We then discuss treatment in practice, and the potential role of psychopharmacology (Section 10.3). The efficacy of the drugs proposed for treating abnormal repetitive behaviour is then evaluated, together with possible side effects (Section 10.4). We end with a summary and highlight topics still requiring research.

10.1.2. A brief history of the science and practice of veterinary behavioural medicine

Early interest in behaviour problems, especially those of 'companion animals' (e.g. pets) was led by Tuber 30 years ago, who proposed 'animal clinical psychology' as a scientific discipline (Tuber *et al.*, 1974): an approach broadly resembling human clinical psychology (*cf.* e.g. Rosenhan and Seligman, 1995) in relying on understanding a condition's behavioural and environmental origins, and using this in treatment rather than drug-use. As ethological approaches and the idea of animal clinical psychology grew in acceptance within the veterinary profession, the potential for interventions involving medicine (Dodman and Shuster, 1998) and surgery (Hart and Hart, 1985) also became realized. However, it was not until 1997 that an international conference on 'veterinary behavioural medicine' was held (Mills *et al.*, 1997), with recent years seeing a better appreciation of the diverse perspectives that have developed in this area. For instance, some authors (e.g. Overall, 1997; Pageat, 1998) have an approach more in line with human psychiatry than psychology, advocating a medical paradigm in which problem behaviour is viewed as having a physical cause within the animal that needs treatment (Donaldson, 1998). This focuses attention on the need for intervention at

Table 10.2. Comparison of the different emphases of behavioural and medical approaches to problem behaviour.

	Behavioural model	Medical model
Conceptualization of the problem	The product of experience	A physical illness
Causal emphasis	The environment	The individual
Treatment focus	Signs of problem, environmental causal factors	Physical correlates of the problem, pathophysiological change
Treatment modalities	Behavioural and cognitive therapy	Physical interventions (medical and surgical)
Typical practitioner	Trainer or psychologist	Veterinary surgeon

the level of the individual patient rather than its environment. Table 10.2 summarizes the key differences between these 'medical' perspectives and behavioural ones.

While appealing to methods familiar to veterinarians, the medical approach has not been widely adopted, and its potential problems have also been highlighted (e.g. Mills, 2003). These stem largely from the concept that the subject animal is somehow diseased, when in fact many behavioural problems are extensions of normal behaviour (see e.g. Chapters 2–4, this volume). As we argue here, a more holistic perspective is needed, recognizing the value of ethological and psychological, as well as medical, perspectives when treating problematic behaviour.

10.1.3. The concept of behaviour problems in veterinary behavioural medicine

The definition of a 'behaviour problem' is subjective, because a behaviour is only a 'problem' in the clinical sense when an animal's owner judges it unacceptable. Sometimes these behaviours are normal and adaptive, for example a fear response following a traumatic experience, whilst in other cases, they are maladaptive, e.g. a phobia (*cf.* Box 1.4, Chapter 1).

A behaviour problem typically arises because of the sub-optimal nature of the dynamic 'system' that includes the animal concerned, its human companions, the physical and biological environment and the relationships between these. The animal's behaviour is a response to that system. When sub-optimal, the system, or change in part of it, produces 'tension' which requires change elsewhere for all components to adapt. Because all individuals will tend to try to optimize the environment for themselves, within their individual limitations, the change initiated by one component may be opposed by another part of it, potentially resulting in further tension. This systems concept can be illustrated with a typical potential problem scenario: an owner buys a second dog. This produces a change in the social environment and the distribution of resources within the home. The original dog responds to this on the basis of its evolutionary (i.e. canid-typical) behavioural tendencies, plus ontogenetic factors (i.e. its individual life-history). This may involve threatening the new dog, e.g. when it approaches certain resources, the new dog perhaps then learning to avoid these. Depending on when these behaviours occur, their form and the owner's knowledge of canine behaviour, the owner may or may not become aware of them. The owner's response (or lack thereof) to these interactions then determines how the system develops further. For example, if the resident dog growls at the new dog when the owner is present and the owner considers this unacceptable, he may punish the dog, leading to further tension, which may ultimately escalate to a more concerning situation.

This systems approach is the conceptual framework that we use in this chapter, and that guides our treatment approaches. It has several implications. First, for a behaviour problem to be fully understood, the roles that all parts of this system play need to be recognized. Second, the problem is only resolved when the person making the complaint is satisfied. This may involve bringing about a change in their own perceptions or behaviour, as much as changing the 'problem animal' itself. Third, it is possible to study objectively both the behaviour itself and the system the animal is part of. Fourth, it highlights how pharmacological interventions cannot be a 'quick fix': drug treatments for behaviour problems can never be considered in isolation of the system's other factors. In Section 10.3, we discuss treatment approaches in more detail, but first, we define different types of repetitive behaviour.

10.2. The Concept of Abnormal Repetitive Behaviour in Veterinary Behavioural Medicine: Stereotypy, Stereotypic Behaviour and Compulsive Disorder

10.2.1. Differentiating similar behaviours

The terms 'stereotypy', 'stereotypic behaviour' and 'compulsive behaviour' are all often used in veterinary behavioural medicine, sometimes as a collective term for a range of repetitive behaviours, sometimes just for particular types with particular features. Here, we categorize repetitive behaviours more formally. We do this for three reasons. First, in the veterinary literature there has been a tendency to describe almost any apparently abnormal repetitive behaviour as a stereotypy and/or as a compulsive disorder, creating confusion both within and beyond the field. For example, practitioners may apply different terms to very similar phenomena (see Box 10.1), and/or group what some would argue are diverse phenomena under a single heading. Overall's (1997), definition of 'obsessive [*sic*] compulsive disorder', for example, includes 'repetitive stereotypic motor [or] locomotor ... behaviours' as well as 'grooming, ingestive or hallucinogenic behaviours that occur out of context'. Second, a common underlying assumption of the veterinary medical approach is

Box 10.1. The Concept of 'Stereotypy' in Veterinary Behavioural Medicine: Terminology in Practice

D. MILLS

Thirty-two veterinary behaviourists from across Europe and North America, chosen for their standing in the field, participated in research examining terminology in practice (data from Sheppard and Mills, unpublished, courtesy of Ceva Sante Animale). Semi-structured interviews were undertaken, with each participant asked to list the five 'key' (i.e. most common) features of the condition we termed 'stereotypy'. Information relating to any preferred terminologies was also gathered. Pairwise comparisons were made between subjects by calculating the ratio of the number of reports (signs or treatment items) on which both participants agreed, to the total number of reports given by both participants.

In relation to the term 'stereotypy', 22 of the 32 participants provided key features. Most of the 10 who declined to do this did so because they considered 'stereotypy' a purely descriptive term for a type of behaviour and not a diagnosis; thus in their opinion stereotypy might be one of the signs of a range of disorders, but not a primary disorder in its own right (see Section 10.2.1). Sixteen of the 22 providing key features agreed these related to 'stereotypy', i.e. were reasonably comfortable with that term; but six instead offered an alternative that they felt was similar or potentially seen as such, but which they also felt was preferable, namely 'compulsive disorder' or 'obsessive–compulsive disorder'. The proportion of subjects referring to each of the following features is given below:

> Repetitive movements – 100%
> Behaviour is out of its normal context – 55%
> Behaviour is performed very frequently or for prolonged periods – 50%
> Caused by stressful environments – 32%
> Behaviour is difficult to interrupt – 32%
> Derived from normal behaviour patterns – 18%
> Other specific features referred to by just 1 or 2 subjects – 10%

In relation to interruption of behaviour, it is worth noting that two subjects stated that the behaviour could not be interrupted, whilst the others said this was possible (but often difficult).

The mean level of agreement on key signs between all 22 participants was not high: 0.34 out of a maximum of 1.0 (0.3 amongst the 16 subjects using the term 'stereotypy', and 0.42 between the 6 preferring 'compulsive disorder'). Sample sizes were too small for analysis, but by inspection, there appeared no outstanding differences between the key signs offered by subjects comfortable with the term 'stereotypy' and those preferring 'obsessive disorder' or 'obsessive–compulsive disorder'.

Participants were then asked to identify the five most important treatment interventions for the problem. Eighteen did so (10 for 'stereotypy', 8 for their preferred alternative term; thus some subjects identified treatments but not key features). The proportion of subjects referring to each of the following treatments is given below:

> Medication – 100%
> > Specifically:
> > Tricyclic antidepressants (mainly clomipramine) – 89%
> > Specific serotonin uptake inhibitors – 28%
> > Monoamine oxidase inhibitors – 28%
> > Anticonvulsants – 11%
> Teach alternative behaviours to identifiable triggers of the problem – 72%

Continued

Box 10.1. *Continued*

Environmental manipulation/enrichment – 50%
Interrupt the behaviour – 39%
Avoid inadvertent owner reinforcement of the behaviour – 33%
Obedience training – 33%
Increase stimulation/activity – 33%
Change to management regime – 17%

NB: Whilst all considered the *potential* value of pharmacological intervention, this does not mean that they felt that pharmacological intervention is always required. The mean level of agreement on key treatments between the 18 participants was 0.46 (0.43 amongst the 10 subjects using the term 'stereotypy', 0.46 between the other 8). Again solely by inspection, subjects preferring the two different terms did not seem to differ consistently in their treatment recommendations.

that the diagnostic label applied to a problem infers something about its underlying biology. However, this clearly has not been the case (e.g. with the term 'stereotypy'), in either the veterinary or ethological literatures, and we therefore suggest a scheme that distinguishes descriptive terms from those implying particular mechanisms. Third, clearer distinctions are important because of potential treatment implications: a significant clinical difference between stereotypy and compulsive behaviours is described in the human literature and needs to be embraced in the animal literature.

Our proposed scheme is shown in Fig. 10.1. Note that in contrast to other authors in this volume, we distinguish between 'stereotypy', and 'stereotypic behaviour'. These terms are frequently used interchangeably to describe repetitive, unvarying behaviour with no obvious function (e.g. Mason, 1991 and all previous chapters). However, in such contexts the terms are typically used merely descriptively, referring to a range of mechanistically varied conditions with superficial phenotypic similarity (see also Mason, 1991). In contrast, we use the term 'stereotypic behaviour' for behaviours that appear repetitive and stereotyped in form, but about which we do not know, or are not concerned with, their mechanism. It is thus merely descriptive (much like clinical terms such as 'dermitis' – a label of evident changes saying nothing about the underlying cause, which may vary between cases). Thus some stereotypic behaviours are not abnormal (e.g. ritual courtship behaviour), and where they are, they may be a sign of psychological changes, or instead of myriad medical diseases such as tumours or post distemper viral encephalitic chorea (e.g. Oliver and Lorenz, 1993; Blackshaw *et al.*, 1994), to name but two (see Luescher, 2002).

We suggest restricting the term 'stereotypy' to a specific subset of stereotypic behaviours with the mechanistic basis outlined by Garner (Chapter 5, this volume; see also Box 10.2 and Chapter 11) , i.e. manifestations of recurrent or continuous perseveration (failures to inhibit all or part of a response, leading to repetition), associated with basal ganglia disruption. Our term 'compulsive disorder' applies to a distinct subset again (although we do suggest some overlap). This, we propose to refer to

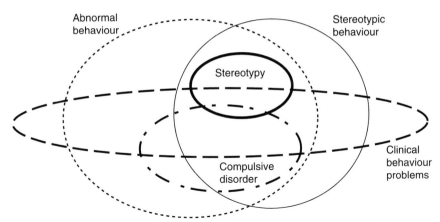

Fig. 10.1. Our schema of the relationship between clinical behaviour problems, abnormal behaviour, stereotypic behaviour, stereotypy and compulsive disorder. This Venn diagram illustrates how abnormal repetitive behaviours may be conceptualized, with 'clinical behaviour problems' and 'stereotypic behaviours' each encompassing both normal and abnormal behaviours; only some stereotypic behaviours being presented as behaviour problems by owners; and 'stereotypy' and 'compulsive disorder' each being specific subsets of abnormal stereotypic behaviour. Note that by 'abnormal' we mean a behavioural pathology *sensu* Box 1.4, Chapter 1. Note too that the areas of the above ellipses are not supposed to represent the relative magnitudes of the respective sets – data are simply not available to allow this. Furthermore, this schema is provisional: thus all lines should be regarded as 'blurred' rather than as firm, well-validated boundaries.

those stereotypic behaviours arising from a failure to switch to more appropriate goal-given environmental cues; and suggest that with future refinement, this definition may include that they represent stuck-in-set perseveration (failures to switch to a more appropriate behavioural set) associated with prefrontal cortex disruption (Chapter 5, this volume; also Box 10.2). However, for now a working definition (adapted from Hewson and Luescher, 1996) might be: goal-directed behaviours originally associated with conflict or frustration, but subsequently shown outside of the initial context, and abnormal-seeming because out of context, repetitive, exaggerated or sustained. Differential responses to certain pharmacological agents may also help to distinguish stereotypy from compulsive disorder, although it is too simplistic to consider this as diagnostic. In particular, given our current state of knowledge, one might suggest stereotypies would respond to intervention with dopaminergic agents, but compulsive disorders, to serotonergic agents (an issue discussed further below).

Note that not all compulsive disorders are stereotypic (see Fig. 10.1). Thus some animals engage in largely static responses (e.g. visual fixations and 'star-gazing' – see images on the book's website), which they may hold for considerable time, and are not repetitive in that bouts are not seen successively. Note too the area of overlap we give between stereotypy and CD (see Fig. 10.1). This is for behaviours that might fulfil both sets of criteria, or that seem intermediate between them – something we discuss further below.

To illustrate the very real importance of separating mechanism-based diagnostic labels from the broader descriptive term 'stereotypic', we list below several diverse reasons why a dog may 'tail-chase' – a not uncommon stereotypic behaviour:

- Acute motivational conflict, i.e. when the animal has two or more opposing but similarly strong motivations.
- Seizure, i.e. a spontaneous burst of electrical activity in the brain.
- Middle ear infection, with inflammation disrupting activity in the adjacent peripheral nervous system controlling balance and movement.
- Play.
- Pain within the tail or lower back, due to damage to the nerves, bones or soft tissues.
- Pain around the region of the tail, including skin inflammation and infections, impacted or infected anal sacs or trauma.

Box 10.2. Implications of Recognizing Mechanistic Differences in Abnormal Repetitive Behaviour

J.P. Garner

Stereotypies and compulsive behaviours seem to reflect dysfunction in different behavioural control systems (see Chapter 5; also Chapter 11, this volume). The differing roles of each system lead to: different forms of 'perseveration' associating with each behaviour; differing diagnostic criteria, proposed below; and possibly also the differing treatment strategies used in humans (where stereotypies are explicitly excluded from, and are insufficient for, a diagnosis of human OCD; and where stereotypies and compulsive behaviours seem to have clinical differences, e.g. in Tourette's syndrome, stereotypies respond well to 'neuroleptics' but poorly to Serotonin Reuptake Inhibitors (SRIs), while impulsive/compulsive behaviours respond well to SRIs but poorly to neuroleptics which are dopamine D2 receptor antagonists, (Cohen *et al.*, 1992; Scahill *et al.*, 1997; Kossoff and Singer, 2001)).

Such differences may help to refine the treatment of animal Abnormal Repetitive Behaviour. To illustrate, a behaviour such as fly-snapping, where the same identical movement can be repeated over and over again, might be diagnosed as a stereotypy, and by analogy to humans might then be expected to respond preferentially to neuroleptics; while a superficially similar behaviour such as an apparent light fixation, where the same inappropriate goal is repeated with flexible behaviour, might be diagnosed as a compulsive behaviour, and by analogy to humans, expected to respond preferentially to SRIs.

If these two classes of behaviour really are different, why in the veterinary literature should they appear to respond similarly to treatment? One potential reason is that a difference in clinical response simply cannot be detected because the classes of behaviours are not clearly distinguished. Another possibility is that, as discussed in this chapter, some drugs used to treat stereotypies and compulsive behaviours in animals may induce a general suppression of all behaviour. For example, a Parkinsonian general inhibition of volitional behaviour was reportedly responsible for treatment effects of both serotonergic and GABAergic drugs in one study in poultry (Kostal and Savory, 1996) where this mechanism was specifically investigated. This indicates the need for future attention to this potential mechanism.

Continued

Box 10.2. *Continued*

Feature	Class of Abnormal Repetitive Behaviour (ARB)	
	Stereotypies	Impulsive/compulsive behaviours
Behavioural control (executive) system	Contention Scheduling System (CSS)	Supervisory Attentional System (SAS)
Biological substrate of the behavioural control system	Basal ganglia motor system	Prefrontal – caudate circuit loops
Role of the behavioural control system (NB: see also Chapter 11)	Selecting and sequencing individual responses, behaviour chains and movements on the basis of stimuli	Selecting and sequencing goals, abstract rules or 'cognitive-attentional sets'; and inhibiting contextually inappropriate responses selected by the CSS
Neuropsychological measure reflecting behavioural control system dysfunction and correlated with ARB class	Recurrent perseveration: the inappropriate repetition of responses (see Boxes 5.2 and 5.3)	Stuck-in-set perseveration: the inappropriate repetition of abstract rules or cognitive-attentional sets with variable responses (see Boxes 5.2 and 5.3)
Diagnostic criteria	A motor pattern is repeated inappropriately • No apparent goal or function • Behaviour is invariant with every repetition	A goal is repeated inappropriately • Repeated inappropriate goal • Behaviour is flexible and goal-directed with each repetition, but may become secondarily invariant
Most effective drugs typically used in humans	Neuroleptics, or atypical neuroleptics. Opiate blocking drugs are sometimes used, particularly in self-injurious forms	SRIs

- A learned response, aimed at attracting owner attention.
- Compulsive disorder (as defined above).
- Stereotypy (as defined above).

Yet despite its potential value, our scheme does have some caveats. First, 'compulsive disorders' apparently share features with our current definition of stereotypy, and at present differentiation at a clinical level may be challenging. Second, the various explanations for a stereotypic behaviour are not mutually exclusive; for instance, a dog that starts tail-chasing for one reason may subsequently learn to do it at other times to gain attention. Third, this distinction between compulsive disorder and stereotypy remains hypothetical (see also Chapter 11): it must be appreciated that the diagnosis of stereotypy in this mechanistic sense has not been validated – although steps have been made to validate the diagnosis of compulsive disorder, as discussed later.

10.2.2. A historical perspective on abnormal repetitive behaviours: from 'seizure' to 'OCD'

Historically, companion animal repetitive behaviours such as 'fly-snapping', tail-chasing or fabric-sucking were often considered symptomatic of seizures, and treatment – usually unsuccessful – attempted with seizure-controlling drugs. These behaviours were also frequently described as stereotypies, with little consideration for the implications of this term. Whilst seizures do remain a possibility (Dodman *et al.*, 1996), in the early 1990s Luescher and colleagues proposed that many of these behaviours were instead more homologous to the stereotypic behaviours of farm and zoo animals (Luescher *et al.*, 1991). At about the same time, researchers in the US National Institute for Health recognized apparent similarities with human obsessive–compulsive disorder (OCD) (Goldberger and Rapoport, 1991), notably in the repetition of a behaviour apparently directed towards a particular goal. OCD is a common human psychiatric disorder (e.g. Karno *et al.*, 1988), and interest therefore developed in the value of companion animal behaviours as models for evaluating potential pharmacological interventions.

10.2.3. Comparisons with humans: from OCD to CD

So how well do companion animal behaviours model human OCD? OCD includes behaviours (compulsions) such as hand washing, checking, arranging/ordering, counting and hoarding, accompanied by intrusive thoughts (obsessions) such as concern of contamination, concern for symmetry or pathological doubt (American Psychiatric Association, 1994). The compulsive behaviours are often reportedly carried out to reduce discomfort, or to prevent a dreaded event (e.g. Anthony *et al.*, 1998), and the intrusive thoughts are often perceived as disturbing. However, in humans, obsessions and compulsions do not closely correspond (e.g. Summerfeldt *et al.*, 1999), and in animals, we cannot determine the occurrence of obsessions. Unlike obsessions, compulsions *can* be measured objectively (e.g. from the strength of their motivation and difficulty in arresting them). The term 'compulsive behaviour' (rather than OCD) is thus generally preferable for such behaviours in non-humans.

However, even the similarities between human and companion animal compulsive conditions are not fully known. Many of the behaviours in companion animals are amenable to the same pharmacological treatment as OCD (Overall, 1992a–c; Rapoport *et al.*, 1992; Luescher, 2002), but differences also exist between human and companion animal conditions. For example, some changes in blood cell count are typical for dogs with compulsive disorder, yet not for humans with OCD (Irimajiri *et al.*, 2003). Furthermore, many human patients find their compulsive behaviours senseless, embarrassing or disgusting – features it would be highly speculative to suggest occur in animals.

In addition to these problems, we must remember too that veterinary behavioural medicine is still young, and some of the behaviours widely considered 'compulsive disorders' in fact better meet our proposed definition of stereotypies. There is currently negligible research to help resolve this issue: it is one of the many gaps in our knowledge. One useful starting point might be to investigate cases of apparent CD, which do not respond to the normal treatment programme described later. Another might be to ascertain a behaviour's developmental history, and/or look at its behavioural correlates. We develop these ideas further in the section below, and also outline the current state of science relating to animal compulsive disorders.

10.2.4. Refining the concept of compulsive disorder in animals

As discussed in Section 10.2.1, compulsive behaviours apparently reflect recurrent expressions of motivated, goal-directed acts, at least initially. The variability of these behaviours varies widely, however, and so differentiation from stereotypy via form alone can be hard. Behaviours such as 'fly-snapping', spinning or 'hind-end-checking' might superficially appear fixed, goalless repetitions of motor patterns and thence stereotypy-like, but closer inspection typically suggests that they are variable in form and appear goal-directed. For example, Miniature Schnauzers that are 'hind-end checkers', do not simply look at their hind-end in a constant fashion, but may turn either way, or get up and check the floor where they have been sitting.

As well as careful observation, sometimes response to treatment may help to differentiate compulsive disorders from stereotypies. For example, feather-picking in parrots often appears compulsive in form, but contrary to what we might then predict (see above), it may not respond well in the long term to serotonergic agents (Mertens, 1997), but can be suppressed with dopamine antagonists (Iglauer and Rasmin, 1993). Furthermore, horses which reduced weaving as a result of environmental intervention with a mirror showed a simultaneous reduction in apparently impulsive threatening behaviour (McAfee *et al.*, 2002). This might suggest that weaving, in at least some horses, is associated with poor abilities to modulate a wider range of goal-directed behaviour and thence is a compulsive disorder, not a stereotypy as frequently implied. Again this supposition has circumstantial support from drug effects: Nurnberg *et al.* (1997) found in one case that weaving was more effectively controlled with an agent used to control OCD in people (a serotonin reuptake inhibitor (SRI)) than with an agent used to control the stereotypies associated with psychoses such as schizophrenia (a dopamine antagonist or neuroleptic). Current work by one of us (DM) aims to investigate this hypothesis more empirically.

It is thus not always 100% clear how compulsive behaviours relate to stereotypies; indeed as discussed below, we believe they may sometimes interrelate. More work is therefore needed to validate the veterinary diagnosis of compulsive disorder (see also Hewson, 1997; and Chapter 11, this volume). In the current absence of a definitive diagnostic clinical test, in practice it is based partly on excluding other behavioural and medical conditions, on historical data, plus close observation. The defining criteria and diagnostic process will, however, undoubtedly be refined in the future. For example, although the value of this clinically remains to be assessed, the work of Garner and colleagues suggests that animals exhibiting either stereotypy or compulsive behaviour are likely to show a range of other, very specific, behavioural traits (see Box 5.2, and Chapter 5).

10.2.5. Developmental considerations: a link between compulsive disorder and stereotypy?

Many repetitive behaviour problems are initially associated with behavioural conflict and frustration, 'conflict' being the presence of two opposing, similarly strong motivations (e.g. approach and withdrawal), and 'frustration', a situation where an animal is motivated to perform a behaviour but prevented from doing so. A variety of behaviours are caused by frustration or conflict (see Box 1.1, Chapter 1). However, with prolonged or repeated conflict, they apparently shift away from this normal, acute response and become generalized to other contexts of high arousal. For example, horses may start to weave before feeding, but over time weave in other situations associated with high arousal, ultimately appearing to weave spontaneously.

As the threshold of eliciting arousal decreases, the number of eliciting contexts increases and the compulsive behaviour becomes more frequent. On the basis of clinical experience, it would thus seem that with repeated or prolonged conflict/frustration, these behaviours potentially become 'emancipated' from their original context; exaggerated, repetitive or sustained; and triggered in a variety of situations by progressively lower levels of arousal. Such a developmental pattern has also been reported in some laboratory rodents (reviewed by Würbel in Chapter 4). This change may be because dopaminergic responses within the mesoaccumbens and/or nigrostriatal systems change in response to stress (e.g. Cabib et al., 1998; Chapter 8, this volume). We speculate that this also reflects a shift from prefrontal to basal ganglionic regulation of the behaviour (Norman and Shallice, 1986; Chapter 8, this volume), and if so, it may be at some point in this process that a compulsive behaviour should actually be considered a stereotypy. This may have clinical relevance, since it has also been observed that with time, environmental manipulation alone becomes less likely to reduce stereotypic behaviour in rodents in the laboratory (reviewed by Lewis *et al.* in Chapter 7), while likewise, in a range of cases of compulsive disorder, treatment outcome was negatively affected by problem duration (Luescher, 1997). Thus compulsive disorders in dogs and cats that had been present for a longer time were less likely to be treated successfully – e.g. via psychopharmacological intervention, typically serotonergic – than were problems with more recent onsets.

Our developmental hypothesis remains speculative, but the coincidence of events deserves scientific investigation, and elucidating the nature of the relationship would represent a significant clinical advance. Current research by one of us (DM) is investigating these factors further in horse stereotypic behaviour.

10.2.6. Conclusion

Our proposed definitions of stereotypic behaviour, stereotypy and compulsive behaviour highlight that:

- not all repetitive or 'stereotypic' behaviour is abnormal, nor obsessive or compulsive;
- 'stereotypic' is often the term best used, since it is descriptive only, implying no knowledge of mechanism;
- stereotypies and compulsive disorders should be considered different categories of stereotypic behaviour; and
- other causes of repetitive behaviour require consideration (for instance sometimes there may be a recurrent physical trigger for the behaviour – such as a poorly fitting bridle as in some forms of head-shaking in the horse (Cook, 1980) – or an internally derived but normal response to pain, e.g. in neuralgic forms of equine head-shaking (Newton *et al.*, 2000)).

We also highlight how, despite our definitions, stereotypy and compulsive disorder can be hard to distinguish in practice, and we hypothesize that in some instances, they represent a developmental continuum.

10.3. Treatment Approaches to Stereotypic Behaviour in Veterinary Behavioural Medicine

Next, let us look at how veterinarians deal with stereotypic behaviour in practice. We start by outlining two aspects of context – the scientific background of the work, and the practical and 'human' constraints affecting treatment – before describing the environmental, behavioural and pharmacological approaches used in treatment.

10.3.1. Scientific basis and associated problems

Veterinary behavioural medicine in practice occurs in the context of a clinic. It tends to focus on individual cases rather than on populations, and demands practical solutions over scientific precision. The development of veterinary behavioural medicine has thus been primarily pragmatic rather than theoretical (see Chapter 9 for similar issues in environment enrichment). Furthermore, research funding opportunities are limited; and the few active researchers often have to exploit clinical case material and so, for ethical reasons, may not utilize untreated controls (e.g. placebo-treated animals), instead comparing treatments that they believe will be effective (again see Chapter 9, for a parallel situation when allocating enrichments). Together, this means that scientific research into any given problem is often rare, and of a variable standard. In addition, information, opinions and ideas are often communicated within the profession by word of mouth, and via non-peer-reviewed or limited-peer-reviewed texts, and so are subject to less rigorous criticism than might occur in other fields.

To some extent this is merely because, as we have seen, the field is young; and it does perhaps explain why terms like 'stereotypy' and 'compulsive' disorder have been used so interchangeably, and/or applied to phenomena whose real underlying causes are unknown (see Section 10.2.1). But it also means that when practitioners are faced with the task of managing stereotypic behaviour, they often have a dearth of scientific support.

10.3.2. Practical considerations and constraints when choosing treatments

As we have seen, veterinary behaviour medicine primarily responds to the concerns of owners. Thus another important aspect of context is that the veterinarian typically treats stereotypic behaviour through the mediation of an animal owner. One implication of this is that the *ideal* treatment programme may be impossible. A veterinarian has no power to enforce a treatment strategy, and the final treatment programme is formulated in consultation with the client. If they object to certain types of drug-use, for example, or demand a rapid response (for instance, to save an animal from being re-homed or euthanased), this shapes how the veterinarian must act.

In some instances, it may also be impractical to address the likely causal factors of the problem; or they may simply be unknown. When the causal factors are unknown, the veterinarian may be left with a range of possibilities ('differential diagnoses'). Furthermore, multiple conditions may be present in a single case. In such instances, it may be necessary to start treatment simply on the basis of the most likely cause of the problem, or using the least specific treatments in the clinician's professional opinion (whilst also ensuring that no harm is done by the treatment). Approaches other than tackling possible causal factors may also be taken. For example, if the causal stimuli *are* known but cannot practicably be eliminated, the animal's perception of these may be modified (e.g. through behaviour therapy) to make them less influential. When control or perception of the causal factors cannot be manipulated, it may instead be possible to redirect the behaviour on to another substrate, for example a horse that repetitively crib-bites may be taught to crib-bite on to a rubber board (a 'cribbing-board'). This clearly does not resolve the ongoing activity, but it reduces the damage otherwise done to the stable and the horse's teeth. Finally, an intractable problem behaviour may, as a last resort, simply be physically prevented, for example by muzzling or placing a 'cribbing-strap' (see book's website for photo) on a horse that crib-bites (though this may cause it additional frustration: see Box 2.3, Chapter 2).

These very different strategies for behaviour problem management clearly have divergent welfare implications for the animal, as we summarize in Table 10.3. This diversity of strategy reflects the needs of veterinarians not just to improve the functioning and welfare of the animal, but also to please the client, prevent damage to property and take public safety seriously (see also Chapter 9 for similar considerations in enrichment use).

As a final potential constraint, note the need for client compliance. For instance, treatment is typically best instigated progressively, primarily because the number of tasks set affects owner compliance – and so ultimately treatment success (Takeuchi *et al.*, 2000). Progress must also be reviewed regularly, with response to treatment being used to assess client compliance, as well as to help evaluate the need for any further investigation into the problem or additional types of intervention.

10.3.3. Treating stereotypic behaviour in companion animals

In Section 10.1.3, we discussed how problem behaviours result from suboptimal systems of which the animal is but a part; and in Section 10.2.1 we outlined the many potential causes of stereotypic behaviour. Above, we illustrated the constraints within which veterinarians often operate. These issues all shape how companion animal stereotypic behaviour is or should be treated in practice.

A medical assessment is essential first, so that relevant medical conditions can be identified and treated. Possible examples include middle ear infection or tail/sacral pain in association with tail-chasing (*cf.* Section 10.2.1); while dermatological investigation and treatment, or the surgical excision of a painful lesion, could be important in a case of self-mutilation. Videotape analysis of the behaviour might also reveal an underlying medical condition, such as a post-ictal phase (a brief period of lethargy and confusion) following a seizure.

When the stereotypic behaviour seems more behavioural/psychological in origin, then the animal's 'behavioural history' is used to diagnose problems and guide treatment. Some properties of the behaviour itself can be

Table 10.3. Broad welfare implications of different strategies aimed at treating problem behaviour.

Treatment strategy	Potential welfare benefits	Potential welfare problems
Preventing performance of the behaviour	Reduces harm of ongoing behaviour	Does not address the underlying cause May restrict the ability of the animal to adapt to its environment (see also Box 2.3, Chapter 2)
Removing causal factors underlying the behaviour	Resolves welfare issues related to the aetiology *and* performance of the behaviour	May reduce the welfare of others affected by the change
Change the perception of stimuli	Resolves welfare issues related to the performance of the behaviour Eliminates perception of a poor environment	Changes may extend beyond the target and so interfere with normal functioning
Redirect the behaviour towards another substrate	If other individuals are affected by the ongoing behaviour, their welfare may be improved	May not improve the welfare of the patient
Encourage other behaviours to compete for expression with the problem behaviour	Resolves welfare issues related to the performance of the behaviour May reduce the impact of threats in the environment	The alternative behaviours may also be a consequence of poor welfare

Broadly speaking, the strategies listed above may be achieved by a variety of methods, from environmental manipulation to chemical intervention. Thus, for example, drugs can operate via several of these mechanisms: heavy sedation will prevent the behaviour; anxiolytics may alter the perception of the environment; and in some cases drugs may correct the causal factor if this is a central disturbance in neurochemistry. Note that every specific treatment has its own specific risks additional to those proposed here (for more details see Cooper and Mills, 1997).

useful. For instance, to help distinguish between play, acute conflict behaviour, compulsive disorder and long-established stereotypy, one would assess what behaviour is shown, including 'body language'/'facial expressions' indicating affect; the severity and frequency of the problem and ideally also use videotape to assess the behaviour's consistency, constancy, potential for disruption, context and the extent to which it is stereotypy-like or instead apparently goal-directed. Such information might then suggest likely pharmacological treatments (see Section 10.2.1).

However, pharmacological treatments should not be used without assessing, first the risks involved (see below), and second, cues and causes within the animal's environment that highlight alternative or additional treatment routes. One example might be a need to control inadvertent reinforcement by the owner. Close examination of what triggers the stereotypic behaviour, people's reactions to it, and how the animal behaves immediately afterwards may all indicate this; for instance, performance *only* in the owner's presence, and certain expressions of owner concern, might suggest conditioning by, say, owner attention or the delivery of a treat. Altering owner–pet interaction would then clearly form a central part of treatment. Another important potential proximate cause in the environment is acute motivational conflict or frustration, perhaps generalized to a range of arousing stimuli. Investigating such factors as when and where the behaviour occurs, and the behaviour's target, can be ways of identifying an association between the behaviour and environmental conflicts, thereby highlighting treatment routes via behavioural modification (e.g. training and psychotherapy to alter the perception of a stimulus so that it becomes less threatening), and/or environmental management (e.g. avoidance/elimination of particular situations).

General stress reduction measures are also frequently warranted: *any* environmental factor resulting in frustration (e.g. no exercise off property for dogs), conflict (e.g. inconsistent interaction with the owner) or stress (e.g. conflict with another individual; separation anxiety; even disease) may potentially contribute to the problem. These frequently arise as a result of confinement, and the frustration of motivations such as those for social interaction or exploration (*cf.* Chapter 9, this volume). For example, repetitive locomotory behaviours in horses are more common amongst socially isolated animals (Luescher *et al.*, 1998; Bachmann *et al.*, 2003), while in dogs, clinical experience indicates that these problems are more common in animals who do not receive sufficient exercise. Lack of predictability and control may also arise from inconsistent owner–animal interactions, via e.g. lack of training to commands and thus inconsistent use of commands, the inappropriate use of punishment, or an inconsistent routine. Casual interaction with the owner may thus need to be avoided, and replaced with highly structured interactions using a command–response–reward format. For example, a Miniature Schnauzer that licked its inner thigh excessively and also frequently stared at the ceiling and froze in this position was successfully treated by avoiding casual interaction; providing interaction only in a command–response–reward format; obedience training; avoidance of punishment; and increased, regular exercise (A. Luescher, personal observation).

Here, it should be recognized that different breeds have specific needs, as reflected by their varying prevalences of stereotypic behaviour. Repetitive locomotory behaviours in horses are particularly common in thoroughbred and warm-blood types (Luescher *et al.*, 1998; Bachmann *et al.*, 2003), for example, while specific repetitive behaviours have recognized breed predispositions amongst household pets (e.g. as listed earlier in Table 10.1). Overall and Dunham (2002) even suggest a possible association between the form of the behaviour and purpose for which a breed was developed: herding breeds often present with tail-chasing, while guarding breeds and dogs bred for focal attention tend to present with apparent hallucinations. Furthermore, individual temperament may also predispose an animal to compulsive problems, because the primary problem is linked to sensitivity to stressors. Thus there is overlap between proximate causes and predisposing factors. In general, the animal should therefore be provided with an environment with an appropriate balance of predictability and control considering not only its species and breed, but also its individual nature (Wiepkema and Koolhaas, 1993).

In cases where causes of stress cannot be removed, it may be possible instead to desensitize the animal to the stressful situation. Such behaviour therapy includes graded exposure to the eliciting situation and response prevention (Abel, 1993; March, 1995). Treatment can also involve pharmacological intervention and behaviour modification. For example, 'response substitution' can also be used as a behaviour modification technique. This involves distracting the animal from the performance of the behaviour, issuing a command for a non-compatible behaviour, and reinforcing the latter. For example in one case (AL), a Border Collie that chewed its right foot compulsively was trained to lie with its head flat on the floor between its paws. Whenever he showed any inclination to chew his foot, the dog was provided with a distraction and given a command for lying down in the described position, and then rewarded for obeying. With time, the reward was delayed more and more so that the dog had to stay in the position for increasing time before receiving the reward. After 1 week of continuous training, the compulsive behaviour had disappeared and did not recur.

If an animal's stress reactivity or other aspects of behaviour seem intrinsically unusual, and/or environmental changes simply cannot be effected, yet other measures still may be taken, especially pharmaceutical approaches. A general behavioural assessment can determine whether the animal is behaving normally in other contexts, further directing treatment choice. For instance, Overall and Dunham (2002) report that nearly 75% of dogs with apparent compulsive disorder had an additional concurrent issue, most commonly attention-seeking, impulsive aggression, or separation anxiety, perhaps reflecting more general problems of behavioural or emotional control. Some individuals may also appear generally anxious, perhaps having physiological predispositions that are less amenable to environmental manipulation but could be helped via medical interventions that reduce the impact of stressors (e.g. pharmacological interventions to reduce anxiety). Table 10.4 summarizes treatment recommendations for compulsive disorder.

10.4. Drugs in the Treatment of Stereotypic Behaviour

From Section 10.3, we can see that drugs are not always needed and, if employed, are an adjunct to environmental and behavioural modification. Furthermore, drugs inevitably involve a risk of side effects (see later) –

Table 10.4. Treatment recommendation for compulsive disorder (based on Luescher, 2002).

- Identify and remove cause of conflict, frustration and specific stressors which trigger the behaviour, or desensitize animal to these stimuli (see also Chapter 9, this volume)
- Reduce general stress within the environment
  ○ Avoid inconsistent interactions, and instigate structured exercises
  ○ Provide animal with opportunities to control aspects of its environment (see also Chapter 9, this volume)
  ○ Provide a consistent, but not monotonous, routine
  ○ Stop all forms of punishment administered by the owners
  ○ Ensure diet and exercise regime are balanced
- Consider potential for pharmacological intervention, and possibly dietary manipulation to increase systemic uptake of tryptophan
- Consider need for response substitution
  ○ May require prevention of compulsive behaviour at first
  ○ Distract and reward animal for other behaviours, when compulsive behaviour is anticipated

risks that differ from the behavioural and environmental strategies discussed above. It should also be recognized that, in some cases, drug therapy may not be treating the true cause but instead stimulating a system which masks the symptoms: thus drugs may sometimes only be treating the signs rather than the cause of the problem.

All this does not, however, mean that chemical interventions are unjustified. For one, rapid behavioural control may be insisted upon by the owner. Some further general guidelines can also be recognized. Medication may be necessary in animals that are emotionally predisposed to the problem (see above); and also in longer standing cases, which otherwise appear to have a poorer prognosis (Luescher, 1997). Furthermore, some agents may also help to improve the response to, and speed of, training, through enhancing memory (especially the serotonergic agents; King *et al.*, 2000) or sensitivity to rewards (especially inhibitors of monoamine oxidase B; Mills and Ledger, 2001). They may therefore be utilized to minimize the time taken for responses to occur with psychological intervention.

The most commonly recommended agents for abnormal repetitive behaviours enhance serotonin levels, modulate dopamine activity, or antagonize the opiate system. The use of such drugs is discussed in the following section. We start with some general caveats about interpreting drug action (see Section 10.4.1).

10.4.1. Drug action: a basic primer

No one transmitter, receptor or drug relates directly to or acts independently on any specific behaviour. Furthermore, as Chapters 5–8 illustrate, the neuroanatomical and neurophysiological links within the brain are complex and interrelated. Whilst many drugs primarily affect one neurotransmitter system or receptor sub-type, they are not always exclusive in effect, and by affecting one system may induce changes in different systems elsewhere, either directly or indirectly as the body compensates. Furthermore, drugs given by mouth or by injection do not just go to the specific regions of the brain associated with the problem, but affect the relevant receptors wherever they are found in the body: one reason why side effects can result. Thus if a drug reduces general activity, for example, through sedation or some other mechanism such as inducing nausea, it might theoretically also reduce repetitive behaviour as a mere consequence. Furthermore, doses used may mean that the consequent tissue levels of neurotransmitter are many times higher than the natural endogenous level (Shankaran and Gudelsky, 1998) and potentially result in effects that extend beyond the specificity of the known transmitter–receptor interaction. Thus drugs may be operating by chemically flooding the system to produce an extreme response that only resembles the desired condition at the phenomenological level (face validity) but not the mechanistic level (construct validity) (Nesse and Berridge, 1997). This potential problem exists when interpreting the results of experiments designed to both generate *and* eliminate stereotypic behaviour pharmacologically (see Box 7.3, Chapter 7). Thus until neurophysiological mechanisms behind a condition are fully elucidated we must consider the possibility that response to a particular medication potentially may be operating at only a phenomenological level (controlling the signs) rather than at a mechanistic one (affecting the cause). These issues are well illustrated by the results of Nurnberg *et al.* (1997) who reported on the use of three different types of psychoactive medication in a single case of weaving in a horse (an opiate antagonist – naltrexone; a sedative antipsychotic – acepromazine and a serotonergic antidepressant – paroxetine). In this case, all reduced weaving to some extent, with paroxetine producing a 95% reduction, acepromazine a 40% reduction and naltrexone a 30% reduction. Whilst it might be argued that these different levels of effect reflect the different roles of the various transmitter systems involved in regulating the behaviour, these results might also reflect varying levels of inhibitory side effects.

Thus although drugs in the next section are grouped according to their main known receptor interaction, they are not necessarily receptor-specific and their effects may actually be due to their action on other receptor systems, or to other unknown interactions.

10.4.2. The pharmacological treatment of stereotypic behaviour – evidence of efficacy

There have been very few relevant controlled studies on drug efficacy (see Box 10.3 for an outline of best practice and the abbreviations used in drug administration). In fact, only serotonergic agents have been shown clinically effective in any form of randomized, controlled studies on any captive animal (Hewson *et al.*, 1998a), the remaining direct evidence coming from

Box 10.3. The Use and Testing of Pharmacological Compounds: Clinical Trials, Experimental Studies and Intervention Studies

A. LUESCHER

Clinical trials assess the effect of an intervention strategy, while keeping other determinants of outcome constant. A clinical pharmacological trial is usually preceded by one study to establish dose range and drug safety, and a pilot study on a few patients demonstrating likely benefits. The clinical trial then tests effectiveness in a large number of patients, in real-world situations, as well as identifying side effects.

Randomized controlled clinical trials are the standard of excellence. A sample of patients is recruited that is as representative of the population of affected individuals as possible. Clear inclusion/exclusion criteria need to be applied (although these limit the population to whom the results can be generalized), e.g. recruiting patients with a specific condition necessitates an accepted method of diagnosis (in the best case, an accepted gold-standard diagnostic test; in the worst case, expert agreement). The sample is then divided randomly (in a formal randomized or matched manner) into two groups with comparable prognoses (e.g. for compulsive disorder, with similar breed, age and sex distributions, and similar clinical signs, severities and problem durations). Both groups are treated identically, with the exception that the treatment group is exposed to the intervention under test, and the control group, to a standard intervention or placebo treatment. (For ethical and legal licensing reasons, the effect of the intervention strategy is often compared to the best-proven or established treatment, and only under certain circumstances to a no-treatment control). In clinical trials involving behavioural drugs, treatment and control groups need to either get no behaviour and/or environmental modification at all, or get the same, standardized (and therefore simple!) behaviour and/or environmental modification. To avoid subjective bias, the experimenter and the animal owner have to be 'blinded' as to which group subjects belong to; knowledge of treatment allocation may otherwise influence initial randomization, interactions with animal owners (inducing a bias in them) and outcome measurement. The outcome also has to be measurable unequivocally and objectively, using validated tools. The 'Gold Standard' procedure for trial conduct is laid out in published international guidelines to good clinical practice (GCP).

Finer details of the trial design depend on how many patients can be recruited and other factors. The most straightforward design is a 'parallel arm study', where after a baseline period without treatment, the two groups are followed for some time, one getting the experimental, the other the control treatment. At the end, the two samples are compared. Another common design is the 'crossover' trial, in which one sample first gets Treatment A, then (potentially after a 'wash-out' period, dependent on the rate of metabolism and excretion of the drug) Treatment B, while the other sample first gets B, then A. Comparison then is generally made *within* subject, comparing its disease state at the end of the first versus the end of the second period; order of treatment potentially being controlled for statistically. This crossover design is useful when large numbers of patients cannot be recruited, since each subject is its own control. (See e.g. Fletcher *et al.*, 1996 for more details.)

> *Dosing and delivery abbreviations*:
>
> P.O. – per os (by mouth)
> i.v. – intravenously
> s.c. subcutaneously
> sid – once a day
> bid – twice a day

Usually drugs are dosed on a live body weight basis (kg), but where there is a narrow safety margin, surface area (m^2) may be used to obtain a dose more consistent with metabolic rate.

case studies. If there are few sound data for companion animals, there are nil for exotic species; and so if drugs are to be used on such animals (see e.g. Box 10.4), it must be recognized that therapeutic doses remain largely unknown, and that dosing regimes extrapolated from other species could carry serious potential risks to the health of the animal.

It also needs to be recognized that without a consensus definition for different forms of abnormal repetitive behaviour, different authors have to date defined similar terms differently and, conversely, may have described different conditions similarly (e.g. see Box 10.1). These caveats must be borne in mind as we discuss the evidence below. See Box 10.3 for dosing and delivery abbreviations.

10.4.3. Drugs acting on serotonin receptors

10.4.3.1. Serotonin Reuptake Inhibitors (SRIs)

Compounds that inhibit the synaptic reuptake of biogenic amines including 5-HT, and thus their degradation by the presynaptic neuron, are believed to potentiate the actions of these molecules. However, their therapeutic effect may be delayed by several weeks. This indicates that their effect is not simply due to the accumulation of neurotransmitter in the synapse, but possibly due to a cascade of other events triggered by the accumulation of neurotransmitter (Baldessarini, 1996b). For example, autoreceptors desensitize as a consequence of chronic administration of an SRI, resulting in greater release of serotonin, while reuptake is still inhibited (Baldessarini, 1996b). The concentration of serotonin in the synapse following treatment with these compounds does not exceed three to five times the normal value (Fuller, 1994). Serotonin is not only released at synapses, but also from axonal swellings on the nerve and acts on many peripheral targets (Sanders-Bush and Mayer, 1996). The increase of serotonin at the serotonergic synapse after application of an SRI therefore might not be as important in the treatment of compulsive disorder as previously believed. The study of reuptake inhibitor action is further complicated by the fact that their effect may differ for different brain regions (Fuller, 1994).

We usually use SRIs for several weeks or more after a satisfactory therapeutic effect has been achieved, before attempting to wean the patient off the drug over 3–4 weeks. 'Weaning' off the compound should then always be attempted, but is not always successful (i.e. clinical signs return), necessitating long-term drug administration (possibly for years). Note too that the use of the drug should always be accompanied by behaviour modification.

Because SRIs indirectly reduce the density and sensitivity of serotonin receptors, a 'rebound effect' (i.e. a period of greatly reduced serotonergic functioning and recurrence of problem behaviours) is possible following abrupt discontinuation of the drug. Such a rebound has been observed in dogs when clomipramine was discontinued abruptly (C.J. Hewson, personal communication, Prince Edward Island; D. Mills,

Box 10.4. Pacing, Prozac and a Polar Bear

E.M.B. Poulsen and G. Campbell Teskey

'Snowball' was an adult female polar bear (*Ursus maritimus*), who spent much of her day pacing. She was born in captivity in 1969, and lived with a conspecific at the Calgary Zoo in an 836 m² concrete pit enclosure. Archives indicate that at the age of 2.5, she began to display the stereotypies common in captive polar bears. Over the years, several interventions were attempted, but all failed to abolish the behaviours. These included an unspecified tranquilizer, enrichment programmes, naturalistic enclosure redesign and diet variation. Snowball also had recurring hair loss that again had never been successfully treated. This involved the loss of the whole hair (the hair was not broken) primarily on the flank region. Therefore, in 1993, 22 years after her pacing first appeared, we decided to try another approach. We reasoned that after two decades of impoverishment, Snowball likely had a neurological disorder that could not simply be reversed by attempts to improve the enclosure. We began by videotaping her during daylight hours (7:00 to 16:00) to quantify her behavioural repertoire. We discovered that she spent ca. 70% of this time pacing, and that while pacing she also had a facial tic and a repetitive vocalization like a 'huff' or cough. These behaviours were very unvarying, occurring in very specific locations within the enclosure to the extent that Snowball always placed her paws in the exact same spot. Since humans can show drastic reductions in compulsive behaviours (which can be highly stereotypic) following the usage of specific serotonin reuptake inhibitors (SSRI), we thought we would try this in Snowball. The Eli-Lilly pharmaceutical company donated quantities of the SSRI Prozac®. In consultation with veterinarians we arrived at a dosage range of 0.62 to 1.32 mg/kg, derived from human dosages and accounting for the polar bear's large fat deposits and a significant scaling factor. After much discussion with zoo staff, we reached a broad-based consensus and began administering the drug in Snowball's food. We continued the videotaping and analysis of her behaviour. The trial was in an A-B-A form, each phase lasting approximately 5, 14 and 24 weeks, respectively.

   Over the next weeks, the amount of time Snowball spent pacing gradually reduced, and she spent more and more time engaged in normal-seeming behaviours like walking about the whole enclosure, swimming, interacting positively with the other polar pear and looking at the zoo visitors. After about 8 to 9 weeks of drug administration the stereotypic behaviours ceased. An unanticipated but positive side effect also emerged: her recurring hair loss vanished. Snowball did not appear to display any deleterious side effects. It was then time to take her off the drug to see if the pacing would return. The drug was withdrawn. For several weeks Snowball remained free of stereotypic behaviours, but then they began to return, and within a few months after cessation of the drug she had returned to her pre-drug level of pacing. Within a year of the completion of this study she was euthanased for a medical problem unrelated to the use of the drug (hip degeneration). However, it is likely that this condition would have been exacerbated by her previous constant locomotion.

   Thus certain forms of pharmacotherapy can provide behavioural and potential welfare benefits to captive animals. However, we urge that such animals should be housed according to their species-specific nutritional, environmental and social needs such that these problems do not emerge in the first place. We also hope that if other zoos use pharmacologic treatments, they do so in conjunction with real improvements in naturalistic enclosure design and environmental enrichment programmes (see also Mills and Luescher, this chapter).

personal communication, 1997). It is therefore recommended that the drug be withdrawn gradually towards the end of treatment, by decreasing dose (but not dose frequency) (Hewson and Luescher, 1996).

SRIs are heterogeneous and include the tricyclic antidepressant, clomipramine, and atypical antidepressants such as the 'specific' serotonin reuptake inhibitors (SSRIs) fluoxetine, paroxetine and sertraline. While clomipramine has side effects in common with other tricyclics and, through its first metabolite chlordesipramine, also affects norepinephrine reuptake, the atypical antidepressants appear to have fewer side effects (Baldessarini, 1996a). These compounds are considered in more detail below.

CLOMIPRAMINE. Clomipramine is the only tricyclic antidepressant with anti-obsessional activity in humans. However, like the other tricyclics, clomipramine has anticholinergic effects. It may thus affect micturition (i.e. urination behaviour), cause gastrointestinal disturbances, and lower the seizure threshold as a side effect. At high doses, it may also affect heart function, producing sinus tachycardia and arrhythmias, and can be mildly hypotensive. Clomipramine also reduces REM sleep and has some sedative properties. Occasionally, irritability and aggressiveness are seen as side effects (Product Monograph, 1996a; D. Mills, personal observation). Clomipramine should not be given with monoamine oxidase inhibitors, anticholinergic agents or antihistamines. Caution has to be exercised if clomipramine is used with SSRIs, neuroleptics, e.g. haloperidol (compounds which are discussed further below), or benzodiazepine anxiolytics, e.g. diazepam, since these drugs increase the plasma concentration of clomipramine (Product Monograph, 1996a).

A randomized, placebo-controlled crossover clinical trial involving 51 dogs diagnosed with a variety of compulsive disorders (such as self-licking and tail-chasing) found that clomipramine was about four times more likely to improve the condition than placebo (Hewson et al., 1998a). As further examples, Goldberger and Rapoport (1991) used clomipramine in an experimental single-blind trial for the treatment of acral lick granuloma (a condition involving repetitive licking of a specific area of a limb) in dogs, with the tricyclic antidepressant desipramine as a control. Six out of nine dogs showed a reduction in licking as rated by the owner. Rapoport et al. (1992) also found only such serotonergic agents to effectively control acral lick granuloma in dogs. A dose of 3 mg/kg P.O. sid was subsequently recommended for treatment of dogs (Marder, 1991). For cats, a dose of 1 mg/kg P.O. sid clomipramine is generally recommended. A significant reduction in three cases of apparently ritualistic stereotypic motor behaviour in dogs was also reported when clomipramine was used at a dose of 3 mg/kg P.O. bid for several months, in combination with counter-conditioning (Overall, 1994). Clomipramine has also been used, with much more limited success, to control feather-picking in birds (Grindlinger and Ransay, 1992). Seibert et al. (2004), in a placebo-controlled study of 11 feather-picking cockatoos (6 'pickers' and 5 'mutilators'), reported that clomipramine at 3 mg/kg P.O. bid was effective in eight of the birds at 6 weeks. Seven of the eight birds began to improve at 3 weeks with significantly greater improvement noted at 6 weeks. However, one clomipramine-treated bird (a self-mutilator) was worse at 6 weeks, and the remaining two treated birds were unchanged. No adverse events were reported during the study period.

OTHER TRICYCLIC ANTIDEPRESSANTS. Other tricyclics have been tried, with one publication suggesting that amitryptiline was associated with improvement in about 60% of cases of cats and dogs with 'OCD' (as defined by the authors), compared to 83% for clomipramine (Overall and Dunham, 2002). Amitryptiline has a much weaker effect on serotonin reuptake and thus this apparent relatively high efficacy is somewhat surprising; however, given the nature of the controls in this particular study, it cannot be concluded that the amitryptiline itself explains the effect.

FLUOXETINE. Fluoxetine is the most commonly used SSRI in companion animal practice and has also been used in captive exotic species (Poulsen *et al.*, 1996; Chitty, 2003; Hugo *et al.*, 2003; e.g. Box 10.4) – although, as with clomipramine, success in controlling feather-picking in birds is limited (Mertens, 1997).

Fluoxetine specifically inhibits the reuptake of serotonin by blocking presynaptic uptake channels, i.e. it makes the serotonin released into the synapse last longer. Because of its specificity, it does not have the anti-cholinergic side effects of the tricyclics. Fluoxetine's first metabolite, norfluoxetine, has similar potency and specificity as a reuptake blocker (Fuller *et al.*, 1991). Fluoxetine and norfluoxetine have a much longer half-life (1 day and 2.1 to 5.4 days, respectively; Product Monograph, 1996b) compared to clomipramine (1.5 to 9 h; Hewson *et al.*, 1998a), which means less frequent dosing is required.

A double-blind, randomized placebo-controlled study of fluoxetine on 62 dogs with acral lick lesions, concluded that the drug was effective in controlling the condition over a 6-week period, and during a 2-month follow-up treatment period both licking behaviour and lesion scores continued to improve (Wynchank and Berk, 1998). Three treatment trials compared clomipramine with desipramine, fluoxetine with fenfluramine and sertraline with placebo, respectively, in the treatment of canine acral lick granuloma (Rapoport *et al.*, 1992). For 5 weeks, 3 mg/kg clomipramine, desipramine or sertraline, or 1 mg/kg fluoxetine or fenfluramine were given. Clomipramine and fluoxetine produced similar clinical effects, with an approximately 40% reduction of licking compared to controls. Sertraline resulted in an approximately 20% reduction in self-licking. However, lethargy, loss of appetite, diarrhoea and growling were reported in the clomipramine group; and lethargy, loss of appetite and hyperactivity with fluoxetine treatment (Rapoport *et al.*, 1992). In another study, one dog treated with fluoxetine, at approximately 0.5 mg/kg P.O. sid, exhibited dilated pupils and developed fibrogingival hyperplasia (a form of growth of the gums of the mouth). After several years of continued treatment, however, the dog did not develop renal or hepatic problems

(Overall, 1995). Successful treatment of compulsive tail-mutilation in a Bichon Frise dog with 1 mg/kg fluoxetine P.O. sid for 3 weeks has been reported (Melman, 1995). The recommended dose of fluoxetine for dogs and cats varies between authors, with some recommending 0.5 to 1 mg/kg P.O. sid (Marder, 1991; Mills and Simpson, 2002), and others twice this dose (Landsberg *et al.*, 2003).

OTHER SPECIFIC SEROTONIN REUPTAKE INHIBITORS. Paroxetine is another SSRI frequently used in practice. It has a shorter half-life than fluoxetine and no active metabolite. It is therefore more quickly eliminated from the system, and so it is particularly important that the dose of paroxetine be 'tapered' gradually at the end of treatment to avoid recurrence of the signs. Paroxetine also has anticholinergic side effects, like the non-specific SRIs. Side effects are dose-dependent. Therefore, it is better to use a low dose and wait (for up to 4 weeks) to evaluate the therapeutic effect, rather than start with a high dose or increase the dose rapidly: the latter procedure produces a more rapid onset of therapeutic effect, but also increases the risk of side effects. Sertraline has a similar action and side effect profile to fluoxetine, with a shorter half-life and less-active metabolite.

10.4.3.2. Serotonin agonists

BUSPIRONE. Buspirone is used for the symptomatic relief of excessive anxiety in human patients with generalized anxiety disorder (Product Monograph, 1993), and therefore has been of interest to those supposing a link between abnormal repetitive behaviours and anxiety. However, buspirone is an agonist at 5HT1A autoreceptors, and so decreases the synthesis and release of serotonin; it also increases dopamine and noradrenaline turnover (Baldessarini, 1996b). It is thus not considered effective for treatment of OCD in humans (Baldessarini, 1996b) and was found ineffective as an adjuvant to fluoxetine in humans (Grady *et al.*, 1993). Buspirone was likewise ineffective in the treatment of an apparently compulsive circling dog, and may even exacerbate the problem (Overall, 1995).

10.4.3.3. Serotonin precursors: tryptophan as a dietary route to treatment?

It may be possible to influence serotonin metabolism with the nutritive amino acid, L-tryptophan (Teff and Young, 1988). L-tryptophan is a serotonin precursor (Fuller, 1994), but competes with other large neutral amino acids for uptake into the brain (Fenstrom and Wurtman, 1972). Its effect on brain serotonin levels is thus dependent on the amino-acid profile of the diet (Hedaya, 1984; Colombo *et al.*, 1992). Its uptake is improved by circulating insulin, as stimulated by dietary sugars and arginine. Thus careful dietary formulation is essential to maximize the effectiveness of supplementation (Clark and Mills, 1997). L-tryptophan at 0.05 and 0.1 mg/kg reduced performance of a stereotypic head twist by about a third in one horse (Bagshaw *et al.*, 1994). L-tryptophan has been recommended for the clinical treatment of equine compulsive behaviour at 2 g/day for a typical horse (McDonnell, 1996), although doses as high as 5 g/day may be necessary (D. Mills, personal observation). L-tryptophan has also been used to control repetitive self-mutilation in captive primates (Weld *et al.*, 1998).

10.4.4. Drugs acting on dopamine receptors

The dopamine antagonist, haloperidol, has been shown to inhibit stereotypic behaviour in voles (Kennes *et al.*, 1988) and pigs (von Borell and Hurnik, 1991), but not compulsive behaviour in dogs. Whilst SRIs, such as clomipramine, may help to reduce OCD symptoms in humans through their additional effect on dopaminergic structures (Kelland *et al.*, 1990; Altemus *et al.*, 1994), dopamine antagonist monotherapy is not usually effective in reducing compulsive rituals in OCD patients (McDougle *et al.*, 1994a). By contrast, dopaminergic agents may perhaps appear the obvious choice for treating stereotypies (Chapters 5–8, this volume). Potentially effective drugs that interact with the dopaminergic system fall into two broad groups: the neuroleptics (anti-schizophrenic or antipsychotic drugs), including haloperidol, pimozide and phenothiazines; and the monoamine oxidase inhibitors.

Typical neuroleptics act mostly on D2 receptors. After oral administration, there is extensive first-pass metabolism by the liver. Neuroleptics tend to be eliminated slowly from the body and the metabolites are generally not active (Baldessarini, 1996a). In practice, however, these agents are not widely used because of serious side effects. In humans, neuroleptics reduce initiative, spontaneous motor activity, interest in the environment, and the manifestation of emotion or affect, although intellectual functions are retained. In rodents, unconditioned escape and avoidance behaviour is not affected, but conditioned avoidance may be reduced (Product Monograph, 1992). Low-potency antipsychotics such as acepromazine have a marked sedative effect. High-potency antipsychotics such as fluphenazine are not sedative at clinical doses, but are more likely to produce extrapyramidal side effects such as tremor, rigidity, hypersalivation, bradykinesia and motor restlessness. With higher doses, seizure risk increases. Hypotension, hyperthermia, sedation, arrhythmias and dermatologic reactions may be further side effects (Baldessarini, 1996a). The role of neuroleptics is also limited because of the risk of behavioural side effects that – ironically – include stereotypy-like behaviours termed tardive dyskinesia. The use of neuroleptics on animals thus has serious potential welfare implications, even if they control the problematic behaviour.

10.4.4.1. Neuroleptics

HALOPERIDOL. Haloperidol is perhaps the best-known neuroleptic, with a recommended dose rate of 20 mg/m^2 in the French veterinary literature, where it appears to be more widely discussed than elsewhere (Pageat, 1998).

The authors' experience in the use of haloperidol is very limited, but a dose of 1–2 mg bid orally invariably resulted in undesirable behavioural effects in the experience of one of us (AL). Haloperidol has been used experimentally in cats, at a dose of 0.1–0.2 mg/kg to counteract dopamine-induced stereotypy (Cools and van Rossum, 1970). Furthermore, a single injection (2 mg/kg i.v.) in a double-blind placebo-controlled study of 20 cats with excessive grooming found evidence of a sustained improvement in this behaviour at least 4 months later (Willemse et al., 1994). In parrots, haloperidol has been used effectively to treat feather picking (Iglauer and Rasmin, 1993; Lennox and Van Der Heyden, 1993, 1999). However, there are important breed differences in sensitivity, with Quaker parakeets, and Umbrella and Moluccan cockatoos, requiring lower dosages (Ritchie and Harrison, 1994). Haloperidol appears to work best in cockatoos and in cases of self-mutilation, further supporting the idea that there are mechanistically different forms of stereotypic behaviour (Welle, 1998).

PHENOTHIAZINES. Phenothiazines are quite non-specific agents: they interact with neurotransmitter systems other than dopamine (e.g. serotonin and norepinephrine), have antihistaminic and anti-tryptaminergic properties and may produce extrapyramidal side effects (Baldessarini, 1996a).

Long-acting formulations of phenothiazines, such as fluphenazine enanthate or decanoate, may be of interest for the treatment of animals that are difficult or dangerous to dose orally. If such preparations are to be used, an appropriate dose rate needs to be established first with a short-acting compound. Recommended therapeutic doses of fluphenazine in companion animals are 50 mg/m^2 (Pageat, 1998) but this has yet to gain wide acceptance and the authors have no experience of this agent for the treatment of compulsive disorder.

OTHER POTENT NEUROLEPTICS. The French literature (Pageat, 1998) refers to other related neuroleptics related to haloperidol, which may be used to control stereotypic behaviour in dogs. These include pimozide (4–12 mg/m^2), sulpiride (100–500 mg/m^2), tiapride (100–600 mg/m^2) and thioridazine (30 mg/m^2). Any of these might theoretically be considered as an augmenting agent in cases that are refractory to SRI monotherapy (McDougle et al., 1994a,b), but the risks and implications of intervention – especially in terms of side effects – are important and must be fully explained to the client.

10.4.4.2. Monoamine oxidase inhibitors

General monoamine oxidase inhibitors block the metabolism of a wide range of monoamine neurotransmitters including dopamine, norepinephrine and serotonin. They may have hypertensive side effects, resulting from an interaction with the amino-acid tyramine in the diet. Selegiline

(L-deprenyl) irreversibly inhibits the degradation of monoamine oxidase B (MAO-B) only, by binding to the enzyme's active site, and so does not have these interactions. Selegiline principally inhibits the reuptake of dopamine and secondarily that of noradrenaline (Knoll, 1981; Paterson *et al.*, 1990). Whilst there have been no peer-reviewed publications, its efficacy has been reported especially in relation to compulsive licking in several small cases series in both cats (Landsberg *et al.*, 2003) and dogs (Pageat, 1998). Response is reported to be rapid and efficacy high (>80%).

10.4.5. Drugs interacting with opioid receptors

Opioid antagonists include naloxone, naltrexone and nalmefene. Their effect is mostly on mu opioid receptors, but to a lesser effect on kappa and delta receptors (Reisine and Pasternak, 1996). Side effects in man include nausea and a general suppression of behaviour, and so their direct role in modulating stereotypic behaviour remains equivocal (see also Box 1.3, Chapter 1, this volume). Nausea might theoretically bring about a general suppression in behaviour, which could also reduce the level of repetitive behaviour, especially oral/ingestive behaviours including oral stereotypies. Careful analysis of the total time budgets of patients receiving such medication might be a useful first step towards trying to determine whether such a phenomenon occurs in captive animals.

10.4.5.1. Naloxone

When given orally, naloxone is almost completely metabolized in the liver (first pass metabolism) before reaching the systemic circulation. If given intravenously, it is also cleared very quickly. Intramuscular administration produces a more prolonged effect. In dogs, a half-life of about 70 min has been established after intravenous administration (Pace *et al.*, 1979).

Naloxone can temporarily inhibit tail-chasing in dogs (Brown *et al.*, 1987). A single dose of 0.2 mg/kg injected subcutaneously stopped a 7-month-old male Bull Terrier from chasing its tail for approximately 3 h. Because of the need for a parenteral route and the short-lived clinical effect, the use of naloxone in dogs is impractical. However, a double-blinded crossover study found that a single injection of naloxone (1 mg/kg s.c.) in 12 cats appeared to suppress excessive grooming for between 2.5 weeks and 6 months (Willemse *et al.*, 1994).

10.4.5.2. Naltrexone

Naltrexone is a more potent opiate antagonist than naloxone, and even when administered orally, has a clinical effect that lasts 24 to 72 h in humans. Oral naltrexone has been recommended for use in companion animals because of a presumed long half-life in these species. However, the half-life appears similar to that of naloxone in the dog (Pace *et al.*,

1979) and is considerably less than that reported for humans (Garret and el-Koussi, 1985). In humans, the first metabolite (6-beta-naltrexol) is also an active opiate antagonist, with a half-life of approximately 13 h. In dogs, however, naltrexone is not metabolized to 6-beta-naltrexol, but conjugated and excreted in urine and bile (Garret and el-Koussi, 1985). The clinical dose of naltrexone for companion animals is between 2.2 (White, 1990) and 4.4 mg/kg/day orally (Marder, 1991). Naltrexone may cause liver damage at higher doses.

In one study, 11 dogs with stereotypic self-licking, chewing or scratching were injected subcutaneously with naltrexone 1 mg/kg, or nalmefene 1 to 4 mg/kg (Dodman et al., 1988b) and these behaviours were reduced for at least 90 min in seven subjects. In an 'open-label' trial, dogs with acral lick dermatitis were given 2.2 mg/kg orally once or twice per day. This resulted in owner-reported cessation or substantial decreases in the amount of licking and re-epithelization of the lesion in 7 of the 11 dogs (White, 1990). One dog treated for compulsive tail-chasing with naltrexone at 2 mg/kg orally every 6 h, developed intense generalized pruritus (itchiness of the skin), which subsided when the dose was reduced to 1 mg/kg every 6 h (Schwartz, 1993).

Opioid antagonists have also been used experimentally to reduce crib-biting in horses (Dodman et al., 1987). Naloxone injected intravenously at 0.02 to 0.04 mg/kg suppressed cribbing for an average duration of 20 min. Three cribbing horses administered naltrexone intravenously at 0.04 to 0.4 mg/kg greatly reduced or suppressed the behaviour completely for 1.5 to 7 h (although their weaving was relatively unaffected). Nalmefene given intramuscularly or subcutaneously to five horses at 0.08 to 0.1 mg/kg was effective for between 1.5 and to almost 7 h. Sustained-release formulations can extend that time to several days in some cases, but are not available commercially. The relatively short duration and parenteral administration route limit the practical application of these drugs.

10.4.5.3. Hydrocodone

The opiate agonist hydrocodone has been used orally at 0.25 mg/kg for the treatment of canine acral lick dermatitis, with complete remission of licking in one, and partial remission in two, of the three treated dogs (Brignac, 1992).

10.5. Summary and Conclusion

The veterinary profession has a range of potential treatments for behaviour problems including psychotherapy, enrichment strategies, surgery and medicine, and through the work of able practitioners, veterinary behavioural medicine has undoubtedly improved the quality of life for many companion animals. However, the management of a given case is currently part science and part art, dependent upon both the therapist's knowledge and judgement. Furthermore, the clinician's 'treatment path' also needs to recognize the particular problem of concern to the owner, as well as the potential animal welfare implications of both the problem behaviour and its treatment.

In recent years there has been growing interest in the similarities between human psychiatric issues (especially OCD) and companion animal stereotypic behaviour, and thence in the potential of psychopharmacology to manage such problems. However, as we have seen, the practical application of the limited scientific information on the effect of drugs on animal stereotypic behaviour is not always straightforward. Many other factors must be considered both when trying to effect a successful treatment, and also when interpreting the effects of a chemical compound. For example, stereotypic behaviour might be suppressed in some cases – especially in species for which there is negligible research, such as exotics – as a result of either the normal side effects of some agents, or even overdosing to the point of sedation. It is thus currently unwise to make simple generalizations about pharmacological treatments for stereotypic behaviour: the clinical use of medicines in any given instance should be considered somewhat experimental, with each patient monitored carefully for potential side effects or adverse reactions, and the results put into the public domain. Psychopharmacological intervention should thus never be used as a 'quick fix' for individuals wishing to keep animals in an unacceptable environment. In these cases, rehoming should be considered. However, if an owner refuses to allow this, then therapeutic intervention may help to alter the patient's perception of the poor environment and thus improve its quality of life. This applies just as much to zoo animals as it does to companion animals. Drugs do not justify keeping animals in unsuitable conditions and should not be considered as an alternative to good husbandry; none the less it could be argued that a 'pill-popping polar' is happier than one that is deprived of medication in the same environment, when rehousing is not possible. Pharmaceutical therapy may also be useful when some fundamental aspect of animal temperament or behavioural functioning seems involved. Thus overall, when used with care, and (importantly) used along with other approaches, drug treatments can play an important role in overall therapy.

So what scientific research is needed to better support practitioners? Unfortunately, there have been as yet no large-scale placebo-controlled studies into the use of psychopharmacology for treating stereotypic behaviour in animals, nor specific investigation into its potential side-effects. This is clearly a major need. Even smaller-scale studies are currently few and far between. It is also ironic that the species humans choose as companions, and apparently greatly care for, are also the ones for which we have the least basic scientific information on the causes of stereotypic behaviour. Clearly there is an urgent need for more research, at both fundamental and applied levels. Partly, this would be valuable for helping clinicians improve the scientific bases of their treatment rationales. One crucial focus here is how to distinguish between different forms of abnormal repetitive behaviour. Recent work helping to objectively differentiate specific forms of stereotypic behaviour (reviewed elsewhere in this volume; see e.g. Chapters 5 and 11) may help in this respect, especially if it can be adapted to clinical tests which help to refine treatment choice. Our own attempted schema gives some preliminary guidelines to veterinarians, but is still very provisional. Fundamental information would also help in preventing stereotypic behaviours before they even emerge. Improvements in management (e.g. environmental enrichment), and owner education all clearly have major roles to play in their aetiology, but more research is needed to further identify the best preventative approaches. In the future, perhaps it would also be fruitful to develop breeding strategies to produce animals better adapted to their environment, and/or, given the apparent efficacy of tryptophan in some cases, to explore 'psychodietetics'.

However, while repetitive behaviour problems in companion animals still occur, the potential for veterinary intervention will remain. Depending on the circumstances of that animal, psychopharmacological intervention may often be justifiable – even with our current state of knowledge – on both welfare and practical grounds.

References

Abel, J.L. (1993) Exposure with response prevention and serotonergic antidepressants in the treatment of obsessive compulsive disorder: a review and implications for interdisciplinary treatment. *Behaviour Research and Therapy* 31, 463–478.

Altemus, M., Swedo, S.E., Leonard, H.L., Richter, D., Rubinow, D.R., Potter, W.Z. and Rapoport, J.L. (1994) Changes in cerebrospinal fluid neurochemistry during treatment of obsessive–compulsive disorder with clomipramine. *Archives of General Psychiatry* 51, 794–803.

American Psychiatric Association (1994) *Diagnostic and Statistical Manual of Mental Disorders*, 4th edn. American Psychiatric Association, Washington, DC.

Anthony, M.M., Downie, F. and Swinson, R.P. (1998) Diagnostic issues and epidemiology in obsessive–compulsive disorder. In: Swinson, P.R., Anthony, M.M., Rachman, S. and Richter, M.A. (eds) *Obsessive Compulsive Disorder. Theory, Research and Treatment.* The Guilford Press, New York, pp. 3–32.

Bachmann, I., Audige, L. and Stauffacher, M. (2003) Risk factors associated with behavioural disorders of crib-biting, weaving and box-walking in Swiss horses. *Equine Veterinary Journal* 35, 158–163.

Bagshaw, C.S., Ralston, S.L. and Fisher, H. (1994) Behavioral and physiological effect of orally administered tryptophan on horses subjected to acute isolation stress. *Applied Animal Behaviour Science* 40, 1–12.

Baldessarini, R.J. (1996a) Drugs and the treatment of psychiatric disorders, psychosis and anxiety. In: Harman, J.G., Goodman-Gilman, A. and Limbird, L.E. (eds) *Goodman & Gilman's the Pharmacological Basis of Therapeutics*, 9th edn. McGraw-Hill, New York, pp. 399–430.

Baldessarini, R.J. (1996b) Drugs and the treatment of psychiatric disorders, depression and mania. In: Harman, J.G., Goodman-Gilman, A. and Limbird, L.E. (eds) *Goodman & Gilman's the Pharmacological Basis of Therapeutics,* 9th edn. McGraw-Hill, New York, pp. 431–459.

Blackshaw, J.K., Sutton, R.H. and Boyhan, M.A. (1994) Tail chasing or circling behavior in dogs. *Canine Practice* 19, 7–11.

Brignac, M.M. (1992) Hydrocodone treatment of acral lick dermatitis. *Proceedings of the Second World Congress of Veterinary Dermatology.* Montreal, Quebec.

Brown, S.A., Crowell-Davis, S., Malcolm, T. and Edwards, P. (1987) Naloxone-responsive compulsive tail-chasing in a dog. *Journal of the American Veterinary Medical Association* 190, 1434.

Cabib, S., Giardino, L., Calza, L., Zanni, M., Mele, A. and Pulisiallegra, S. (1998) Stress promotes major changes in dopamine receptor densities within the mesoaccumbens and nigrostriatal systems. *Neuroscience* 84, 193–200.

Chitty, J. (2003) Feather-plucking in psittacine birds. 2. Social environmental and behavioural considerations. *In Practice* 25, 550–555.

Clark, J. and Mills, D.S. (1997) Design consideration for the evaluation of tryptophan supplementation in the modification of equine behaviour. In: Mills, D.S., Heath, S.E. and Harrington, L.J. (eds) *Proceedings of the First International Meeting Veterinary Behavioural Medicine.* UFAW, Potters Bar, UK, p. 220.

Cohen, D.J., Riddle, M.A. and Leckman, J.F. (1992) Pharmacotherapy of Tourette's syndrome and associated disorders. *Psychiatric Clinics of North America* 15, 109–129.

Colombo, J.P., Cervantes, H., Kokoroyic, M., Pfister, U. and Perritaz, R. (1992) Effect of different protein diets on the distribution of amino acids in plasma, liver, and brain in the rat. *Annals of Nutrition and Metabolism* 36, 23–33.

Cook, W.R. (1980) Head-shaking in horses. Part 3. Special diagnostic procedures. *Equine Practice* 2, 31–40.

Cools, A.R. and van Rossum, J.M. (1970) Caudal dopamine and stereotype behaviour of cats. *Archives Internationales de Pharmacodynamie et de Therapie* 187, 163–173.

Cooper, J.J. and Mills, D.S. (1997) Welfare considerations relevant to behaviour modifications in domestic animals. In: *Proceedings of the First International Meeting Veterinary Behavioural Medi-*

cine. UFAW, Potters Bar, UK, pp. 164–173.

Dodman, N.H. and Shuster, L. (1998) *Psychopharmacology of Animal Behavior Disorders.* Blackwell Science, Malden, Massachusetts.

Dodman, N.H., Shuster, L., Court, M.H. and Dixon, R. (1987) Investigation into the use of narcotic antagonists in the treatment of a stereotypic behavior pattern (crib-biting) in the horse. *American Journal of Veterinary Research* 48, 311–319.

Dodman, N.H., Shuster, L., Court, M.H. and Patel, J. (1988a) Use of a narcotic antagonist (nalmefene) to suppress self-mutilative behavior in a stallion. *Journal of the American Veterinary Medical Association* 192, 1585–1587.

Dodman, N.H., Shuster, L., White, S.D., Court, M.H., Parker, D. and Dixon, R. (1988b) Use of narcotic antagonists to modify stereotypic self-licking, self-chewing and scratching behavior in dogs. *Journal of the American Veterinary Medical Association* 193, 815–819.

Dodman, N.H., Knowles, K.E., Shuster, L., Moon-Fanelli, A.A., Tidwell, A.S. and Keen, C.L. (1996) Behavioral changes associated with suspected complex partial seizures in Bull Terriers. *Journal of the American Veterinary Medical Association* 208, 688–691.

Donaldson, D. (1998) *Psychiatric Disorders with a Biochemical Basis.* Parthenon, Carnforth, UK.

Fernstrom, J.D. and Wurtman, R.J. (1972) Brain serotonin content: physiological regulation by plasma neutral amino acids. *Science* 178, 414–416.

Fletcher, R.H., Fletcher, S.W. and Wagner, E.H. (1996) *Clinical Epidemiology, the Essentials,* 3rd edn. Lippincott Williams and Wilkins, Philadelphia, 276 pp.

Fuller, R.W. (1994) Mini-review: uptake inhibitors increase extracellular serotonin concentration measured by brain microdialysis. *Life Sciences* 55, 163–167.

Fuller, R.W., Wong, D.T. and Robertson, D.W. (1991) Fluoxetine, a selective inhibitor of serotonin uptake. *Medical Research Reviews* 11, 017–034.

Garrett, E.R. and el-Koussi, A.E.A. (1985) Pharmacokinetics of morphine and its surrogates. V. Naltrexone and naltrexone conjugate pharmacokinetics in the dog as a function of dose. *Journal of Pharmaceutical Sciences* 74, 50–56.

Goldberger, E. and Rapoport, J.L. (1990) Canine acral lick dermatitis: response to the anti-obsessional drug clomipramine. *Journal of the American Animal Hospital Association* 27, 179–182.

Grady, T.A., Pigott, T.A., L'Heureux, F., Hill, J.L., Bernston, S.E. and Murphy, D.L. (1993) Double-blind study of adjuvant buspirone for fluofetine-treated patients with obsessive compulsive disorder. *American Journal of Psychiatry*, 50, 819–821.

Grindlinger, M.H. and Ransay, E. (1992) Treatment of feather-picking with clomipramine. *Proceedings of the Annual Meeting of the Association of Avian Veterinarians.*

Hart, B.L. and Hart, L.A. (1985) *Canine and Feline Behavioral Therapy.* Lea and Febiger, Philadelphia, 275 pp.

Hartigan, P.J. (2000) Compulsive tail chasing in the dog: a mini-review. *Irish Veterinary Journal* 53, 261–264.

Hedaya, R.J. (1984) Pharmacokinetic factors in the clinical use of tryptophan. *Journal of Clinical Psychopharmacology* 4, 347–348.

Hewson, C.J. (1997) Clomipramine in dogs: pharmacokinetics, neurochemical effects, and efficacy in compulsive disorder. Ph.D. thesis, Ontario Veterinary College, Guelph, Ontario, Canada.

Hewson, C.J. and Luescher, U.A. (1996) Compulsive disorder in dogs. In: Voith, V.L. and Borchelt, P.L. (eds) *Readings in Companion Animal Behavior.* Veterinary Learning Systems, Trenton, New Jersey, pp. 153–158.

Hewson, C.J., Parent, J.M., Conlon, P.D., Luescher, U.A. and Ball, R.O. (1998a) Efficacy of clomipramine in the treatment of canine compulsive disorder: a randomized, placebo-controlled, double blind clinical trial. *Journal of the American Veterinary Medical Association* 213, 1760–1766.

Hewson, C.J., Conlon, P.D., Luescher U.A. and Ball R.O. (1998b) The pharmacokinetics of clomipramine and desmethylclomipramine in dogs: parameter estimates following a single oral dose and 28 consecutive daily oral doses of clomipramine. *Journal of Veterinary Pharmacology and Therapeutics* 21, 214–222.

Hugo, C., Seier, J., Mdhluli, C., Daniels, W., Harvey, B.H., Du Toit, D., Wolfe-Coote, S., Nel, D. and Stein D.J. (2003) Fluoxetine decreases stereotypic behaviour in primates. *Progress in Neuropsychopharmacology and Biological Psychiatry* 27, 639–643.

Iglauer, F. and Rasmin, R. (1993) Treatment of chronic feather picking in psittacine birds with a dopamine antagonist. *Journal of Small Animal Practice* 54, 564–566.

Irimajiri, M., Jay, E., Luescher, U.A. and Glickman, L. (2003) Mild polycytaemia in canine compulsive disorder. Presented at *American Veterinary Society of Animal Behavior Annual Scientific Meeting*, Denver, 2003.

Karno, M., Golding, J.M., Sorensen, S.B. and Burnam, M.A. (1988) The epidemiology of obsessive–compulsive disorder in five US communities. *Archives of General Psychiatry* 45, 1094–1099.

Kelland, M.D., Freeman, A.S. and Chiodo, L.A. (1990) Serotonergic afferent regulation of the basic physiology and pharmacological responsiveness of nigrostriatal dopamine neurons. *Journal of Pharmacology and Experimental Therapeutics* 253, 803–811.

Kennes, D., Odberg, F.O., Bouquet, Y. and De Rycke, P.H. (1988) Changes in naloxone and haloperidol effects during the development of captivity induced jumping stereotypy in bank voles. *Journal of Pharmacology* 153, 19–24.

King, J.N., Simpson, B.S., Overall, K.L., Appleby, D., Pageat, P., Ross, C., Chaurand, J.P., Heath, S., Beata, C., Weiss, A.B., Muller, G., Paris, T., Bataille, B.G., Parker, J., Petit, S. and Wren, J. (2000) Treatment of separation anxiety in dogs with clomipramine: results from a prospective, randomised, double-blind, placebo-controlled, parallel-group, multicenter clinical trial. *Applied Animal Behaviour Science* 67, 255–275.

Knoll, J. (1981) The pharmacology of selective MAO inhibitors. In: Youdim, M.B.H. and Paykel, E.S. (eds) *Monoamine Oxidase Inhibitors – The State of the Art*. John Wiley & Sons, Chichester, UK.

Kossoff, E.H. and Singer, H.S. (2001) Tourette syndrome: clinical characteristics and current management strategies. *Paediatric Drugs* 3, 355–363.

Kostal, L. and Savory, C.J. (1996) Behavioral responses of restricted-fed fowls to pharmacological manipulation of 5-HT and GABA receptor subtypes. *Pharmacology, Biochemistry and Behavior* 53, 995–1004.

Landsberg, G., Hunthausen, W. and Ackerman, L. (2003) *Handbook of Behaviour Problems of the Dog and Cat*, 2nd edn. Saunders, Oxford, UK, 545 pp.

Lennox, A.M. and Van Der Heyden, N. (1993) Haloperidol for use in treatment of psittacine self-mutilation and feather plucking. In: *Proceedings of the Annual Conference of the Association of Avian Veterinarians 1993*, pp.119–120.

Lennox, A.M. and Van Der Heyden, N. (1999) Long-term use of haloperidol in two parrots. In: *Proceedings of the Annual Conference of the Association of Avian Veterinarians 1999*, pp. 133–137.

Luescher, U.A. (1997) Factors affecting the outcome of behavioral treatment. *Paper presented at Meeting of the American Animal Hospital Association*, March 1997, San Diego, California.

Luescher, U.A. (2002) Compulsive behaviour. In: Horwitz D.F., Mills D.S. and Heath, S.E. (eds) *BSAVA Manual of Canine and Feline Behavioural Medicine*. BSAVA, Gloucester, pp. 229–236.

Luescher, U.A., McKeown, D.B. and Halip, J. (1991) Stereotypic and obsessive–compulsive disorders in dogs and cats. *Veterinary Clinics of North America (Small Animal Practice)* 21, 401–413.

Luescher, U.A., McKeown, D.B. and Dean, H. (1998) A cross-sectional study on compulsive behaviour (stable vices) in horses. *Equine Veterinary Journal Supplement* 27, 14–18.

March, J.S. (1995) Cognitive-behavioral psychotherapy for children and adolescents with OCD: a review and recommendations for treatment. *Journal of the American Academy of Child and Adolescent Psychiatry* 34, 7–18.

Marder, A.R. (1991) Psychotropic drugs and behavioral therapy. *Veterinary Clinics of North America (Small Animal Practice)* 21, 329–342.

Mason, G.J. (1991) Stereotypies: a critical review. *Animal Behaviour* 41, 1015–1037.

McAfee, L.M., Mills, D.S. and Cooper, J.J. (2002) The use of mirrors for the control of stereotypic weaving behaviour in the stabled horse. *Applied Animal Behaviour Science* 78, 159–173.

McDonnell, S.M. (1996) Pharmacological aids to behavior modification in horses. In: *Meeting of the Swiss Society for Equine Medicine*, June 1996, Basel, Switzerland.

McDougle, C.J., Goodman, W.K. and Price, L.H. (1994a) Dopamine antagonists in tic-related and psychotic spectrum obsessive compulsive disorder. *Journal of Clinical Psychiatry* 55, 24–31.

McDougle, C.J., Goodman, W.K., Leckman, J.F., Lee, N.C., Heninger, G.R. and Price L.H. (1994b) Haloperidol addition in fluvoxamine-refractory obsessive–compulsive disorder – a double blind placebo-controlled study in patients with and without tics. *Archives of General Psychiatry* 51, 302–308.

Melman, S.A. (1995) Case report: a tale of (tail) mutilation. *American Veterinary Society of Animal Behavior Newsletter* 17, 7–8.

Mertens, P.A. (1997) Pharmacological treatment of feather-picking in pet birds. In: Mills, D.S., Heath, S.E. and Harrington, L.J. (eds) *Proceedings of the First International Meeting Veterinary Behavioural Medicine*. UFAW, Potters Bar, UK, pp. 209–211.

Mills, D.S. (2003) Medical paradigms for the study of problem behaviour: a critical review. *Applied Animal Behaviour Science* 81, 265–277.

Mills, D.S. and Davenport K. (2002) The effect of a neighbouring conspecific versus the use of a mirror for the control of stereotypic weaving behaviour in the stabled horse. *Animal Science* 74, 95–101.

Mills, D.S. and Ledger, R. (2001) The effect of oral selegiline hydrochloride on learning and training in the dog: a psychobiological interpretation. *Progress in Neuropsychopharmacology and Biological Psychiatry* 25, 1597–1613.

Mills, D.S. and Simpson, B.S. (2002) Psychotropic agents. In: Horwitz, D.F., Mills, D.S. and Heath, S.E. (eds) *BSAVA Manual of Canine and Feline Behavioural Medicine.* BSAVA, Gloucester, pp. 237–248.

Mills, D.S., Heath, S.E. and Harrington, L.J. (1997) In: *Proceedings of the First International Meeting Veterinary behavioural Medicine.* UFAW, Potters Bar, UK.

Nesse, R.M. and Berridge, K. (1997) Psychoactive drug use in evolutionary perspective. *Science* 277, 63–65.

Newton, S.A., Knottenbelt, D.C. and Eldridge, P.R. (2000) Headshaking in horses: possible aetiopathogenesis suggested by the results of diagnostic tests and several treatment regimes in 20 cases. *Equine Veterinary Journal* 32, 208–216.

Norman, D.A. and Shallice, T. (1986) Attention to action: willed and automatic control of behaviour. In: Davidson, R.J., Schwartz, G.E. and Shapiro, D. (eds) *Consciousness and Self-regulation: Advances in Research and Theory.* Plenum Press. New York, pp. 1–18.

Nurnberg, H.G., Keith, S.J. and Paxton, D.M. (1997) Consideration of the relevance of ethological animal models for human repetitive behavioural spectrum disorders. *Biological Psychiatry* 41, 226–229.

Oliver, J.E. and Lorenz, M.D. (1993) *Handbook of Veterinary Neurology,* 2nd edn. WB Saunders, Philadelphia, pp. 9–10.

Overall, K.L. (1992a) Recognition, diagnosis and management of obsessive–compulsive disorders. Part I. *Canine Practice* 17, 40–44.

Overall, K.L. (1992b) Recognition, diagnosis, and management of obsessive–compulsive disorders. Part II. *Canine Practice* 17, 25–27.

Overall, K.L. (1992c) Practical pharmacological approaches to behavior problems. In: *Purina Specialty Review: Behavioral Problems in Small Animals.* Ralston Purina Company, pp. 36–51.

Overall, K.L. (1994) Use of clomipramine to treat ritualistic stereotypic motor behavior in three dogs. *Journal of the American Veterinary Medical Association* 205, 1733–1741.

Overall, K.L. (1995) Animal behavior case of the month. *Journal of the American Veterinary Medical Association* 206, 629–632.

Overall, K.L. (1997) *Clinical Behavioral Medicine for Small Animals.* Mosby, St Louis, p. 515.

Overall, K.L. and Dunham, A.E. (2002) Clinical features and outcome in dogs and cats with obsessive–compulsive disorder: 126 cases (1989–2000). *Journal of the American Veterinary Medical Association* 221, 1445–1452.

Pace, N.L., Parrish, R.G., Lieberman, M.M., Wong, K.C. and Blatnick, R.A. (1979) Pharmacokinetics of naloxone and naltrexone in the dog. *Journal of Pharmacology and Experimental Therapeutics* 208, 254–256.

Pageat, P. (1998) *Pathologie du Comportement du Chien,* 2nd edn. Editions du Point Veterinaire, Maisons-Alfort.

Paterson, I.A., Juorio, A.V. and Boulton, A.A. (1990) 2-phenylethylamine – a modulator of catecholamine transmission in the mammalian central nervous system. *Journal of Neurochemistry* 55, 1827–1837.

Poulsen, E.M.B., Honeyman, V., Valentine, P.A. and Teskey, G.C. (1996) Use of fluoxetine for the treatment of stereotypical pacing behaviour in a captive polar bear. *Journal of the American Veterinary Medical Association* 209, 1470.

Product Monograph (1992) Haldol, haloperidol tablets, solution and injection USP, antipsychotic agent. McNeil Pharmaceutical Ltd., Don Mills, Ontario.

Product Monograph (1993) BuSpar (buspirone hydrochloride) tablets, anxiolytic. Montreal, Quebec: Bristol.

Product Monograph (1996a) Anafranil (clomipramine hydrochloride), antidepressant antiobsessional. Ciba-Geigy Pharmaceuticals, Mississauga, Ontario.

Product Monograph (1996b) Prozac (fluoxetine hydrochloride) capsules and oral solution, antidepressant/antiobsessional/antibulimic. Eli Lilly, Scarborough, Ontario.

Rapoport, J.L., Ryland, D.H. and Kriete, M. (1992) Drug treatment of canine acral lick: an animal model of obsessive compulsive disorder. *Archives of General Psychiatry* 49, 517–521.

Reisine, T. and Pasternak, G. (1996) Opioid analgesics and antagonists. In: Harman, J.G., Goodman-Gilman, A. and Limbird, L.E. (eds) *Goodman & Gilman's the Pharmacological Basis of Therapeutics,* 9th edn. McGraw-Hill, New York, pp. 521–555.

Ritchie, B.W. and Harrison, G.J. (1994) Formulary. In: Ritchie, B.W., Harrison, G.J., Harrison, L.R. (eds) *Avian Medicine: Principles and Application.* Wingers Publishing, Lake Worth, Florida, pp. 457–478.

Rosenhan, D.L. and Seligman, M.E.P. (1995) *Abnormal Psychology,* 3rd edn. W.W. Norton and Company, New York.

Sanders-Bush, E. and Mayer, S.E. (1996) 5 hydroxytryptamine (serotonin) receptor agonists and antagonists. In: Harman, J.G., Goodman-Gilman, A. and Limbird, L.E. (eds) *Goodman & Gilman's the Pharmacological Basis of Therapeutics*, 9th edn. McGraw-Hill, New York, pp. 249–263.

Scahill, L., Riddle, M.A., King, R.A., Hardin, M.T., Rasmusson, A., Makuch, R.W. and Leckman, J.F. (1997) Fluoxetine has no marked effect on tic symptoms in patients with Tourette's syndrome: a double-blind placebo-controlled study. *Journal of Child and Adolescent Psychopharmacology* 7, 75–85.

Schwartz, S. (1993) Naltrexone-induced pruritus in a dog with tail-chasing behavior. *Journal of the American Veterinary Medical Association* 202, 278–280.

Seibert, L.M., Crowell-Davis, S.L., Wilson, G.H. and Ritchie, B.W. (2004) Placebo-controlled clomipramine trial for the treatment of feather-picking disorder in Cockatoos. *Journal of the American Animal Hospital Association* 40, 261–269.

Shankaran, M. and Gudelsky, G.A. (1998) Effect of 3,4-methylenedioxymethamphetamine (MDMA) on hippocampal dopamine and serotonin. *Pharmacology, Biochemistry and Behaviour* 61, 361–366.

Summerfeldt, L., Richter, M.A., Anthony, M.M. and Swinson, R.P. (1999) Symptom structure in obsessive compulsive disorder: a confirmatory analytic study. *Behavior Research and Therapy* 37, 297–311.

Takeuchi, Y., Houpt, K.A. and Scarlett, J.M. (2000) Evaluation of treatments for separation anxiety in dogs. *Journal of the American Veterinary Medical Association* 217, 342–345.

Teff, K.L. and Young, S.N. (1988) Effects of carbohydrate and protein administration on rat tryptophan and 5-hydroxytryptamine: differential effects on the brain, intestine, pineal, and pancreas. *Canadian Journal of Physiology and Pharmacology* 66, 683–688.

Tuber, D.S., Hothersall, D. and Voith, V.L. (1974) Animal clinical psychology: a modest proposal. *American Psychologist* 29, 762–766.

von Borell, E. and Hurnik, J.F. (1991) The effect of haloperidol on the performance of stereotyped behavior in sows. *Life Sciences* 49, 309–314.

Weld, K.P., Mench, J.A., Woodward, R.A., Bolesta, M.S., Suomi, S.J. and Higley, J.D. (1998) Effect of tryptophan treatment on self biting and central nervous system serotonin metabolism in rhesus monkeys. *Neuropsychopharmacology* 19, 314–322.

Welle, K.R. (1998) A review of psychotropic drug therapy. In: *Proceedings of the Annual Conference of the Association of Avian Veterinarians 1998*, pp. 121–123.

White, S.D. (1990) Naltrexone for treatment of acral lick dermatitis in dogs. *Journal of the American Veterinary Medical Association* 196, 1073–1076.

Wiepkema, P.R. and Koolhaas, J.M. (1993) Stress and animal welfare. *Animal Welfare* 2, 195–218.

Willemse, T., Mudde, M., Josephy, M. and Spruijt, B.M. (1994) The effect of halo-peridol and naloxone on excessive grooming behaviour of cats. *European Neuropsychopharmacology* 4, 39–45.

Wynchank, D. and Berk, M. (1998) Behavioural changes in dogs with acral lick dermatitis during a 2 month extension phase of fluoxetine treatment. *Human Psychopharmacology – Clinical and Experimental* 13, 435–437.

11 Stereotypic Behaviour in Captive Animals: Fundamentals and Implications for Welfare and Beyond

G. Mason

Department of Animal and Poultry Science, University of Guelph, Guelph, Ontario, N1G 2W1, Canada

Editorial Introduction

To end, I synthesize the previous chapters plus some additional literature. I also outline outstanding research questions for future work – and for the next edition of this book.

First, I look at the fundamental causes of stereotypic behaviour. I show that the three main causes of repetition are disinhibition (as implicated in perseveration), reinforcement, and sustained elicitation by internal or external stimuli; the three main causes of predictability are environmental constancy, routine-formation, and the repeated elicitation of very specific action patterns; and the three main types of 'source behaviour' are escape attempts, surrogates for natural behaviour patterns, and a third category of more puzzling, heterogeneous forms (some of which probably reflect dysfunction). I discuss two mechanisms in more detail than previous chapters: the processes underlying normal routine-formation; and the various types of perseveration stemming from the (mal)function of different brain regions. I particularly focus on the system targeted by Cabib in Chapter 8, showing that this could underlie forms of 'affective' perseveration, and, correspondingly, distinct stereotypic behaviours – additional to the 'motor' and 'stuck-in-set' dichotomy proposed in Chapters 5 and 10. I then summarize how captivity induces the repetition inherent in stereotypic behaviour: by causing frustration; and/or by altering CNS functioning through stress and/or through impeding normal development.

Next I consider the practical and ethical implications of stereotypic behaviour. I argue that environments that induce it typically also reduce animal welfare. However, at the individual level, 'coping', and the 'scar-like' effects of routine-formation and early experience, may eliminate close correspondence between the behaviour and underlying stress and frustration. Indeed paradoxically, highly stereotypic individuals often fare better in these inadequate environments than their less active peers: patterns that could reflect coping, or perhaps instead the activity-reducing effects of some other psychological or physical conditions. Last,

I discuss the extent to which stereotypic behaviour indicates brain malfunction. I argue that while some forms do not (e.g. those of normal humans or free-living wild animals), some definitely do (e.g. many of those in Chapter 6). As for farm, laboratory and zoo animals, whether some or all of their stereotypic behaviours indicate CNS malfunction is an intriguing, disturbing possibility that needs more research.

Throughout, I use the term 'stereotypic behaviour' in the broad sense recommended in Chapter 10. However, I end by questioning the value of classifying a behaviour solely according to ill-defined aspects of phenotype, and suggest that neither this nor the standard definition (*repetitive, unvarying with no apparent goal of function*) actually reflect how people use terms like 'stereotypic behaviour' in practice. I therefore propose a new definition centred on the causal mechanisms of repetition: *that stereotypic behaviours are repetitive behaviours induced by frustration, repeated attempts to cope, and/or CNS dysfunction*. Where such causal factors are unknown, I recommend the blander term, 'Abnormal Repetitive Behaviour' (ARB; *cf.* Chapter 5). I also suggest that stereotypic behaviours should be sub-categorized into frustration-induced and malfunction-induced forms, the former being maladaptive but readily reversible responses of normal animals to abnormal environments, and the latter, a spectrum of pathologies – some of which warrant the more precise label 'stereotypy' (*cf.* Chapters 5 and 10) – evidenced by various forms of abnormal perseveration and/or direct signs of impaired in CNS functioning.

GM

11.1. Introduction

11.1.1. Outline of this chapter

In this final chapter, I argue that stereotypic behaviour involves diverse mechanisms which are differentially involved across different forms, which explain different aspects of the behaviour, and which likely represent a spectrum from the responses of a normal animal to an abnormal environment, through to the pathological signs of profound brain dysfunction. I begin by synthesizing the previous chapters plus additional literature, first asking some fundamental questions (Section 11.2): overall, what behavioural processes account for the repetitive nature of stereotypic behaviours, and for the very predictable, unvarying nature of some? And what determines the form displayed, be it pacing, rocking or chewing? For each issue, three main processes emerge as most important, from which I select two to discuss in more detail than in the preceding chapters. First, I look at the different types of perseveration that could correlate with repetition, elaborating on the form hinted at in Chapter 8 by Cabib, and comparing this with those reviewed by Garner and colleagues in Chapters 5 and 10. Second, I look at motor learning processes that could render an initially variable behaviour more predictable and routine-like with time and repetition. I end with an overview of the likely routes by which captivity induces stereotypic behaviour.

In Section 11.3, I move on to the ethical and practical implications of stereotypic behaviour. Overall, what does it say about captive animals' welfare, and do stereotypic individuals always have poorer welfare than non-stereotypic ones? Do some forms of stereotypic behaviour indicate brain malfunction in captive animals? And if at least some forms do, does this have practical implications? In the final section, as discussed further below, I propose new definitions for 'stereotypic behaviour' and 'stereotypy'. In this, as in all the sections, I also highlight outstanding research questions.

11.1.2. Stereotypic behaviour: scope and definitions

As we have seen in this volume and its website, captive animals show diverse forms of repetitive behaviour which baffle, intrigue or worry us. Many broadly fit the classic, decades-old definition of 'stereotypy', in being *'unvarying and repetitive ... with no apparent goal or proximate function'* (see previous chapters). However, different cases meet this description to very different extents. Some are highly unvarying: route-tracing Amazon parrots and polar bears, for instance, may place their feet in exactly the same location each time they repeat a circuit (e.g. Wechsler, 1991; Garner *et al.*, 2003b); but in others, in contrast, a variety of postures and movements are employed (as in self-biting or hair-plucking, *cf.* e.g. Chapters 4–6), animals seeming to have an inflexibility of goal rather than an inflexibility of action pattern. A similar spectrum occurs in the degree of repetition: at one extreme are long, continuous bouts of repeated movements (each bout of route-tracing or spot-pecking by caged canaries, for instance, typically involves 15–100 reiterations; Keiper, 1969); while at the other extreme are cases like the pet dog that keeps staring at a light – its stance is still each time, and this behaviour occurs intermittently, and yet it is recurrent day after day, week after week (*cf.* e.g. Mills and Luescher, Chapter 10). Furthermore, assessment of that final defining feature of 'stereotypy', its apparent lack of goal or function, is typically very subjective – the caveat *apparent* probably there just to save us from being precise about something so hard to assess!

So, given this descriptive, imprecise definition, it is unsurprising that very diverse behaviour patterns have been pooled under this label. It is equally unsurprising that when certain repetitive behaviours are *not* termed 'stereotypies', this typically is not for clear, objective reasons. Thus some behaviours are quite arbitrarily given alternative labels instead, such as 'compulsive behaviour' in some companion animal cases, 'redirected behaviour', e.g. belly-nosing in piglets, and even 'exercise' for wheel-running (e.g. Box 4.2). The quandary of definition has been raised repeatedly in the preceding chapters (especially Chapters 2, 4, 5 and 10). In Section 11.4, I therefore revisit this issue, asking whether it is useful to have a classification based purely on phenotype, and even whether *unvarying and repetitive ... with no apparent goal or proximate function* accurately captures all that people mean when using the terms 'stereotypy' or 'stereotypic behaviour' in practice.

In the following sections, however, rather than worry too much about this or be sidetracked by questions like 'how unvarying *is* "unvarying"?', I am going to follow Mills and Luescher's pragmatic recommendation (see Chapter 10), and use their broad, heterogeneous catch-all term 'stereotypic behaviour' for all apparently functionless, repetitive behaviours. I thus use this to encompass cases from the most clockwork-like, rhythmic forms of pacing and nodding, through to the more flexible wood-chewing of Chapter 2, hair-plucking of Chapter 5 and self-biting of Chapter 6. This term, encompassing a broader group than Garner's 'Abnormal Repetitive Behaviour' (see Chapter 5), is convenient because it does not imply any known or unitary cause. I will also avoid the term 'stereotypy', until my final section (11.4), because of recent calls to restrict it to cases with a known and specific aetiology (see Chapters 5 and 10).

11.2. Fundamentals: the Behavioural Processes Involved in Stereotypic Behaviour

11.2.1. What behavioural processes account for the repetitive nature of stereotypic behaviours?

Overall, the repetitive nature of stereotypic behaviour seems to arise for three main reasons. These are not mutually exclusive, and so potentially may act in concert. The first involves sustained or recurrent eliciting stimuli in the animal's internal or external environment; the second, reward and reinforcement (e.g. 'coping'); and the third, perseveration or its correlates.

11.2.1.1. Sustained elicitation

The first explanation for sustained repetition is that motivationally salient factors (e.g. releasing stimuli in the environment) elicit the prolonged and/or recurrent performance of specific normal behaviour patterns. Thus for instance, sustained nutritional deficits and/or the appetite-enhancing effects of small amounts of food were hypothesized to elicit repeated foraging behaviours in captive ungulates (e.g. food-deprived sows; see Bergeron *et al.*, Chapter 2); the absence of a suitably tunnel-like den, perhaps combined with the concave shape of cage corners, to trigger stereotypic digging in caged gerbils; and the aversiveness of the cage environment combined with the salience of odours outside, to repeatedly elicit escape-attempts in laboratory mice (both discussed by Würbel in Chapter 4).

That the behavioural responses to these stimuli do not habituate or extinguish could suggest that they are reinforced, continuing because of correlations between their performance and some positive outcome (as discussed further below). Alternatively, some 'constraint on learning' (sensu e.g. Shettleworth, 1972) could mean that the responses cannot be suppressed despite failing to be beneficial. To illustrate with examples from research on learning, natural foraging behaviours often spontan- eously appear in animals trained to expect food, and persist despite delaying obtaining the reward ('misbehaviour'; e.g. Timberlake and Lucas, 1989); while the punishment of escape responses often fails to suppress them, sometimes even enhancing them further (e.g. Mackintosh, 1974). This inappropriate type of stimulus-induced response may be even harder to suppress if elicited by 'super-normal' stimuli: oyster- catchers, for instance, presented with very large fake eggs will attempt to roll them into their nest, and even choose them over normal eggs despite being far too big to actually brood (e.g. Tinbergen, 1951). Thus stimulus-response links that are evolutionarily reliable or important can be tricked by artificial situations into triggering bizarre, counterproduc- tive actions.

Relevant to stereotypic behaviour are those stimulus–response links perhaps involved in ungulate foraging, at least as envisaged by the 'frus- trated food search' and 'inflexible time budget' hypotheses (see Chapter 2); and in the sustained escape attempts or shelter-seeking efforts of caged rodents (Chapter 4) – which Würbel suggests may have evolved to simply persist until successful. However, we still do not fully understand the role of such effects in stereotypic behaviour, and they remain an open topic for future research. Could normal motivational mechanisms really account for activities that are repeated thousands of times a day, or can they only explain less extreme forms of stereotypic behaviour? And are some spe- cific natural behaviours inherently likely to be elicited in a sustained manner, due to the way their control mechanisms have been shaped by natural selection? So far, we really do not know.

11.2.1.2. Reward and reinforcement

Despite superficially seeming functionless, some stereotypic behaviour may have reinforcing consequences. Thus in this volume we have seen that wheel-running is a reinforcer for rodents (Box 4.2); that bouts of self- injurious biting in primates correlate with reductions in physiological stress (Chapter 6); and that the oral stereotypic behaviours of several ungulates are linked with short-term reductions in heart rate, while crib-biting by horses shows 'rebound' performance if prevented for a time, and non-nutritive sucking by calves has physiological effects likely contributing to satiety (reviewed in Chapter 2). We have even seen that pet animals may perform repetitive behaviours to obtain attention or treats from their owners (see Chapter 10).

Furthermore, there are numerous verbal reports that stereotypic behaviours can be satisfying, calming and/or reinforcing for humans (reviewed by Mason and Latham, 2004); numerous papers showing that the performance of certain natural behaviour patterns is inherently reinforcing for animals (e.g. Mason *et al.*, 2001); and numerous accounts, again from humans, of repetitive activities like chanting, exercise and dancing being stress-relieving (e.g. reviewed by Mason and Latham, 2004). Thus it could well be that some specific natural behaviour patterns are self-reinforcing ('do-it-yourself enrichments': Mason and Latham, 2004); that repetition *per se* has emergent benefits ('mantra effects': Mason and Latham, 2004); or that stereotypic behaviour has some other positive effects; see Chapter 2 for further possibilities). It could even be that such effects have yet further benefits still, through giving animals more control over their state (*cf.* the classic papers by Weiss *et al.*, cited in Box 1.3; plus work reviewed in Chapter 9 and by Berkson, 1996).

This general idea – that some stereotypic behaviour helps animals to cope – is a long-standing one, and although certain hypothesized mechanisms for it have now been discredited (see Box 1.3), it clearly remains a live issue today. Nevertheless, more evidence is still needed to ascertain its true role in stereotypic behaviour: so far, there is not a single case in which we know for sure, both that beneficial consequences arise from the stereotypic behaviour (rather than merely correlating with it), *and* that this causes repetition via reinforcement. This is a fascinating area for future research.

11.2.1.3. Behavioural disinhibition

The third and final main explanation for repetition is behavioural disinhibition, as manifest for instance in 'perseveration': the generalized tendency to inappropriately repeat recently performed or otherwise prepotent behaviours (e.g. as reviewed in Chapter 5). Such tendencies vary naturally between normal individuals (e.g. see Chapter 5), change with age (e.g. Hauser, 1999; Ridderinkhof *et al.*, 2002, Tapp *et al.*, 2003), and also increase with acute stress (e.g. reviewed Mason and Latham, 2004) or, more profoundly, with early social deprivation (see Chapter 6). So far, in every case investigated to date, captive animals with high levels of stereotypic behaviour have proved more generally perseverative, e.g. taking longer to extinguish a learned response that is made unrewarding. Examples from stereotypic rodents, birds and bears were reviewed by Garner in Chapter 5, and more recently this has also been shown in rhesus monkeys with self-injurious biting (Lutz *et al.*, 2004), and in stereotypic horses (Hemmings *et al.*, 2006). Strictly speaking, of course, such correlational findings do not show that altered behavioural control *causes* stereotypic behaviour (although this seems most parsimonious; and circumstantial evidence comes from mice, whose extinction scores predict how stereotypic they are several weeks later: Latham, 2005). Furthermore, questions also remain as to whether such effects hold for all stereotypic behaviour; whether this necessarily indicates CNS malfunction (see Section 11.3.2); and the precise mechanisms involved. Overall, however, perseveration and its correlates appear important in stereotypic behaviour, and look a fruitful topic for further research. I discuss the different potential mechanisms below, in Section 11.2.4.

11.2.2. What behavioural processes account for the predictable, unvarying nature of some stereotypic behaviours?

Some apparently functionless behaviours are repetitive (e.g. a dog chasing and rechasing a ball), and yet not so 'unvarying' that anyone would term them stereotypic. In contrast, the most extreme examples of stereotypic behaviour are extraordinarily predictable from one performance to the next. So what could account for this sort of predictability? Overall, there seem to be three main reasons, again not mutually exclusive. The first is the repeated production of similar actions for reasons endogenous to the animal; the second, an unvarying environment that simply does not require behaviours to be modulated; and the third, a developmental shift in the way behaviour is controlled, low variability emerging through 'routine-formation' (e.g. procedural learning).

11.2.2.1. The reiteration of particular actions

Some action patterns are always produced in a similar way, as if instructions for them are coded in the central nervous system. Examples include certain grooming movements in infant rodents, and the patterned stepping evident even *in utero* in many mammals (reviewed by Berridge, 1994); some bird calls (e.g. the 'cuck-oo' of the cuckoo); and many consummatory behaviours (see e.g. Box 1.1). Such 'Fixed Action Patterns' were the focus of many ethological texts of the 1950s–1970s; and are still studied as part of some current work on birdsong, invertebrate escape responses, central pattern generators, etc. (see Chapter 1). These, if elicited in a continuous sequence, would obviously generate very predictable stereotypic behaviours. However, individual movements – learnt ones, not just innate – can also be reiterated similarly like this, if the brain regions responsible for action selection call up the same movement repeatedly. This occurs in forms of motor perseveration (as discussed further in Section 11.2.4). Thus overall, in this first type of process, predictability is an inherent by-product of the mechanistic causes of the behaviour's repetition.

11.2.2.2. Environmental predictability

The second cause of predictability is, in contrast, exogenous to the animal: an unvarying environment. After all, as Clubb and Vickery comment in Chapter 3, 'there are only so many ways an animal can walk around in a small square cage'. That stereotypic behaviour might be predictable simply because there is no great reason to vary it, was first suggested decades ago by zoo biologists like Hediger and Morris (reviewed in Mason, 1993). More recently, this also formed the core of Lawrence and Terlouw's 'channelling' hypothesis (Lawrence and Terlouw, 1993; see Chapters 2 and 4): focusing on the oral behaviours of pigs, they suggested that 'strongly motivated behaviour is highly modified or *channelled* by the environment into the few simple behavioural elements allowed by the available incentives', and 'thus the behavioural variability ... of foraging sequences should reflect the variability of the foraging environment'. This hypothesis is supported with anecdotal evidence, both from pigs in environments differing in complexity (see Chapter 2) and also from carnivores. Thus the rhythmic pacing of caged mink or bears can appear extremely unvarying, but if stimuli in and beyond the cage change, so too do these animals' behaviours – quickly shifting in location to follow the sounds of a passing feeding-machine, or instantaneously changing in form so as not to collide with a moving cage-mate or a new obstacle placed in the animal's way (Mason, 1993; Vickery, 2003). Predictability of form could therefore simply reflect the lack of environmental change or external stimuli needing responding to; and thus be quite independent of the causes of the behaviour's repetition. This intuitive and sensible idea has not, however, yet been formally and quantitatively investigated.

11.2.2.3. Routine-formation

The third likely process to explain predictability involves stereotypic behaviour decreasing in variability over time due to progressive changes in the behaviour's control. This idea crops up repeatedly in the literature (e.g. reviewed by Mason, 1991a,b; Mason and Latham, 2004; and Chapter 4), with possible examples including the pre-feeding locomotion of mink (Mason, 1993) and escape movements of early weaned mice (see Chapter 4). Such changes may be due to the normal processes, perhaps arising through repetition, underlying phenomena like the 'crystallization' of song in maturing young birds or development of 'routines' (e.g. the well-used pathways of rodents, or habits and motor skills of humans; see Section 11.2.5). However, although the idea is oft-repeated in the literature on stereotypic behaviour, so far it is little backed with good quality data. Only a few studies on pigs (Cronin, 1985), bears (Mason and Vickery, 2004) and mink (Mason, 1993) show statistical changes in predictability with age or repetition, and, as far as I know, only the latter reports data that are both quantitative and longitudinal, i.e. following individuals over time. In addition, there has been no serious investigation of possible mechanisms (something I return to in 11.2.5); little research into whether observed increases in predictability parallel the many other changes said to occur during 'establishment' (as reviewed in Chapters 3 and 4); and most importantly, no work into whether effects are really caused by the repetition of the behaviour, or instead are mere correlates of it (with both repetition and predictability being, say, products of increased time in captivity and/or increased age at assessment, *cf.* Chapter 7). Here, like the previous potential explanation, predictability is again a secondary property of the stereotypic behaviour, quite unrelated to its primary cause of repetition (although it may then increase bout number; see

Section 11.2.5): an issue that will be important when we discuss different types of perseveration (Section 11.2.4).

11.2.3. What determines the form of a stereotypic behaviour?

This book and its website depict repetitive chewing, tail-chasing, pacing, self-biting, swimming, tongue-rolling, fur-plucking and many other stereotypic behaviours that differ, not just in repetition and rigidity, but also in their basic unit of repetition. This aspect of form is often related to taxonomic group (see Figure 1.3), and also to timing (with, across a whole range of species, pre-feeding stereotypic behaviour typically being locomotory, but post-feeding, oral; see Chapters 2 and 3). So what determines the action that is repeated, i.e. a stereotypic behaviour's 'source behaviour' (*cf.* Mason, 1991b)? Once again I present three broad explanations, although my third category is not a very tidy one.

11.2.3.1. Surrogates for natural activities

The first group of source behaviours comprises activities like vacuum or redirected movements (see Box 1.1) which resemble a specific natural behaviour pattern (albeit one constrained by captivity). For example, finches deprived of nesting material may stereotypically pluck and carry their own feathers (see work by Hinde, reviewed by Mason, 1993). Similar likely examples in this volume include, once again, the foraging-like oral movements of ungulates, induced by unnatural dietary regimes (Chapter 2), and the stereotypic digging of gerbils deprived of a suitably tunnel-like den (Chapter 4); as well as the digit-sucking, self-clasping and body-rocking of young primates denied normal maternal contact (Chapter 6; Berkson, 1996), plus the object-sucking often seen in other newly weaned young mammals (see Box 6.2). Thus here, the cause of repetition can be inferred from the behaviour's very form: animals repeat X' because deprived of X, where X is a natural behaviour pattern and X' its surrogate. Some cases may simply represent stimulus-induced responses, performed as normally as they can be in the constraints of captivity; while others may well be accompanied by motivational frustration (which perhaps is then partially alleviated if X' has motivational consequences that help redress states caused by the lack of X; see Section 11.2.1).

11.2.3.2. Escape attempts

The second broad group of source behaviours consists of escape attempts. One of the earliest accounts, from Meyer-Holzapfel on the pacing of a dingo separated from its pack, was illustrated in Chapter 3 (Box 3.1). Later came experimental work from Duncan and Wood-Gush in the early 1970s, showing how food-frustration led hens to pace against the doors of their cages (reviewed in Mason, 1993), and Chapter 4 highlights more recent, elegant experimental work on the bar-mouthing of laboratory mice. Further likely examples from the preceding chapters include the pacing of some young mammals when separated from their mothers (see Novak *et al.*, Chapter 6, plus Box 6.2). Clubb and Vickery (Chapter 3) even suggest that escape attempts underlie all the pacing typical of captive carnivores – since this behaviour is often directed at enclosure boundaries, and is increased by a multitude of factors that make the enclosed area aversive or regions outside it attractive. Thus here, animals deprived of X do not try and replace it with X', but instead attempt to remove themselves from the frustrating situation.

11.2.3.3. And the rest . . .

Finally, we have a third group whose origins are more problematic because they are neither obvious surrogates for particular thwarted natural activities, nor attempts to escape. Here, either the deprivation of X leads to A, B and C (to pursue the notation above), or X' is repeated *despite* no deprivation of X. For example, frustrated ranging may underlie the pacing of a polar bear, perhaps by enhancing motivations to escape (as reviewed in Chapter 3), but it cannot explain the repetitive 'huffing' noises these animals sometimes make as they pace (e.g. Box 10.4). The deprivation of maternal contact may motivate compensatory digit-sucking, self-clasping and body-rocking, but how it leads to eye-poking, or placing a hind-leg behind the head, is much harder to explain (see Chapter 6). Furthermore, animals that groom themselves or conspecifics to excess (see Chapters 4–5), have clearly *not* been deprived of the chance to groom in a more naturalistic manner, any more than humans who tooth-grind at night (e.g. Pingitore *et al.*, 1991) have been deprived of normal chewing. Thus in this last group, the source behaviour seems to reveal little about the behavioural or environmental deficit responsible for repetition.

We could perhaps call this third group of puzzling actions 'displacement activities', to reflect their apparent irrelevance (*cf.* Box 1, Chapter 1), but this would merely be a label, not an explanation. So what *could* explain them? One idea raised in Chapter 8 by Cabib, and by Spruijt and van den Bos in accompanying Box 8.1, is that some stereotypic responses are exaggerated appetitive behaviours resulting from excessive responsiveness to any and all cues predicting reward (a suggestion I develop further in Section 11.2.4). Another hypothesis from Spruijt and van den Bos is that intrinsically rewarding activities are performed repeatedly when animals are chronically stressed, almost as a means of self-comfort (Box 8.1). This idea somewhat resembles Swaisgood and Shepherdson's broader proposal (albeit referring to enrichment-use) that for animals in very barren environments, just doing 'something, anything', may be better than nothing (see Chapter 9). Mason and Latham (2004) in turn hypothesized that the *repetition* of simple actions could be rewarding via 'mantra effects', the form of action again being quite arbitrary. In other instances still, arbitrary actions might be reinforced by outside events, conditioned by the adventitious arrival of food (see e.g. Mackintosh, 1975 and Timberlake and Lucas, 1989 on 'superstitious responses' in pigeons) or by a distressed owner's attempts to control a pet's behaviour (see above and Chapter 10). Some actions may even self-stimulate acupuncture sites (see Chapter 6). Finally, some forms may instead best be explained through the mechanics of CNS dysfunction: just as different amphetamine effects reflect different sites of action (e.g. Box 7.3), and the distinctive choreas of Huntingdon's disease arise from quite particular basal ganglial pathologies, so too may some captive animals' stereotypic behaviours simply be by-products of specific malfunctions in particular circuits.

11.2.4. Where repetition correlates with perseveration, what different processes could be involved?

As Chapters 5 and 7 discuss, the initiation, termination and sequencing of behaviour patterns depends on loops within the forebrain, which run from the cortex and back again through the basal ganglia. These allow cortical information to be processed by the basal ganglia before being relayed to further cortical areas important in producing behaviour. The basal ganglia have thus been said to 'translate intention into action' (e.g. Graybiel, 1998, quoting James Parkinson). These loops, sometimes given different names by different authors, include the motor/skeletomotor/ sensorimotor loop; the prefrontal/cognitive loop; the limbic/motive loop; and the oculomotor loop (e.g. Rolls, 1999; Haber, 2003; Columbia University Medical Center, 2005; see also Fig. 7.2 and Box 7.2). As Chapters 5 and 7 describe, all involve an indirect pathway which is inhibitory, plus a direct pathway which is excitatory; thus inhibition of the indirect pathway or stimulation of the direct pathway both activate movement, although this can occur in a variety of ways and, depending on the exact mechanism and loop involved, have a variety of behavioural effects.

Although usually functioning in parallel, these loops have somewhat dissociable functions and effects (see Chapter 5). However, importantly, they are not closed nor completely independent. They influence each other (e.g. Kalivas and Nakamura, 1999; Rolls, 1999; Haber *et al.*, 2000; Haber, 2003), something little touched on in this volume (though see Chapter 8). They are also, as we have seen, influenced by other pathways, especially midbrain inputs (e.g. the nigrostriatal pathway to the motor loop: Chapters 5 and 7; and the mesoaccumbens pathway to the limbic loop: Chapter 8), with cortical dopamine also modulating the descending projections from the cortex: important during sensitization to stereotypy-inducing drugs, and in stress-induced behavioural disinhibition (e.g. Karler *et al.*, 1998, McFarland *et al.*, 2004). Furthermore, each loop can be functionally and anatomically subdivided (e.g. Rolls, 1999, Chudasama *et al.*, 2003). Thus despite the dichotomy proposed in Chapters 5 and 10, there are *several* forebrain loops, and furthermore, they are neither discrete nor indivisible (e.g. Chudasama *et al.*, 2003): their complexities and distributed functions are a topic of much ongoing research, and the mapping of functions and dysfunctions onto anatomy is a work in progress. These caveats aside, however, the altered functioning of these different loops often seems to affect the properties of behaviour in different ways (e.g. yielding different types of perseveration), and they may even ultimately underlie different classes of stereotypic behaviour – a hypothesis first raised for captive animals by Garner (see Chapter 5 and Box 10.2). Below, I therefore look at these possible effects in more detail, starting with the loops discussed by Chapter 5, 7 and 10, before presenting the brain regions focused on by Chapter 8 in a similarly 'systems level' way.

11.2.4.1. The motor loop and stereotypic behaviour

The motor loop arises from various parts of the sensory and motor cortex, enters the basal ganglia at the putamen of the dorsal striatum, and returns to regions of the cortex involved in motor/premotor control: see Chapters 5 and 7. As these chapters review, this system matches stimuli to suitable responses, functioning to 'call up' and thence generate appropriate behavioural actions. It is thus directly important for the selection of specific motor 'programs' and thence the control of skeletal musculature. When it malfunctions, it causes problems with initiating individual movements (e.g. Parkinson's disease) or with suppressing them (e.g. amphetamine stereotypy). As Chapter 5 reviews, dysfunctions in this pathway also cause particular forms of motor or response perseveration, in which individual actions are repeatedly performed.

How to assess the role of this system in captive animals' stereotypic behaviour? Perhaps a first screen is observation: any stereotypic behaviour whose movements vary from one repetition to the next cannot be related to this type of perseveration (though similarity from one repetition to the next is, in contrast, insufficient evidence on its own, since as Section 11.2.2 shows, such predictability could arise in other ways). The first experiment to address this question examined how stereotypers responded when learned responses were put into extinction (Garner and Mason, 2002). With hindsight, this was rather naïve, since motor perseverations are not the only forms to attenuate extinction (as we will see below), but such approaches could be made more relevant by observing if repeated responses made in extinction are always similar (e.g. always using the nose, left paw, etc.): necessary – if again not sufficient – to infer motor perseveration. A more elegant approach, however, is suggested by Garner (see Chapter 5): to use tests that probe animals' spontaneous tendencies to generate repeated responses by asking them to 'gamble' for rewards by, say, pressing one of several, arbitrary, operant levers for a randomly delivered treat. Such tests have shown that highly stereotypic blue tits, Amazon parrots and mice do indeed spontaneously generate more predictable sequences of responses (Garner *et al.*, 2003a,b; see also Chapter 5). More invasive tests of this hypothesis could probe animals' sensitivities to the stereotypy-inducing effects of amphetamine and similar, to see if levels of captivity-induced stereotypic behaviour predict levels of stimulant-induced stereotypy (*cf.* e.g. Box 7.3); use the techniques employed by Lewis *et al.* (see Chapter 7), focusing on the putamen; or look at gene expression in the putamen, as correlates with cocaine-induced stereotypies in monkeys (Saka *et al.*, 2004).

11.2.4.2. The prefrontal loop and stereotypic behaviour

Also discussed in this volume is the prefrontal loop (see Chapters 5 and 10) which arises from the parietal cortex and other regions, inputs the basal ganglia through the head of the caudate (again part of the dorsal striatum), and then projects to the lateral/doroslateral prefrontal cortex. Functionally, this has been described as a 'supervisory attentional system' key in planning, impulse control and other high-level, organizational aspects of behaviour. Thus as Chapters 5 and 10 review, damage or alteration to this loop can impair abilities to plan, such that for example complex behaviours are not sequenced appropriately or to completion. In tests of perseveration, such subjects also show a particular form called 'stuck-in-set', characterized by difficulties in changing the rules or 'attentional sets' used to guide behaviour (such as transferring a learned skill to a new situation, or altering what is attended to if the type of stimuli that need to be monitored are changed, e.g. Wallis *et al.*, 2001). This system has been implicated in some obsessive–compulsive disorders in humans (e.g. Harris and Dinn, 2003), but what role does it play in captive animals' stereotypic behaviour? So far we have just one, though very neat, piece of evidence: using tests for stuck-in-set perseveration, mice which excessively overgroom and pluck the fur of other animals were shown to have greater difficulties in these tasks than control animals (see Chapter 5).

11.2.4.3. The limbic loop and stereotypic behaviour

The limbic loop is Cabib's focus in her discussion of stress–sensitization (Chapter 8). It arises in the temporal lobes, anterior cingulate cortex, hippocampal formation and orbitofrontal part of the prefrontal cortex, loops into the ventral striatum (e.g. the nucleus accumbens), and returns to input on the anterior cingulate and the medial/orbitofrontal parts of the pre-frontal cortex. As Chapter 8 mentions, it regulates motivational aspects of behavioural control, such as responding to cues learnt to predict reward, putting effort (e.g. lever pressing) into obtaining reward, and responding to novelty (see references cited in Box 8.1; plus Kalivas and Nakamura, 1999; Rolls, 1999; Robinson and Berridge, 2003; Salamone *et al.*, 2005). Thus it is important in appetitive behaviour, changes in its functioning particularly affecting the motivational control of these activities. For example, animals become persistent in extinction tests (e.g.

Reading et al., 1991, reviewed by Rolls, 1999), but we also see more specific changes too. For instance, amphetamine injected into the nucleus accumbens of rats quadruples the lever-pressing they show when a cue predicting sugar is presented. It does this despite not increasing their baseline lever pressing for reward, nor the apparent pleasure they get from sugar, and thus seems to enhance the effect that motivationally relevant cues have on behaviour (reviewed by Rolls, 1999; Wyvell and Berridge, 2000). Nucleus accumbens lesions also lead to impulsive choices by rats, in which they cannot resist the lure of a small but immediate reward, even in preference to a large but delayed one (Cardinal et al., 2001). Likewise, marmosets with lesions to the orbitofrontal cortex become impaired in their ability to perform a 'detour task' where they have to reach around a transparent partition to gain food. Instead, despite the barrier, they try to grab directly at the treat (Wallis et al., 2001). This type of response, where motivationally important cues elicit relevant, yet impulsive and unsuccessful, behaviour has been termed 'affective persev-eration' (e.g. Hauser, 1999). This is quite distinct from the stuck-in-set behaviour emphasized in the previous section; thus orbitofrontally lesioned marmosets also become impaired in 'reversal learning' (e.g. A is paired with a treat and B not; but the situation is then reversed so that A should now be ignored), without being hampered in their abilities to shift attention to new tasks or stimulus-types (Clarke et al., 2005; see also Dias et al., 1996 and Hauser, 1999).

In terms of stereotypic behaviour, this system (especially the nucleus accumbens) underlies the locomotor responses shown to stimulant drugs (e.g. reviewed in Box 7.3; and by Rolls, 1999), while the orbitofrontal cortex is also implicated in some forms of human obsessive compulsive disorder (OCD) (e.g. Harris and Dinn, 2003; Szechtman and Woody, 2004; Mataix-Cols et al., 2004). Fuchs et al., (2004) also implicate it in internally driven compulsions, such as compulsive drug-taking (with Chapter 8 giving further references), as well as in impulsive responding to external cues. So is this loop involved in captive animals' stereotypic behaviours? As we have seen, Chapter 8 presents a case for its role in the stereotypic cage-climbing of stressed, food-restricted DBA mice, behaviour that Cabib argues is an exaggerated response to the cues offered by the (now empty) food hopper. Furthermore, individual frequencies of oral stereotypies (specifically chain-chewing) in tethered pigs positively correlate with the degree of locomotion they show if treated with amphetamine (Ter-louw et al., 1992), just as predicted if the former are accumbens-mediated. Most recently, crib-biting horses have also been found to have around double the accumbens D1 and D2 receptor densities of non-stereotypic controls (McBride and Hemmings, 2005). But how to investigate this possibility further, and in other ways? Aside from drug responses, or the types of detailed neurophysiological measures of Chapters 7 and 8, non-invasive behavioural tests could investigate whether stereotypic in-dividuals show exaggerated responses to multiple different types of pre-dicted rewards (cued food, cued mating opportunities, cued enrichment delivery, and so on); impulsivity when faced with reward cues, even when delays or detours would be more beneficial; and difficulties in suppressing learnt responses when previously rewarded cues are still present (as in reversal learning; or in extinction tests in which a 'reward light' is left on). This is an exciting, so far unexplored, area for future research.

11.2.4.4. The oculomotor loop and stereotypic behaviour

This final loop controls the eye movements used for looking in different directions, e.g. following moving targets ('saccades'). It is thus unlikely to be involved in most stereotypic behaviours, and was not mentioned by the previous authors. However, its damage or dysfunction can cause oculomotor perseverations. Schizophrenics, for instance, show poor abilities to suppress certain unwanted or unnecessary eye movements under test (e.g. Muller *et al.*, 1999; Barton *et al.*, 2005). So could severe alteration in this loop lead to oculomotor stereotypic behaviour? Distinctive repetitive eye-rolling has been observed in veal calves (see Broom and Leaver, 1978; Fraser and Broom, 1990), and it could be revealing to screen them in the type of tests used to investigate abnormal saccades in humans. In the future, it might also be worth looking more closely at smaller captive animals (e.g. rodents), so as not to overlook any abnormal eye movements that they might be displaying.

11.2.5. Where predictability increases with repetition and/or length of time in captivity, what mechanisms are involved?

Although several earlier chapters alluded to 'establishment' – a change in the nature of stereotypic behaviour with time or repetition – none discussed how this might come about. Here, I therefore review some relevant processes and suggest how they could be investigated. With repetition, normal behaviour patterns can shift into routines with forms of automatic processing (Mason and Turner, 1993; Toates, 2001) loosely known as 'central control' (Fentress, 1976; Martiniuk, 1976), e.g. procedural learning (e.g. Graybiel, 1998; Jog *et al.*, 1999; du Lac, 1999; Marsh *et al.*, 2005). These changes enable individuals to execute regularly performed or fast movements with minimal cognitive processing or need for sensory feedback (Fentress, 1973, 1976), speed touch-typers providing one good example.

Several processes are implicated in these changes. For example, at a fine motor level, individual actions become 'ballistic' or 'open loop', i.e. executed without the need for feedback (e.g. proprioception). The brain simply generates fixed motor instructions, which are then executed without sensory guidance (e.g. as in the normal pecking of pigeons, during which the eyes are reflexly closed; Wohlschlager *et al.*, 1993), and even, in studies of humans, despite instructions to cease moving (e.g. Salt-house, 1985). Typists, for example, asked to stop cannot do so in the middle of typing the short and common word 'the' (Kerzel and Prinz, 2003). A second process (or, more likely, processes: see e.g. Marsh *et al.*, 2005) allows longer, more complex behaviours to become 'automated'. With repetition (e.g. practice, in the case of human skills), behaviour patterns come to need less cognitive monitoring. As a result, the individual can perform the sequence faster, and do other activities simultaneously (e.g. play the piano while talking: Mechner, 1995; or, in mice, groom while monitoring a novel environment: Fentress, 1976). This seems to be because each component of a sequence becomes dependent on cues from the preceding component, instead of on external cues (Mechner, 1995): thus each action simply triggers the next, a process sometimes termed 'chunking' (e.g. Graybiel, 1998). The term 'habit' as used by experimental psychologists applies to similar changes that also give recurring behaviours a rigid quality; actions performed repeatedly to gain a reward become less and less modifiable by changes in the quality of that reward (see e.g. Dickinson, 1985[N1]; Killcross and Coutureau, 2003). Brain regions important in these motor changes with repetition are the cerebellum and striatum (e.g. Graybiel, 1998; Jog *et al.*, 1999; de Luc, 1999; Passingham, 1996).

So could motor or procedural learning be involved in stereotypic behaviour? 'Central control' has long been invoked as a cause of 'establishment', but despite its plausibility, and growing understanding of how repeated actions become skills or habits, this hypothesis really has not been systematically tested. Evidence for this idea would include the following changes to the behaviour pattern with repetition (in addition to increasing predictability): increased speeds of performance; improved abilities to attend to external events without ceasing to perform the behaviour; and some odd 'side effects'. First, if interrupted in the middle, a routine-like sequence may need restarting from the beginning. This can be observed in some musicians (Mechner, 1995); in greylag geese egg-rolling with their bills (Tinbergen, 1951); and in rats in a choice maze which, if disturbed half-way down a run, may return to the 'start box' before repeating their choice (Lashley, 1921). Second, an action with similar characteristics to a component of a sequence may trigger the rest of the sequence, even if quite inappropriate; as Norman (1981) put it, 'pass too near a well-formed habit and it will capture your behaviour' (see also Mechner, 1995; and note the difference from motor perseveration, where similar actions should inhibit each other – see Chapter 5). Third, with decreased attention paid to the control of the behaviour itself, one might see what has been observed in a pacing hunting dog (Fentress, 1976), and in rodents running along habitual trails (Fentress, 1976; Berdoy, 2003; Latham and Mason, 2004): collisions with obstacles newly placed in the way. Note too that although that some individuals may have pre-existing general tendencies to routine-formation (e.g. Benus *et al.*, 1987, 1990) – once again, probably detectable via extinction tests – the developmental changes in a stereotypic behaviour should be *specific* to that sequence: *not* simultaneously evident across a suite of behaviours (which would instead indicate more generalized changes, e.g. disinhibition, occurring with age, stress, and/or time in captivity; *cf.* Chapters 5, 7 and 8; and Section 11.2.4).

11.2.6. So overall, why do captive animals perform stereotypic behaviour?

We can see from the preceding chapters and the synthesis above that overall, captive animals perform stereotypic behaviour for the following, non-mutually exclusive, reasons:

1. Internal states induced by the captive environment, and/or cues external to the animal, persistently trigger or motivate a specific behavioural response;

   and/or

2. The environment creates a state of sustained stress which affects how the cortical-basal ganglia loops elicit and sequence behaviour, resulting in abnormal behavioural disinhibition;

   and/or

3. A past, early rearing environment has affected CNS development, again resulting in abnormal behavioural sequencing, with effects evident long past infancy.

Speculatively, future work might even reveal other effects of captivity too, such as diets or stress levels which exacerbate how aging impairs the brain (see e.g. Vallée *et al.*, 1999; Milgram *et al.*, 2004), or increased risks of the streptococcal infections that can trigger anti-basal ganglia auto-immune disease (e.g. Edwards *et al.*, 2004, Snider and Swedo, 2004) – just two possible further topics for future research.

Processes (1)–(3) above explain how captive environments induce behavioural repetition, but do they fully account for the sustained occurrence of stereotypic activities day after day, week after week, year after year? Perhaps they do (we do not know), but in some cases this might, as we have seen, be further promoted by endogenous effects such as reinforcing consequences from the behaviours. Note too that in some cases, the motivational or behavioural control effects of captivity would additionally cause, not just repetition, but also, inherently similar behaviours to be reiterated with little variability. Alternatively, in other cases, as we have seen, such predictability may be superficial, either being a mere artefact of the predictability of the environment or instead just emerging naturally through repetition.

This overview should help us as we now turn to the ethical and practical implications of stereotypic behaviour in captive animals.

11.3. Implications for Animal Welfare and Animal Normalcy

11.3.1. What does stereotypic behaviour say about animal welfare?

Throughout this book (especially Chapters 2–8), the theme has been that stereotypic behaviours emerge when a large discrepancy exists between the conditions offered by captivity and an animal's preferred and/or naturalistic state. The brief of the authors was to discuss *how* these have their effects, not to consider animal welfare *per se*. However, looking at how such discrepancies act does give fundamental insights into stereotypic behaviours' likely links with welfare (a term I use to refer to an animal's subjective affective or emotional state). I therefore start with this framework, before briefly reviewing the empirical data linking welfare and stereotypic behaviour.

When they are the product of thwarted motivations to perform species-typical behaviours or to escape, then stereotypic behaviours are very likely to reflect aversive mental states, since preventing highly motivated behaviours often causes stress (see any animal welfare text). Importantly, as Chapter 2 emphasizes, this may be true even when the behaviours elicited do not look predictable or stereotyped (*cf.* the stone-chewing of hungry pigs housed outdoors). In some instances, however, such frustration-induced stress may be somewhat rectified, if the behaviour itself reduces the underlying motivation. Such effects could well complicate links between stereotypic behaviours and welfare (e.g. Mason and Latham, 2004). Furthermore, if they become routine-like and triggered by a growing range of cues, stereotypic behaviours might track underlying motivational states less closely. Again, this would blur the correspondence between the degree of stereotypic behaviour performed and an individual animal's stress levels (*cf.* Dantzer, 1986; Mason and Latham, 2004).

If a stereotypic behaviour is instead the product of ongoing unavoidable stress which induces changes in the CNS, then this too is also likely to be accompanied by poor welfare. Furthermore, when the limbic loop is affected by such changes (as suggested in Chapter 8; see also 11.2.4), then the resulting stereotypic behaviours could perhaps correlate with the types of strong, persistent feeling of 'something being wrong' linked to limbic loop dysfunction in some human OCDs (Robinson and Berridge, 2003, Szechtman and Woody, 2004; Maltby *et al.*, 2005).

Finally, if early rearing environments have caused lasting CNS changes, then this has more complex implications for welfare. Such early environments may well have caused very poor welfare at the time that they impinged on normal CNS development (see e.g. Chapter 6). Furthermore, like the limbic loop effect suggested above, the malfunctions induced might be accompanied by lasting fearfulness into adulthood (see e.g. Chapter 6; plus review by Mason and Latham, 2004), such that stereotypic behaviours correlate with poor welfare long after the original insult to development. However, it is also possible that lasting effects of early experience are mere behavioural 'scars', products of past stress but no longer reflecting poor welfare (*cf.* e.g. Dantzer, 1986; Mason, 1991b; Mason and Latham, 2004). It could even be that certain early environments impair brain development without ever being stressful or aversive (just as, say, hypoxia is not aversive to many animals, despite being very detrimental to functioning). In this instance, an animal may be rendered dysfunctional and stereotypic without ever having experienced poor welfare.

Small wonder, then, that although stereotypic behaviour is generally held to indicate poor welfare, empirically its links with other welfare measures (e.g. HPA functioning, reproductive success, etc.) are not always clear-cut. For instance, it has long been known that some aversive environments – e.g. very cold ones, or ones involving unpredictable electric shock – do not induce stereotypic behaviour, instead eliciting huddling or crouching (e.g. reviewed Mason, 1991a, 1993; Mason and Latham, 2004). In this book, we have also seen that wild-caught or enriched-reared animals placed in barren environments as adults typically show far less stereotypic behaviour than do animals raised in such conditions all their lives (see Chapter 7) – despite presumably finding these conditions far more frustrating. Furthermore, sometimes animals given enrichments show *enhanced* stereotypic behaviour (reviewed Mason and Latham, 2004), perhaps because habit-like forms (see Section 11.2.5) are 'slipped into' when similar actions are performed, even when those actions involve enrichment-use.

However, despite these counter-examples, typically environments or treatments leading to stereotypic behaviour *are* empirically linked with greater signs of poor welfare (see e.g. Figure 2a, Chapter 1). Indeed many known or likely examples of this have been given in this volume. These include exposing pigs to chronic calorie-restriction (stereotypic behaviours increasing with the degree of dietary restriction; Chapter 2), and food-restricting mice (Chapter 8); plus depriving ungulates of fibre (Chapter 2). Indeed in ungulates, stereotypic behaviours could well reflect discomfort or even pain from gastro-intestinal acidosis (see Chapter 2). Further cases where poor welfare may be directly implicated in the emergence of stereotypic behaviours include exposing animals to inescapable aggression (Chapters 3 and 4); the pacing of wild canids faced with noisy human crowds (see Figure 5, Chapter 3) and of other carnivores prevented from ranging (see Chapter 3); the responses of maternally-deprived infants in a range of species, often greatest the younger they are when deprived (Chapter 6); the self-injurious behaviours of laboratory primates, as predicted by the number of previous 'blood-draws' and other aversive events (Chapter 6); and perhaps the effects of housing gerbils without a naturalistic den (Chapter 4). If all these stereotypic behaviours do indeed represent aversive states like stress, fear and pain, then this is of enormous welfare concern, especially considering the many millions of individuals affected worldwide (see Chapters 1–2).

Within such environments, however, one often finds some individuals with very high levels of stereotypic behaviour, and others with little or none. Do these differences reflect differential adaptation to the captive environment? Should we selectively breed from such non-stereotyping individuals, in order to reduce welfare problems? It would seem not, since paradoxically, individuals that spontaneously develop high levels of stereotypic behaviour often seem to fare *better* than identically treated but non-stereotypic conspecifics (see Figure 1.2b). Two recent examples illustrate this with reproductive success: farmed mink with spontaneously high levels of pacing and similar have the greatest litter sizes and lowest infant mortality rates (Jeppesen *et al.*, 2004; see also Mason and Latham, 2004), as do caged African striped mice with high levels of jumping and looping (van Lierop, 2005). If stereotypic behaviour has beneficial psychological consequences (see Section 11.2.2), or physical ones such as improved gut health (Chapter 2) or better physical fitness (*cf.* e.g. Jeppesen *et al.*, 2004), then such effects are perhaps not surprising.

Unfortunately, however, there is an alternative explanation for such findings: that rather than stereotypic behaviour having benefits, *not* being stereotypic indicates an even more detrimental state. For instance, Novak and colleagues, and Cabib, suggest that depression-like states can be an alternative to stereotypic behaviour: the former describe primate infants newly separated from their mothers as protesting and pacing for a day or two, before lapsing into 'despair' (Chapter 6), while Cabib presents a more neurobiological account of how sustained, uncontrollable stress can, in some mouse strains, reduce tendencies to stereotype and instead promote 'learned helplessness' (Chapter 8). This also fits with some accounts of intensively farmed sows: Cronin (1985) found that the less-stereotypic females were less responsive to novel stimuli, and he judged them as 'less normal' than stereotypers (Cronin, 1985; though *cf.* Schouten and Wiepkema, 1991); these sows also typically proved to be the individuals which had 'protested' most when first tethered (Schouten and Wiepkema, 1991). Other illnesses or forms of physical capacity might also reduce stereotypic behaviour too: hypo-kinetic diseases like Parkinsonism (*cf.* some drug-induced forms; see Chapter 10), for instance, would likely reduce active, whole-body stereotypic behaviours, as would conditions like arthritis or muscle fatigue.

Together with the findings of Lewis and colleagues that *barren-housed* non-stereotypic deer mice have brains quite different from *enriched-housed* non-stereotypers (see Chapter 7), this indicates diverging reasons for a lack of stereotypic behaviour: some positive (when enrichment-related), but some decidedly negative for welfare. This is an important topic to investigate further. It also suggests that Table 1.1 might better reflect the true extent of global animal welfare problems by emphasizing the total numbers of individuals housed in stereotypic behaviour-inducing conditions, not just the proportion developing this behaviour. Such findings also have two practical implications, in suggesting that selecting against stereotypic behaviour, or physically preventing animals from performing it (*cf.* Box 2.3, and Table 10.3), are both likely to be counter-productive for animal welfare. If we want to reduce stereotypic behaviour and improve welfare simultaneously, improving captive conditions will usually be the key (Mason *et al.*, 2006).

11.3.2. Does all stereotypic behaviour indicate pathology?

Chapters 5–8 and 10 present stereotypic behaviour as at least sometimes the product of CNS pathology. Could it *always* indicate this type of malfunction? This issue could help us understand the relationships between stereotypic behaviour and welfare, as discussed above, and could also help us better evaluate environmental enrichments for captive animals, since dysfunctional animals may well take a long time to respond (reviewed by Mason and Latham, 2004). It could also help improve the validity of some rodent research models (see Chapter 5), and perhaps even increase the reintroduction success of captive-bred animals in conservation projects (Vickery and Mason, 2003, 2005).

Broadly speaking the term 'pathology' has been used in two ways in this volume. Würbel (Box 1.4) recommends using it in normative way, comparing the feature in question with the equivalent in some well-defined control group (e.g. wild or free-living animals). Lewis and colleagues, in Chapter 7, arguably do this too, when citing the 'law of stereotypy' as used in some human medicine. However, Würbel's recommendation is not that we solely compare outward phenomema, but instead compare underlying mechanisms. This is to enable us to distinguish truly malfunctional changes from 'maladaptive' ones (sensu Mills, 2003) in which normal animals produce unusual responses when placed in unusual environments (*cf.* our fake-egg-preferring oyster catcher of Section 11.2.1). Nevertheless it is still not clear how this approach would distinguish adaptive long-term responses to challenging environments (*cf.* e.g. phenotypic plasticity) from non-adaptive ones. Novak and colleagues instead adopt a consequentialist approach, describing stereotypic behaviour as pathological if taking up excessive time or causing self-harm (Chapter 6). However, Würbel's concern with this is that something like fever might then be termed pathological. Overall, then, perhaps the best definition combines both approaches: thence stereotypic behaviour is pathological if caused by neurological, neurophysiological or behavioural differences from those of wild, free-living or very enriched-housed animals, and if such changes also have no functional value *in any context* or have demonstrably have harmful consequences. We might add that such changes should be hard to rectify (*cf.* examples in Chapter 6 and 7): a trait instantly reversible with enrichment is, intuitively, perhaps not one we would call pathological.

Stereotypic behaviours are clearly not always pathologies. Perhaps the best evidence comes from normal, healthy humans, who show stereotypic behaviours from babyhood right through to adulthood (e.g. Thelen, 1979; Rago and Case, 1978); indeed at least three of this book's authors (myself included) regularly finger-chew, knee-joggle and pace! Such activities are not malfunctional even where they correlate with perseveration (see e.g. Happaney and Zelazo, 2004 on the compulsive-like behaviours of normal human children; also Zohar *et al.*, 1995 as cited in Chapter 5), since some degree of persistence or perseveration is itself a perfectly adaptive feature of normal behaviour (e.g. Benus *et al.*, 1987, 1990; Hauser, 1999). Stereotypic behaviours have also been seen in free-living wild animals with no other signs of impaired function, e.g. stone-playing and wrist-biting in Japanese macaques (Grewal, 1981; Huffman, 1984), circling by hedgehogs (e.g. Boys-Smith, 1967), tongue-playing by wild giraffes (Veasey *et al.*, 1996), and transient pacing by polar bears (Ames, 1993). At the other extreme, in contrast, some stereotypic behaviours clearly *are* pathological, such as those of deprivation-reared monkeys, described in Chapter 6 as being caused by changes in brain functioning, accompanied by cognitive deficits, potentially resulting in self-injury, and very hard to reverse. A range of less clear-cut cases occurs between these extremes. The stereotypic behaviour of barren-reared deer mice, for example, reflects CNS development differing from that of enriched-reared animals, yet seems reversible, at least up until a certain age (Chapter 7); while the behaviourally activating effects of stress-induced sensitization (Chapter 8) could perhaps be adaptive responses to challenge, at least for free-living mice.

While we may not yet know the full picture in these – and many other – instances, it is worth questioning whether pathological changes underlie the stereotypic behaviours so evident across zoo, farm and laboratory animals. For instance, while sham-chewing, chain-mouthing farmed sows were considered by Chapter 2 primarily as normal animals seeking foraging outlets in a energy-deficit, physically restrictive world, it is perhaps telling that like Chapter 8's mice, their stereotypic behaviour does not emerge straight away, but instead appears gradually, after a few weeks of tethering (Cronin, 1985), with its correlation with amphetamine-responsiveness not evident until 3 months of this treatment (Terlouw *et al.*, 1992). Could the frustration of food restriction, the anxiogenic effects of acidosis (Hanstock *et al.*, 2004) and/or inescapable gastric discomfort, cause progressive CNS stress sensitization? Furthermore, these animals would have been removed far earlier from their mothers than would happen naturally, and we know that extremely early-weaning (at least) does affect piglets' dopaminergic (Fry *et al.*, 1981, Sharman *et al.*, 1982) and serotinergic (Sumner *et al.*, 2002) systems (see further discussion in Chapter 2 and Box 6.2). The circumstantial evidence is thus rather compelling, and fully assessing the degree to which captive animals have CNS dysfunction should thus be a major future research goal.

11.4. Redefining and Classifying Stereotypic Behaviours

I introduced this chapter by describing the standard definition of stereo-typies – *repetitive, unvarying with no apparent goal of function* – as bland, and potentially not very useful. Here, I argue that this definition pools too many diverse phenomena; puts the focus on traits which are either secondary to the key issues (how unvarying a stereotypic behaviour is) or so hard to measure as to be useless (whether or not there is a goal or a function); while omitting features typically implicit whenever people discuss these behaviour patterns. I therefore propose that we replace it with a new definition based on biological causal factors. I also suggest how we might classify and subdivide these behaviours in the future.

11.4.1. Redefining stereotypic behaviour

I have often been asked 'Is such-and-such a stereotypy?', with such-and-such variously being a dog chasing a ball, a toddler sucking its thumb, a piglet performing belly-nosing, a rodent wheel-running, and so forth. However, despite the formal definition given above and elsewhere, the people posing this question were *not* asking 'Is such-and-such repetitive, unvarying, and with no apparent goal of function?' After all, they could assess that perfectly well for themselves! Instead, they were really asking: 'Does this individual have a problem?' Veasey *et al.* (1996) were some of the few to make this explicit: in their paper on giraffe behaviour, they expressly did not class the tongue-playing of wild individuals as stereotypic because it was not linked with food restriction nor with enclosure. I suggest that, like these authors, we acknowledge the implicit baggage usually attached to the terms 'stereotypy' and 'stereotypic behaviour', rendering it explicit in a new definition: *stereotypic behaviour is repetitive behaviour induced by frustration, repeated attempts to cope, and/or CNS dysfunction.* In captive animals, these stem from a deficit in housing or husbandry, where a deficit means something that the animal would change if it could (e.g. a motivational deficit linked with frustration; a health deficit linked with nausea or pain; or a safety deficit causing fear), or that causes a pathological change. Where we simply do not know the biological cause (as, for example, is often the case for zoo animals – see Chapter 9), a better term may instead be the Abnormal Repetitive Behaviour (ARB) of Chapter 5.

This definition reflects how most people use the term in practice, turns the question 'Is such-and-such a stereotypic behaviour?' into an interesting and relevant one, and indeed reflects why researchers find such behaviours worthwhile topics for study. It means we can stop worrying about whether a foetus sucking its thumb *in utero* (see Mason, 1993) has a stereotypy, or whether a dog happily chasing a ball and a cat kneading a pillow while purring loudly are stereotypic: they are not. The behavioural mechanisms involved in these cases might help a full understanding of true stereotypic behaviour, but they are benign examples of rhythmic behaviour that we should simply not be side-tracked by.

Furthermore, this new definition now focuses on the behaviours' mechanistic causes rather than on their phenotypes. My proposed definition thus omits reference to how unvarying the behaviour is. If it is useful to exclude the kneading cat or thumb-sucking foetus, so too do I believe it useful *not* to exclude the hungry pig sporadically stone-chewing in a muddy pasture, or animals plucking out their own or others' pelage in a variable manner. Some (similar) behaviours could look more predictable than this merely because they are performed in a very stable environment, or because they have been rendered habit-like with repetition – and I therefore suggest that 'predictability' is simply too trivial to be a defining characteristic. I also omit 'lack of apparent goal or function' from the new definition. This is partly because this is hard to quantify (at what point does drinking become 'polydipsia'?), but also because if future research shows the tongue-rolling of cows to be an effective way of reducing foregut acidity, for example (*cf.* Chapter 2), I would like to term it 'stereotypic' nonetheless, since it is a response to an environmental deficit, plus closer to the arguably less functional oral stereotypic behaviour of horses than to anything from a cow's normal repertoire.

11.4.2. Categorizing stereotypic behaviours

From this framework, I now build on the suggestions of Garner, Mills and Luescher (Chapters 5 and 10), and of Latham and Würbel (Box 4.2) that stereotypic behaviours should be sub-categorized according to their underlying causes. Like them, I suggest that the causes of *repetition* are key, i.e. the primary distinguishing features of different stereotypic behaviours. Other mechanisms involved in stereotypic behaviours (the extent to which they are reinforced; the extent to which they have become habits over time; their origins in terms of source behaviours) then cut across this basic categorization, further providing useful ways to describe, explain and compare the features of individual cases.

Rather like the old division between 'cage stereotypies' and 'deprivation stereotypies' (*cf.* Mason, 1991a, 1993; Novak *et al.*, Chapter 6), but focusing on biological mechanism rather than environmental cause, I therefore suggest the following:

1. The term *'frustration-induced stereotypic behaviour'* should be used for repetitive behaviours driven directly by motivational frustration, fear or physical discomfort. These behaviours need not be the product of any underlying abnormality (they are maladaptive, not malfunctional). Their source behaviours reflect this deficit, deriving from rapidly

emerging attempts to find a surrogate for a missing normal behaviour, to escape from confinement, or to otherwise alleviate a problem; and they are abolished immediately by a specific change in husbandry that successfully rectifies this underlying deficit. Furthermore, such changes should reduce them in a 'dose-dependent' way – the most strongly preferred treatments being most effective in reducing the stereotypic behaviour.

2. The term *'malfunction-induced stereotypic behaviour'* should be used when early rearing or chronic stress impairs brain functioning. These behaviours are the product of pathology; co-occur with a suite of other effects (e.g. quantifiable changes in CNS physiology/anatomy; specific forms of perseveration, etc.); and may involve source behaviours that do not closely reflect the original cause of repetition. In a suboptimal environment, they emerge slowly or in a discontinuous manner; and are correspondingly slow or difficult to reverse, as well as being potentially reversible by any one of a ranges of factors that helped the underlying deficit, even if not the original cause of the problem.

Expanding and overriding the bimodal scheme proposed in Chapters 5 and 10, I also hypothesize that malfunction-induced stereotypic behaviours should be subdivided according to the brain system most affected – or perhaps more usefully, since the precise anatomical localization of dissociable executive processes is sometimes difficult or controversial (e.g. Chudasama *et al.*, 2003), according to the type of behavioural dysfunction implicated, as reviewed in 11.2.4. Thus I hypothesize that malfunction-induced stereotypic behaviours might be subdivided into:

- motor stereotypic behaviours (or what Chapters 5 and 10 propose to be 'true stereotypies': akin to psychostimulant-induced stereotypies; correlating with motor perseveration; and modulated by the motor loop);
- cognitive-loop-related stereotypic behaviours (akin to some human OCDs; and what Chapters 5 and 10 propose to correlate with stuck-in-set perseveration);
- affective or limbic stereotypic behaviours (akin to other human OCDs, and also to amphetamine-induced hyper-locomotion; should correlate with impulsivity/impaired reversal learning; perhaps most amenable to treatment with serotinergic agents [see e.g. Clarke *et al.*, 2005]);

and, in some special cases:

- oculomotor stereotypies, (caused by malfunction of the oculomotor loop).

Note that each of my two broad causes of repetition (frustration and malfunction) is continuous across a spectrum: thus animals may be not, mildly, or highly frustrated; not, slightly or extremely malfunctional, and so on. Furthermore, these causes can potentially co-occur in syndromes within the same individual (e.g. as different forms of perseveration seem to in schizophrenia; Harris and Dinn, 2003). In some cases, they may then combine to determine the overall levels of stereotypic behaviour. For instance, in barren-housed isolated adult mice, perseverative responding in extinction only significantly correlates with stereotypic behaviour if motivational factors (individual differences in the animals' motivations to escape their barren cages and reach enrichments) are statistically controlled for (Latham, 2005) – suggesting that both frustration and malfunction are at work in these animals. In other cases, different causal factors may predict different aspects of stereotypic behaviour (e.g. bout-length versus bout-repetitiveness, or form versus overall frequency; e.g. as suggested in Chapter 5); or even result in different, dissociable forms of stereotypic behaviour within the same individual or group of individuals, as occurs for instance in laboratory mice and autistic children (reviewed in Chapter 5). The various causal factors may also vary in their relative importance over the lifetime of an individual, or even over a short time span if acute stress or attentional demands temporarily increase the effects of perseveration or central control (see Fentress on murine grooming, e.g. in Mason, 1993; Gimpel, 2005). Thus the categories I propose are not mutually exclusive, and also draw somewhat artificial boundaries across continua. Nevertheless, they may help us classify, and even treat, stereotypic behaviours in a more biologically meaningful way, and according to the primary cause underlying repetition – something argued valuable by Chapters 5 and 10.

11.5. In Conclusion

Overall, this book has given us a range of explanations for captive animals' stereotypic behaviours. Some have been based on an understanding of a species' natural behavioural repertoires, social and sensory worlds, ecological niche, and the effects captivity has on frustration or specific aspects of physiological functioning. These help explain why captive animals may repeatedly perform behaviours that resemble attempts to escape, or thwarted elements of the natural behavioural repertoire. Other explanations have been based on the effects of captivity on the neurotransmitter levels, receptor densities, metabolic activity and synaptic connectivities of specific parts of the mid- and forebrains. These account for the similarities between some forms of captivity-induced stereotypic behaviour and behaviours induced by certain drugs, or evident in some human clinical conditions, and they help explain why some forms are so persistent or accompanied by, say, propensities to self-harm.

Reviewing the evidence presented in this volume and elsewhere, I have proposed a scheme whereby stereotypic behaviours are classified according to the mechanisms underlying their repetition, spanning a spectrum from frustration-induced to malfunction-induced. I have also suggested other mechanisms for investigation, to fully explain all the various properties (predictability, form, etc.) of stereotypic behaviours; plus I have suggested ways of subdividing my basic classificatory scheme according to e.g. the nature of associated perseveration. Testing this scheme, for its usefulness in guiding research, and in developing more effective means of treating these behaviours and improving welfare, will rely on the continued combined inputs of ethology, neuroscience and veterinary medicine. By the next edition, perhaps we will be able to judge its success.

Acknowledgements

My first set of acknowledgements is to the friends and colleagues who fielded 11th hour phone calls and e-mails during the writing of this last chapter: Ros Clubb, Joe Garner, Ronald Keiper, Naomi Latham, Mark Lewis, Danny Mills, Christel Moons, Christine Nicol, Frank Ödberg, Trevor Robbins, Willem Schouten, Sophie Vickery and Tina Widowski. My second set is to the researchers who directly or indirectly inspired and helped during my first incursion into the field of stereotypic behaviour nearly two decades ago: Pat Bateson, Gershon Berkson, Robert Dantzer, John Fentress, Ton Groothuis, Ronald Keiper, Robert Hinde, Alasdair Lawrence, Mike Mendl, Frank Ödberg, Anne-Marie de Passillé, Trevor Robbins, Jeff Rushen, Willem Schouten, Claudia Terlouw, Michelle Turner and Piet Wiepkema. (And if you are in both lists, you get recurrent thanks – or is it stuck in set?)

References

Ames, A. (1993) *The Behaviour of Captive Polar Bears.* UFAW, Potters Bar, UK

Barton J.J., Cherkasova M.V., Lindgren K.A., Goff D.C. and Manoach D.S. (2005) What is perseverated in schizophrenia? Evidence of abnormal response plasticity in the saccadic system. *Journal of Abnormal Psychology* 114, 75–84.

Benus, R.F., Koolhaas, J.M., Van Oortmerssen, G.A (1987) Individual differences in behavioral reaction to a changing environment in mice and rats. *Behaviour* 100, 105–122.

Benus, R.F., Dendaas, S., Koolhaas, J.M., Van Oortmerssen, G.A (1990) Routine formation and flexibility in social and non-social behavior of aggressive and non-aggressive male mice. *Behaviour* 112, 176–193.

Berdoy, M. (2003) *The Laboratory Rat: A Natural History.* http://www.ratlife.org (accessed Jun 2005)

Berkson, G. (1996) Feedback and control in the development of abnormal stereotyped behaviours. In: R.L. Sprague and K.M. Newell (eds) *Stereotyped Movements – Brain and Behavior Relationships.* American Psychological Association, Washington DC, pp. 3–15.

Berridge, K.C. (1994) The development of action patterns. In: J.A. Hogan and J.J. Bolhuis (eds) *Causal Mechanism of Behavioural Development.* Cambridge

University Press, Cambridge, pp. 147–180.

Boys-Smith, J.S. (1967) Behaviour of a hedgehog *Erinaceous europaeus. Journal of Zoology, London* 153, 564–566.

Broom, D.M. and Leaver, J.D. (1978) The effects of group-rearing or partial isolation on later social behaviour of calves. *Animal Behaviour* 26, 1255–1263.

Cardinal, R.N., Pennicott, D.R., Lakmali, S. and Robbins T.W. (2001) Impulsive choice induced by rats by lesions of the nucleus accumbens core. *Science* 292, 2499–2501.

Chudasama, Y., Passetti, F., Rhodes, S.E. V., Lopian, D., Desia, A. and Robbins, T. W. (2003) Dissociable aspects of performance on the 5-choice serial reaction time task following lesions of the dorsal anterior cingulated, infralimbic and orbitofrontal cortex in the rat: differential effects on selectivity, impulsivity and compulsivity. *Behavioural Brain Research* 146, 105–119.

Clarke, H.F., Walker, S.C. , Crofts, H.S., Dalley J. W., Robbins, T. W. and Roberts A. C. (2005) Prefrontal serotonin depletion affects reversal learning but not attentional set shifting. *Journal of Neuroscience* 25, 532–538

Columbia University Medical Centre (2005) Course notes: Lab 11–Basal Ganglia http://healthsciences.columbia.edu/ dept/ps/2007/firstyear/ ns/2006/neuroanat_lab11.pdf (accessed May 2005)

Cronin G.M. (1985) The development and significance of abnormal stereotyped behaviours in tethered sows. Ph.D. dissertation, Wageningen Agricultural University, Wageningen.

Dantzer, R. (1986) Behavioural, physiological and functional aspects of stereotypic behaviour: a review and reinterpretation. *Journal of Animal Science* 62, 1776–1786.

Dias, R., Robbins, T.W. and Roberts, A.C. (1996) Dissociation in prefrontal cortex in attentional and affective shifts. *Nature* 380, 69–72.

Dickinson A. (1985) Actions and habits: the development of behavioural autonomy.

Philosophical Transactions of the Royal Society, Series B 308, 67–78.

Edwards M.J., Dale R.C., Church A.J., Trikouli E., Quinn N.P., Lees A.J., Giovannoni G. and Bhatia K.P. (2004) Adult-onset tic disorder, motor stereotypies, and behavioural disturbance associated with antibasal ganglia antibodies. *Movement Disorders* 19, 1190–1196.

Fentress J.C (1973) Specific and nonspecific factors in the causation of behavior. In: Bateson P.P.G. and Klopfer P.H. (eds), *Perspectives in Ethology 1*. Plenum Press, London, pp. 155–218.

Fentress J.C. (1976) Dynamic boundaries of patterned behaviour: interaction and self-organization. In: Bateson P.P.G., Hinde R.A. (eds.) *Growing points in ethology*. Cambridge University Press, Cambridge, pp. 135–169.

Fraser A.F. and Broom D.M. (1990) *Farm Animal Behaviour and Welfare*. 3rd ed. Baillière Tindall, London.

Fry J.P., Sharman D.F., and Stephens D.B. (1981) Cerebral dopamine, apomorphine and oral activity in the neonatal pig. *Journal of Veterinary Pharmacology and Therapeutics* 4, 193–207.

Fuchs, R.A., Evans, K.A., Parker, M.P., and Ronald, E. (2004) Differential involvement of orbitofrontal cortex subregions in conditioned cue-induced and cocaine-primed reinstatement of cocaine seeking in rats. *Journal of Neuroscience* 24, 6600–6610

Garner, J.P. and Mason, G.J. (2002) Evidence for a relationship between cage stereotypies and behavioural disinhibition in laboratory rodents. *Behavioural Brain Research* 136, 83–92.

Garner, J.P., Mason, G. and Smith, R. (2003a) Stereotypic route-tracing in experimentally-caged songbirds correlates with general behavioural disinhibition. *Animal Behaviour* 66, 711–727

Garner, J.P., Meehan, C.L. and Mench, J.A. (2003b) Stereotypies in caged parrots, schizophrenia and autism: evidence for a common mechanism. *Behavioural Brain Research* 145, 125–134.

Gimpel, J. (2005) 'The effects of housing and husbandry on the welfare of laboratory primates. PhD thesis, University of Oxford, Oxford

Graybiel, A.M. (1998) The basal ganglia and chunking of action repertoires. *Neurobiology of Learning and Memory* 70, 119–136

Grewal, B.S. (1981) Self-wrist-biting in Arashiyama-B troop of Japanese monkeys (*Macaca macaca fuscata*). *Primates* 22, 277–280.

Haber, S.N., Fudge, J.L., and McFarland, N. R. (2000) Striatonigral pathways in primates form an ascending spiral from the shell to the dorsallateral striatum. *Journal of Neuroscience* 20, 2369–2382.

Haber, S. (2003) The primate basal ganglia: parallel and integrative networks. *Journal of Chemical Neuroanatomy* 26: 317–330.

Hanstock T.L., Clayton E.H., Li K.M., and *Mallet P.E.* (2004) Anxiety and aggression associated with the fermentation of carbohydrates in the hindgut of rats. *Physiology and Behavior* 82, 357–368.

Happaney K. and Zelazo P.D. (2004) Resistance to extinction: A measure of orbitofrontal function suitable for children? *Brain and Cognition* 55, 171–184

Harris, C.L. and Dinn, W.M. (2003) Subtyping obsessive-compulsive disorder: Neuropsychological correlates. *Behavioral Neurology* 14, 75–87.

Hauser, M (1999) Perseveration, inhibition and the prefrontal cortex: a new look. *Current Opinion in Neurobiology* 9, 214–222.

Huffman, M.A. (1984) Stone-play of *Macaca fuscata* in Arashiyama B troop: Transmission of a non-adaptive behaviour. *Journal of Human Evolution* 13, 725–735.

Jeppesen, L.L., Heller, K.E. and Bildsøe, A. (2004) Stereotypies in female farm mink (*Mustela vison*) may be genetically transmitted and associated with higher fertility due to effects on body weight. *Applied Animal Behaviour Science* 86,137–143.

Jog, M.S., Kubota, Y., Connolly, C.I., Hillegaart, V. and Graybiel. A.M. (1999). Building neural representations of habits. *Science* 286, 1745–1749

Kalivas, P.W. and Nakamura, M. (1999). Neural systems for behavioural activation and reward. *Current Opinion in Neurobiology* 9, 223–227.

Karler R., Bedingfield J.B., Thai D.K. and Calder L.D. (1997) The role of the frontal cortex in the mouse in behavioral sensitization to amphetamine. *Brain Research* 757, 228–235.

Keiper, R.R. (1969) Drug effects on canary stereotypies. *Psychopharmacologia* 16, 16–24

Kerzel D. and Prinz W. (2003) Performance. In: *Encyclopedia of Cognitive Science.* Nature Publishing Group, pp. 560–565.

Killcross S. and Coutureau E. (2003) Coordination of actions and habits in the medial prefrontal cortex. *Cerebral Cortex* 13, 400–408

Jog M.S., Kubota, Y., Connolly, C.I., Hillegaart, V. and Graybiel, A.M. 1999. Building neural representations of habits. *Science* 286, 286:1745–1749

Lac, S. du (1999). Motor learning. In R. A. Wilson and F. C. Kiel (eds), *The MIT Encyclopedia of the Cognitive Sciences,* pp. 571–572. MIT press, Cambridge, Massachusetts.

Lashley K.S. (1921) Studies of cerebral function in learning II: the effects of long-continued practice upon cerebral localisation. *Journal of Comparative Psychology* 1, 453–468.

Latham, N. (2005) Refining the role of stereotypic behaviour in the assessment of welfare: Stress, general motor persistence and early environment in the development of abnormal behaviour. PhD thesis, Oxford University, Oxford.

Lawrence, A.B. and Terlouw, E.M.C. (1993) A review of behavioral factors involved in the development and continued performance of stereotypic behaviors in pigs. *Journal of Animal Science* 71, 2815–2825.

Lierop, M. van (2005) *Stereotypical behaviours in the striped mouse (Rhabdomys* pumilio): evaluating the coping hypothesis. MSc. Thesis, University of the Witwatersrand, Johannesburg.

Lutz, C., Tiefenbacher, S., Meyer, J. and Novak, M. (2004) Extinction deficits in male rhesus macaques with a history of self-injurious behaviour. *American Journal of Primatology* 63, 41–48

Mackintosh, N.J. (1974) *The Psychology of Animal Learning*. Academic Press, London.

Maltby, N., Tolin, D.F., Worhunsky, P., O'Keefe, T.M., and Kiehl, K.A. (2005) Dysfunctional action monitoring hyperactivates frontal-striatal circuits in obsessive-compulsive disorder: an event-related fMRI sudy. *Neuroimage* 24, 495–503.

Marsh R., Alexander G.M., Packard M.G., Zhu H., Peterson B.S. (2005) Perceptual-motor skill learning in Gilles de la Tourette syndrome. Evidence for multiple procedural learning and memory systems. *Neuropsychologia* 43, 1456–1465.

Martiniuk R. (1976) *Information Processing in Motor Skills*. Holt, Rinehart and Winston, New York.

Mason, G.J. (1991a) Stereotypies: a critical review. *Animal Behaviour* 41, 1015–1037.

Mason G.J. (1991b). Stereotypies and suffering. *Behavioural Processes* 25, 103–115.

Mason G.J. (1993). Forms of stereotypic behaviour. In: Lawrence, A.B. and Rushen, J. (eds.), *Stereotypic Animal Behaviour: Fundamentals and Applications to Welfare*. CAB International Wallingford, UK, pp. 7–40.

Mason G.J., Turner M.A. 1993. Mechanisms involved in the development and control of stereotypies. In: Bateson P.P.G., Klopfer P.H., Thompson N.S. (eds), *Perspectives in Ethology*. Plenum Press, London, pp. 53–85.

Mason, G. Cooper J. and Clarebrough C. (2001) The welfare of fur-farmed mink. *Nature* 410, 35–36.

Mason, G.J. and Latham, N.R. (2004) Can't stop, won't stop: is stereotypy a reliable animal welfare indicator? *Animal Welfare* 13 (Supplement), S57–S69.

Mason, G., Clubb, R., Latham, N. and Vickery, S. (2006) Why and how should we use environmental enrichments to tackle stereotypic behaviour? *Applied Animal Behaviour Science* (Available on-line Aug. 2nd 2006)

Mataix-Cols, D., Wooderson, S., Larence, N. Brammer, M.J., Speckens, A. and Phillips, M.L. (2004) Distinct neural correlate of washing, checking and hoarding symptom dimensions in obsessive-compulsive disorder. *Archives of General Psychiatry* 61, 564–576.

McBride, S. and Hemmings, A. (2005). Altered meso-accumbens and nigro-striatal dopamine physiology is associated with stereotypy development in a non-rodent species. *Behavioural Brain Research* 159, 113–118.

Hemmings A, McBride SD, and Hale C. (2006) Perseverative responding and the aetiology of equine oral stereotypy. *Applied Animal Behaviour Science* (Available on-line June 5th 2006)

McFarland, K., Davidge, S.B., Lapish, C.C. and Kalivas, P. W. (2004) Limbic and motor circuitry underlying reinstatemtn of cocaine-seeking behaviour. *Journal of Neuroscience* 24, 1551–1560.

Mechner F. (1995) *Learning and practicing skilled performance*. The Mechner Foundation, New York:

Milgram N.W., Head E., Zicker S.C., Ikeda-Douglas C., Murphey H., Muggenberg B., Siwak C., Tapp P.D., Lowry S.R. and Cotman C.W. (2004) Long-term treatment with antioxidants and a program of behavioral enrichment reduces age-dependent impairment in discrimination and reversal learning in beagle dogs. *Experimental Gerontology* 39, 753–765.

Mills, D.S. (2003) Medical paradigms for the study of problem behaviour: a critical review. *Applied Animal Behaviour Science* 81, 265–277.

Muller N., Riedel M., Eggert T., and Straube A. (1999) Internally and externally guided voluntary saccades in unmedicated and medicated schizophrenic patients. Part II. Saccadic latency, gain, and fixation suppression errors. *European*

Archives of Psychiatry and Clinical Neuroscience 249, 7–14.

Norman D.A. (1981) The categorization of action slips. *Psychological Review* 89, 1–55.

Passingham, R.E. (1996) Attention to action. *Philosophical Transactions of the Royal Society, Series B*, 1473–1479

Pingitore, G., Chrobak, V. and Petrie, J. (1991) The social and psychological factors of bruxism. *Journal of Prosthetic Dentistry* 65, 443–446.

Rago, W.V. and Case, J.C. (1978) Stereotyped behavior in special education teachers. *Exceptional Children* 44, 342–4.

Reading, P.J., Dunnett, S.B. and Robbins, T. W. (1991) Dissocciable roles of the ventral, medial and lateral striatum on the acquisition and performance of a complex visual stimulus-response habit. *Behavioural Brain Research* 45, 147–161.

Ridderinkhof K.R., Span M.M., and van der Molen M.I.W. (2002) Perseverative behavior and adaptive control in older adults: Performance monitoring, rule induction, and set shifting. *Brain and Cognition* 49, 382–401.

Robinson, T.E. and Berridge, K.C. (2003) Addiction. *Annual Review of Psychology* 54, 25–53.

Rolls, E.T. (1999) *The Brain and Emotion* Oxford University Press, Oxford

Saka, E., Goodrich, C., Harlan, P., Madras, B.K., and Graybiel, A.M. (2004) Repetitive behaviours in monkeys are linked to specific striatal activation patterns. *Journal of Neuroscience* 24, 7557–7565.

Salamone, J.D., Correa, M., Mingote, S.M. and Weber, S. M. (2005) Beyond the reward hypothesis alternative functions of nucleus accumbens dopamine. *Current Opinion in Pharmacology* 5, 34–41.

Salthouse T.A. (1985) Anticipatory processes in transcription typing. *Developmental Psychopathology* 13, 419–449.

Schouten, W.G.P. and Wiepkema, P.R. (1991) Coping styles of tethered sows. *Behavioural Processes* 25, 125–132.

Sharman, D.F., Mann, S.P., Fry, J.P., Banns, H. and Stephens, D. B. (1982) Cerebreal

dopamine metabolism and stereotyped behaviour in early-weaned piglets. *Neuroscience* 7, 1937–1944.

Shettleworth, S.J. (1972). Constraints on learning. In: D.S. Lehrman, R.A. Hinde and E. Shaw (eds.) *Advances in the Study of Behavior, Vol 4.* Academic Press, New York.

Snider, L.A. and Swedo, S.E. (2004) PANDAS: Current status and directions for research. *Molecular Psychiatry* 9, 900–907.

Sumner, B.E.H, Lawrence, A.B., Jarvis, S., Calvert, S.K., Stevenson, J., Farnworth, M.J., Croy, Douglas, A.T., Russell, J.A. and Seckl, J.R. (2002) Enduring behavioural and neural effects of early versus late weaning in pigs. *Stress* 5, supplement (September 2002), p. 105.

Szechtman, H. and Woody, E (2004), Obsessive-compulsive disorder as a disturbance of security motivation. *Psychological Review* 111, 111–127

Tapp P.D., Siwak C.T., Estrada J., Head E., Muggenburg B.A., Cotman C.W., Milgram N.W. (2003) Size and reversal learning in the beagle dog as a measure of executive function and inhibitory control in aging. *Learning and Memory* 10, 64–73.

Terlouw, E.M., Lawrence, A.B. and Illius, A.W. (1992) Relationship between amphetamine and environmentally induced stereotypies in pigs. *Pharmacology, Biochemistry and Behavior* 43, 347–355.

Thelen E. (1979) Rhythmical stereotypies in normal human infants. *Animal Behaviour* 27, 699–715.

Timberlake, W., and Lucas, G.A. (1989) Behavior systems and learning: From misbehavior to general principles. In S.B. Klein and R.R. Mowrer (eds), *Contemporary learning theories: Instrumental conditioning theory and the impact of biological constraints on learning.* Erlbaum, Hillsdale, New Jersey, pp. 237–275

Tinbergen, N. (1951) *The study of instinct.* Oxford University Press, Oxford.

Toates F. (2001) *Biological Psychology.* Pearson Education Limited Harlow, UK.

Vallée M., Maccari S., Dellu F., Simon H., Le Moal M., and Mayo W. (1999) Long-term effects of prenatal stress and postnatal handling on age-related glucocorticoid secretion and cognitive performance: a longitudinal study in the rat *European Journal of Neuroscience* 11, 2906–2916

Veasey, J.S., Waran, N.K. and Young, R.J. (1996) On comparing the behaviour of zoo housed animals with wild conspecifics as a welfare indicator, using the giraffe (*Giraffa camelopardalis*) as a model. *Animal Welfare* 5, 139–153.

Vickery, S.S. (2003) *Stereotypy in Caged Bears: Individual and Husbandry Factors.* Ph. D. Thesis, University of Oxford, Oxford.

Vickery, S.S. and G.J. Mason (2003) Behavioral persistence in captive bears: implications for reintroduction. *Ursus* 14, 35–43.

Vickery, S. and G. Mason. (2004) Stereotypic behavior in Asiatic black and Malayan sun bears. *Zoo Biology* 23, 409–430.

Vickery, S. and Mason, G.J. (2005). Behavioural persistence in captive bears: a reply to Criswell and Galbreath. *Ursus* 16, 274–279.

Wallis J.D., Dias R., Robbins T.W., and Roberts A.C. (2001) Dissociable contributions of the orbitofrontal and lateral prefrontal cortex of the marmoset to performance on a detour reaching task. *European Journal of Neuroscience* 1797–1808.

Wechsler, B. (1991) Stereotypies in polar bears. *Zoo Biology* 10, 177–188.

Wohlschlafer A., Jager R., and Delius J.D. (1993) Head and eye-movements in unrestrained pigeons (*Columbia livia*). *Journal of Comparative Psychology* 107, 313–319.

Wyvell CL. and Berridge KC (2000) Intranucleus accumbens amphetamine increases the conditioned incentive salience of sucrose reward: enhancement of reward "wanting" without enhanced "liking" or response reinforcement. *Journal of Neuroscience* 20, 8122–8130

Index